Communications in Computer and Information Science 452

T0202995

More information about this series at http://www.springer.com/series/7899

Mireya Fernández-Chimeno · Pedro L. Fernandes
Sergio Alvarez · Deborah Stacey
Jordi Solé-Casals · Ana Fred
Hugo Gamboa (Eds.)

Biomedical Engineering Systems and Technologies

6th International Joint Conference,
BIOSTEC 2013
Barcelona, Spain, February 11–14, 2013
Revised Selected Papers

 Springer

Editors
Mireya Fernández-Chimeno
Universitat Politècnica de Catalunya
Barcelona
Spain

Jordi Solé-Casals
University of Vic
Victoria
Spain

Pedro L. Fernandes
Instituto Gulbenkian de Ciência
Oeiras
Portugal

Ana Fred
Technical University of Lisbon
Lisbon
Portugal

Sergio Alvarez
Boston College
Chestnut Hill, MA
USA

Hugo Gamboa
New University of Lisbon
Lisboa
Portugal

Deborah Stacey
University of Guelph
Guelph
Canada

ISSN 1865-0929   ISSN 1865-0937 (electronic)
ISBN 978-3-662-44484-9   ISBN 978-3-662-44485-6   (eBook)
DOI 10.1007/978-3-662-44485-6

Library of Congress Control Number: 2014953274

Springer Heidelberg New York Dordrecht London

Printed on acid-free paper

Springer is part of Springer Science+Business Media (www.springer.com)

Preface

The present book includes extended and revised versions of a set of selected papers from the Sixth International Joint Conference on Biomedical Engineering Systems and Technologies (BIOSTEC 2013), held in Barcelona, Spain from February 11 to 14, 2013.

BIOSTEC was sponsored by the Institute for Systems and Technologies of Information, Control and Communication (INSTICC), in collaboration with the University of Vic.

BIOSTEC 2013 was held in cooperation with the Association for the Advancement of Artificial Intelligence (AAAI) and technically co-sponsored by the Biomedical Engineering Society (BMES) and European Society for Engineering and Medicine (ESEM).

The purpose of the International Joint Conference on Biomedical Engineering Systems and Technologies is to bring together researchers and practitioners interested in both theoretical advances and applications of information systems, artificial intelligence, signal processing, electronics, and other engineering tools in knowledge areas related to biology and medicine.

BIOSTEC is composed of four complementary and co-located conferences, each specialized in at least one of the aforementioned main knowledge areas. Namely:

- International Conference on Biomedical Electronics and Devices – BIODEVICES;
- International Conference on Bioinformatics Models, Methods, and Algorithms – BIOINFORMATICS;
- International Conference on Bio-inspired Systems and Signal Processing – BIOSIGNALS;
- International Conference on Health Informatics – HEALTHINF.

The purpose of the International Conference on Biomedical Electronics and Devices (BIODEVICES) is to bring together professionals from electronics and mechanical engineering, interested in studying and using models, equipment, and materials inspired from biological systems and/or addressing biological requirements. Monitoring devices, instrumentation sensors and systems, biorobotics, micro-nanotechnologies, and biomaterials are some of the technologies addressed at this conference.

The International Conference on Bioinformatics Models, Methods, and Algorithms (BIOINFORMATICS) intends to provide a forum for discussion to researchers and practitioners interested in the application of computational systems and information technologies to the field of molecular biology, including for example the use of statistics and algorithms to understanding biological processes and systems, with a focus on new developments in genome bioinformatics and computational biology. Areas of interest for this community include sequence analysis, biostatistics, image analysis, scientific data management and data mining, machine learning, pattern recognition, computational evolutionary biology, computational genomics, and other related fields.

The goal of the International Conference on Bio-inspired Systems and Signal Processing (BIOSIGNALS) is to bring together researchers and practitioners from multiple areas of knowledge, including biology, medicine, engineering, and other physical sciences interested in studying and using models and techniques inspired from or applied to biological systems. A diversity of signal types can be found in this area, including image, audio, and other biological sources of information. The analysis and use of these signals is a multidisciplinary area including signal processing, pattern recognition, and computational intelligence techniques, among others.

The International Conference on Health Informatics (HEALTHINF) aims to be a major meeting point for those interested in understanding the human and social implications of technology, not only in healthcare systems but in other aspects of human–machine interaction such as accessibility issues and the specialized support to persons with special needs.

The joint conference, BIOSTEC, received 392 paper submissions from 57 countries in all continents. To evaluate each submission, a double-blind paper review was performed by the Program Committee. After a stringent selection process, 56 papers were published and presented as full papers, i.e., completed work (10 pages/30 minute oral presentation), 99 papers reflecting work-in-progress or position papers were accepted for short presentation, and another 88 contributions were accepted for poster presentation. These numbers, leading to a full-paper acceptance ratio of about 14 % and a total oral paper presentations acceptance ratio close to 40 %, show the intention of preserving a high quality forum for the next editions of this conference.

BIOSTEC's program includes panels and six invited talks delivered by internationally distinguished speakers, namely: Pedro Gómez Vilda (Universidad Politécnica de Madrid, Spain), Christian Jutten (GIPSA-lab, France), Adam Kampff (Champalimaud Foundation, Portugal), Richard Reilly (Trinity College Dublin, Ireland), Vladimir Devyatkov (Bauman Moscow State Technical University, Russian Federation), and Pietro Liò (University of Cambridge, UK).

We would like to thank the authors, whose research and development efforts are recorded here for future generations.

December 2013

Mireya Fernández-Chimeno
Pedro L. Fernandes
Sergio Alvarez
Deborah Stacey
Jordi Solé-Casals
Ana Fred
Hugo Gamboa

Organization

Conference Co-chairs

Jordi Solé-Casals — University of Vic, Spain
Ana Fred — Instituto de Telecomunicações, Instituto Superior Técnico, Technical University of Lisbon, Portugal
Hugo Gamboa — CEFITEC/FCT – New University of Lisbon, Portugal

Program Co-chairs

BIODEVICES

Mireya Fernández-Chimeno — Universitat Politècnica de Catalunya, Spain

BIOINFORMATICS

Pedro L. Fernandes — Instituto Gulbenkian de Ciência, Portugal

BIOSIGNALS

Sergio Alvarez — Boston College, USA

HEALTHINF

Deborah Stacey — University of Guelph, Canada

Organizing Committee

Marina Carvalho — INSTICC, Portugal
Helder Coelhas — INSTICC, Portugal
Vera Coelho — INSTICC, Portugal
Andreia Costa — INSTICC, Portugal
Bruno Encarnação — INSTICC, Portugal

Ana Guerreiro	INSTICC, Portugal
André Lista	INSTICC, Portugal
Carla Mota	INSTICC, Portugal
Raquel Pedrosa	INSTICC, Portugal
Vitor Pedrosa	INSTICC, Portugal
Cláudia Pinto	INSTICC, Portugal
Cátia Pires	INSTICC, Portugal
Susana Ribeiro	INSTICC, Portugal
Sara Santiago	INSTICC, Portugal
Margarida Sorribas	INSTICC, Portugal
José Varela	INSTICC, Portugal
Pedro Varela	INSTICC, Portugal

BIODEVICES Program Committee

Farid Amirouche	University of illinois at Chicago, USA
Dinesh Bhatia	Biomedical Engineering Department, Deenbandhu Chhotu Ram University of Science and Technology, Murthal, India
Martin Bogdan	Universität Leipzig, Germany
Luciano Boquete	Alcala University, Spain
Egon L. van den Broek	Human-Centered Computing Consultancy/ University of Twente/Radboud UMC Nijmegen, The Netherlands
Juan Jose Ramos Castro	Universitat Politècnica de Catalunya, Spain
Mireya Fernández-Chimeno	Universitat Politècnica de Catalunya, Spain
Fernando Cruz	College of Technology of Setubal/Polytechnic Institute of Setubal, Portugal
Maeve Duffy	NUI Galway, Ireland
G.S. Dulikravich	Florida International University, USA
Michele Folgheraiter	German Research Center for Artificial Intelligence, Germany
Juan Carlos Garcia Garcia	Universidad de Alcala, Spain
Roman Genov	University of Toronto, Canada
Miguel Angel García Gonzalez	Universitat Politècnica de Catalunya, Spain
Clemens Heitzinger	University of Cambridge, UK
Toshiyuki Horiuchi	Tokyo Denki University, Japan
Leonid Hrebien	Drexel University, USA
Takuji Ishikawa	Tohoku University, Japan
Sandeep K. Jha	Banasthali University, India
Bozena Kaminska	Simon Fraser University, Canada
Walter Karlen	UBC, Canada
Frank Kirchner	DFKI, Germany
Anton Koeck	Austrian Institute of Technology (AIT), Austria
Ondrej Krejcar	University of Hradec Kralove, Czech Republic

Hongen Liao	The University of Tokyo, Japan
Xiao Liu	Brunel University, UK
Mai S. Mabrouk	Misr University for Science and Technology, Egypt
Jordi Madrenas	Universitat Politècnica de Catalunya, Spain
Ratko Magjarevic	Faculty of Electrical Engineering and Computing, Croatia
Jarmo Malinen	Aalto University, Finland
Dan Mandru	Technical University of Cluj Napoca, Romania
Raimes Moraes	Universidade Federal de Santa Catarina, Brazil
Raul Morais	ECT-UTAD/CITAB, Portugal
Umberto Morbiducci	Politecnico di Torino, Italy
Toshiro Ohashi	Hokkaido University, Japan
Kazuhiro Oiwa	National Institute of Information and Communications Technology, Japan
Mónica Oliveira	University of Strathclyde, UK
Abraham Otero	Universidad San Pablo CEU, Spain
Danilo Pani	University of Cagliari, Italy
Sofia Panteliou	University of Patras, Greece
Nathalia Peixoto	Neural Engineering Lab, George Mason University, USA
Mark A. Reed	Yale University, USA
Sang-Hoon Rhee	Seoul National University, Korea, Republic of
Wim L.C. Rutten	University of Twente, The Netherlands
Seonghan Ryu	Hannam University, Korea, Republic of
Mario Sarcinelli-Filho	Federal University of Espirito Santo, Brazil
Chutham Sawigun	Mahanakorn University of Technology, Thailand
Michael J. Schöning	Institute of Thin Film and Ion Technology, Research Centre Jülich GmbH, Germany
Mauro Serpelloni	University of Brescia, Italy
Alcimar Barbosa Soares	Universidade Federal de Uberlândia, Brazil
Federico Vicentini	National Research Council of Italy (CNR), Italy
Bruno Wacogne	FEMTO-ST UMR CNRS 6174, France
Shigeo Wada	Osaka University, Japan
Sen Xu	Bristol-Myers Squibb, Co., USA
Jing Yang	Qingdao Institute of Bioenergy and Bioprocess Technology, China
Aladin Zayegh	Victoria University, Australia

BIODEVICES Auxiliary Reviewers

Rodrigo Andreão	Ifes, Brazil
Gianluca Barabino	University of Cagliari, Italy
Mario Cifrek	University of Zagreb Faculty of Electrical Engineering and Computing, Croatia

Alessia Dessì	University of Cagliari, Italy
Ainara Garde	UBC and Childrens's Hospital, Canada
Igor Lackovic	University of Zagreb, Croatia
Dinko Oletic	University of Zagreb, Faculty of Electrical Engineering and Computing, Croatia
Antonio Petošic	Faculty of Electrical Engineering and Computing, Croatia

BIOINFORMATICS Program Committee

Mohamed Abouelhoda	Nile and Cairo University, Egypt
Tatsuya Akutsu	Kyoto University, Japan
Mar Albà	UPF/IMIM, Spain
Charles Auffray	CNRS Institute of Biological Sciences, France
Rolf Backofen	Albert-Ludwigs-Universität, Germany
Tim Beissbarth	University of Göttingen, Germany
Ugur Bilge	Akdeniz University, Turkey
Inanc Birol	BC Cancer Agency, Canada
Carlos Brizuela	Centro de Investigación Científica y de Educación Superior de Ensenada, Baja California, Mexico
Chris Bystroff	Rensselaer Polytechnic Institute, USA
João Carriço	Universidade de Lisboa, Portugal
Ferran Casals	Universitat Pompeu Fabra, Spain
Kun-Mao Chao	National Taiwan University, Taiwan
Francis Y.L. Chin	The University of Hong Kong, Hong Kong
Antoine Danchin	AMAbiotics SAS, France
Thomas Dandekar	University of Würzburg, Germany
Sérgio Deusdado	Instituto Politecnico de Bragança, Portugal
Eytan Domany	Weizmann Institute of Science, Israel
Richard Edwards	University of Southampton, UK
George Eleftherakis	CITY College, International Faculty of the University of Sheffield, Greece
André Falcão	Universidade de Lisboa, Portugal
Pedro Fernandes	Instituto Gulbenkian de Ciência, Portugal
Fabrizio Ferre	University of Rome Tor Vergata, Italy
António Ferreira	University of Lisbon, Portugal
Elisa Ficarra	Politecnico di Torino, Italy
Liliana Florea	Johns Hopkins University, USA
Gianluigi Folino	Institute for High Performance Computing and Networking, National Research Council, Italy
Andrew French	University of Nottingham, UK
Bruno Gaëta	University of New South Wales, Australia
Max H. Garzon	The University of Memphis, USA
Julian Gough	University of Bristol, UK
Reinhard Guthke	Hans Knoell Institute, Germany

Joerg Hakenberg	Hoffmann-La Roche, Inc., USA
Emiliano Barreto Hernandez	Universidad Nacional de Colombia, Colombia
Hailiang Huang	Mass General Hospital, USA
Bo Jin	Sigma-aldrich, USA
Giuseppe Jurman	Fondazione Bruno Kessler, Italy
Inyoung Kim	Virginia Tech, USA
Jiří Kléma	Czech Technical University in Prague, Faculty of Electrical Engineering, Czech Republic
Andrzej Kloczkowski	Ohio State University, USA
Sophia Kossida	Academy of Athens, Greece
Bohumil Kovár	Institute of Information Theory and Automation of the ASCR, Czech Republic
Lukasz Kurgan	University of Alberta, Canada
Yinglei Lai	George Washington University, USA
Matej Lexa	Masaryk University, Czech Republic
Xiaoli Li	Nanyang Technological University, Singapore
Michal Linial	The Hebrew University of Jerusalem, Israel
Pedro Lopes	Universidade de Aveiro, Portugal
Shuangge Ma	Yale University, USA
Xizeng Mao	University of Georgia, Athens, USA
Elena Marchiori	Radboud University, The Netherlands
Majid Masso	George Mason University, USA
Pavel Matula	Faculty of Informatics, Masaryk University, Czech Republic
Petr Matula	Faculty of Informatics, Masaryk University, Czech Republic
Tommaso Mazza	Casa Sollievo della Sofferenza - Mendel, Italy
Nuno Mendes	Instituto de Gulbenkian de Ciência, Portugal
Imtraud Meyer	University of British Columbia, Canada
Luciano Milanesi	ITB-CNR, Italy
Catarina Moita	Instituto de Medicina Molecular, Portugal
Pedro Tiago Monteiro	Instituto de Gulbenkian de Ciência, Portugal
Shinichi Morishita	University of Tokyo, Japan
Hunter Moseley	University of Louisville, USA
Vincent Moulton	University of East Anglia, UK
Nicola Mulder	University of Cape Town, South Africa
Chad Myers	University of Minnesota, USA
Radhakrishnan Nagarajan	University of Kentucky, USA
Jean-Christophe Nebel	Kingston University, UK
José Luis Oliveira	Universidade de Aveiro, Portugal
Patricia Palagi	Swiss Institute of Bioinformatics, Switzerland
Florencio Pazos	National Centre for Biotechnology, Spain
Matteo Pellegrini	University of California, Los Angeles, USA
Horacio Pérez-Sánchez	Parallel Computer Architecture Group, Spain
Guy Perrière	Université Claude Bernard - Lyon 1, France
Francisco Pinto	Universidade de Lisboa, Portugal

Miguel Rocha	University of Minho, Portugal
Paolo Romano	IRCCS AOU San Martino - IST National Cancer Research Institute, Italy
Simona E. Rombo	DEIS, Università della Calabria, Italy
Juho Rousu	Aalto University, Finland
Derek Ruths	McGill University, Canada
Yvan Saeys	Ghent University, Belgium
J. Cristian Salgado	University of Chile, Chile
Armindo Salvador	Universidade de Coimbra, Portugal
Mark Segal	University of California, San Francisco, USA
Joao C. Setubal	Universidade de São Paulo, Brazil
Hamid Reza Shahbazkia	Universidade do Algarve, Portugal
Christine Sinoquet	LINA/UMR CNRS 6241, University of Nantes, France
Pavel Smrz	Brno University of Technology, Czech Republic
Gordon Smyth	Walter and Eliza Hall Institute of Medical Research, Australia
Peter F. Stadler	Universität Leipzig, Germany
Yanni Sun	Michigan State University, USA
Sandor Szedmak	University of Innsbruck, Austria
Li Teng	The University of Iowa, USA
Silvio C.E. Tosatto	Università di Padova, Italy
Alexander Tsouknidas	Fredereick University, Nicosia, Cyprus, Cyprus
Jose Ramon Valverde	Spanish Research Council, Spain
Massimo Vergassola	Physics of Biological Systems, Institut Pasteur, France
Juris Viksna	University of Latvia, Latvia
Susana Vinga	INESC-ID, Portugal
Gert Vriend	CMBI, The Netherlands
Yufeng Wu	University of Connecticut, USA
Dong Xu	University of Missouri, USA
Tangsheng Yi	University of California at San Francisco, USA
Yanbin Yin	University of Georgia, USA
Jingkai Yu	Institute of Process Engineering, Chinese Academy of Sciences, China
Qingfeng Yu	Stowers Institute for Medical Research, USA
Erliang Zeng	University of Notre Dame, USA
Jie Zheng	Nanyang Technological University, Singapore
Leming Zhou	University of Pittsburgh, USA

BIOINFORMATICS Auxiliary Reviewers

Stefano Castellana	IRCCS Casa Sollievo della Sofferenza - Mendel, Italy
Santa Di Cataldo	Politecnico di Torino, Italy

Marie-Dominique Devignes	LORIA CNRS-UMR 7503, France
Alessandro Guffanti	Genomnia srl, Italy
Andreas Leha	University Medical Center, Germany
Qin Ma	Computational Systems Biology Lab, USA
Martin Mann	University Freiburg, Germany
Sérgio Matos	University of Aveiro, Portugal
Giuseppe Tradigo	University of Calabria, Italy
Vitaly Volpert	CNRS, France
Jiayin Wang	University of Connecticut, USA
Jin Zhang	University of Connecticut, USA
Min Zhao	Vanderbilt Medical Center, USA

BIOSIGNALS Program Committee

Jean-Marie Aerts	Division Measure, Model and Manage Bioresponses (M3-BIORES), Katholieke Universitëit Leuven, Belgium
Jesús B. Alonso	Universidad de Las Palmas de Gran Canaria, Spain
Fernando Alonso-Fernandez	Halmstad University, Sweden
Sergio Alvarez	Boston College, USA
Oliver Amft	TU Eindhoven, The Netherlands
Peter Bentley	UCL, UK
Jovan Brankov	Illinois Institute of Technology, USA
Tolga Can	Middle East Technical University, Turkey
Francis Castanie	TeSA, France
M. Emre Celebi	Louisiana State University in Shreveport, USA
Joselito Chua	Monash University, Australia
Jan Cornelis	VUB, Belgium
Dimitrios Fotiadis	University of Ioannina, Greece
Esteve Gallego-Jutglà	University of Vic, Spain
Juan I. Godino-Llorente	Dept. Ingeniería de Circuitos y Sistemas, Universidad Politécnica de Madrid, Spain
Aaron Golden	Albert Einstein College of Medicine, USA
Thomas Hinze	Friedrich-Schiller University Jena, Germany
Bart Jansen	Vrije Universiteit Brussel, Belgium
Christian Jutten	GIPSA-lab, France
Visakan Kadirkamanathan	The University of Sheffield, UK
Shohei Kato	Nagoya Institute of Technology, Japan
Jonghwa Kim	University of Augsburg, Germany
Georgios Kontaxakis	Universidad Politecnica de Madrid, Spain
Vinod Kumar	IIT Roorkee, India
KiYoung Lee	AJOU University School of Medicine, Korea, Republic of
Lenka Lhotska	Czech Technical University in Prague, Faculty of Electrical Engineering, Czech Republic

Carlos Ortiz de Solorzano	Center of Applied Medical Research of the University of Navarra, Spain
Olga Sourina	Nanyang Technological University, Singapore
Hiroki Takada	University of Fukui, Japan
Asser Tantawi	IBM, USA
Wallapak Tavanapong	Iowa State University, USA
Gianluca Tempesti	University of York, UK
Petr Tichavsky	Institute of Information Theory and Automation, Czech Republic
Anna Tonazzini	CNR, Italy
Duygu Tosun	UCSF, USA
Carlos M. Travieso	University of Las Palmas de Gran Canaria, Spain
Mahdi Triki	Philips Research Laboratories, The Netherlands
Bart Vanrumste	Katholieke Hogeschool Kempen/Katholieke Universiteit Leuven, Belgium
Aniket Vartak	Qualcomm Incorporated, USA
Michal Vavrecka	Czech Technical University, Czech Republic
Eric Wade	USC Division of Biokinesiology and Physical Therapy, USA
Yuanyuan Wang	Fudan University, China
Takashi Watanabe	Tohoku University, Japan
Quan Wen	University of Electronic Science and Technology of China, China
Nicholas Wickstrom	Halmstad University, Sweden
Didier Wolf	Research Centre for Automatic Control - CRAN CNRS UMR 7039, France
Pew-Thian Yap	University of North Carolina at Chapel Hill, USA
Li Zhuo	Beijing University of Technology, China

BIOSIGNALS Auxiliary Reviewer

| Ashwin Belle | Virginia Commonwealth University, USA |

HEALTHINF Program Committee

Sergio Alvarez	Boston College, USA
Francois Andry	Optum / UnitedHealth Group Inc., USA
Philip Azariadis	University of the Aegean, Greece
Adrian Barb	Penn State University, USA
Gillian Bartlett	McGill University, Canada
Rémi Bastide	Jean-Francois Champollion University, France
Ronald Batenburg	Utrecht University, The Netherlands
Bert-Jan van Beijnum	University of Twente, The Netherlands
Riccardo Bellazzi	Universita di Pavia, Italy

Elia Biganzoli	Università degli Studi di Milano, Italy
Egon L. van den Broek	Human-Centered Computing Consultancy/ University of Twente/Radboud UMC Nijmegen, The Netherlands
Edward Brown	Memorial University of Newfoundland, Department of Computer Science, Canada
David Buckeridge	McGill University, Canada
Eric Campo	LAAS CNRS, France
James Cimino	NIH Clinical Center, USA
Miguel Coimbra	Faculdade de Ciências da Universidade do Porto, Portugal
Carlos Costa	Universidade de Aveiro, Portugal
Ricardo João Cruz-Correia	Faculdade de Medicina da Universidade do Porto, Portugal
Julian Dorado	University of Coruña, Spain
Stephan Dreiseitl	Upper Austria University of Applied Sciences at Hagenberg, Austria
Christoph M. Friedrich	University of Applied Science and Arts Dortmund, Germany
Ioannis Fudos	University of Ioannina, Greece
Angelo Gargantini	University of Bergamo, Italy
Alfredo Goñi	University of the Basque Country, Spain
David Greenhalgh	University of Strathclyde, UK
Andrew Hamilton-Wright	Mount Allison University, Canada
Jesse Hoey	University of Waterloo, Canada
Chun-Hsi Huang	University of Connecticut, USA
Ivan Evgeniev Ivanov	Technical University Sofia, Bulgaria
Edward Jones	Florida A&M University, USA
Stavros Karkanis	Technologial Educational Institute of Lamia, Greece
Anastasia Kastania	Athens University of Economics and Business, Greece
Andreas Kerren	Linnaeus University, Sweden
Georgios Kontaxakis	Universidad Politecnica de Madrid, Spain
Baoxin Li	Arizona State University, USA
Giuseppe Liotta	University of Perugia, Italy
Daniel Lizotte	University of Waterloo, Canada
Guillaume Lopez	The University of Tokyo, Japan
Martin Lopez-Nores	University of Vigo, Spain
Michele Luglio	University of Rome "Tor Vergata", Italy
Emilio Luque	University Autonoma of Barcelona (UAB), Spain
José Luis Martínez	Universidade Carlos III de Madrid, Spain
Paloma Martínez	Universidad Carlos III de Madrid, Spain
Alice Maynard	Future Inclusion, UK
Sally Mcclean	University of Ulster, UK
Gianluigi Me	Università degli Studi di Roma "tor Vergata", Italy

Rob van der Mei   CWI Amsterdam, The Netherlands
Gerrit Meixner   German Research Center for Artificial Intelligence
                         (DFKI), Germany
Mohyuddin Mohyuddin   King Abdullah International Medical Research
                         Center (KAIMRC), Saudi Arabia
Christo El Morr   York University, Canada
Sai Moturu   Massachusetts Institute of Technology, USA
Erik Van Mulligen   Erasmus Medical Center, The Netherlands
Radhakrishnan Nagarajan University of Kentucky, USA
Hammadi Nait-Charif   Bournemouth University, UK
Goran Nenadic   University of Manchester, UK
Shane O'Hanlon   Graduate Entry Medical School, University
                         of Limerick, Ireland
José Luis Oliveira   Universidade de Aveiro, Portugal
Rui Pedro Paiva   University of Coimbra, Portugal
Chaoyi Pang   The Australian e-Health Research Centre,
                         CSIRO, Australia
Danilo Pani   University of Cagliari, Italy
José J. Pazos-arias   University of Vigo, Spain
Carlos Eduardo Pereira   Federal University of Rio Grande Do Sul -
                         UFRGS, Brazil
Rosario Pugliese   Universita' di Firenze, Italy
Juha Puustjärvi   University of Helsinki, Finland
Arkalgud Ramaprasad   University of Miami, USA
Marcos Rodrigues   Sheffield Hallam University, UK
Valter Roesler   Federal University of Rio Grande do Sul, Brazil
Stefan G. Ruehm   UCLA (University of California, Los Angeles),
                         USA
George Sakellaropoulos   University of Patras, Greece
Ovidio Salvetti   National Research Council of Italy - CNR,
                         Italy
Akio Sashima   AIST, Japan
Bettina Schnor   Potsdam University, Germany
Kulwinder Singh   University of Calgary, Canada
Paolo Spagnoletti   LUISS Guido Carli University, Italy
Irena Spasic   University of Cardiff, UK
Deborah Stacey   University of Guelph, Canada
Jan Stage   Aalborg University, Denmark
Kåre Synnes   Luleå University of Technology, Sweden
Abdel-Rahman Tawil   University of East London, UK
Francesco Tiezzi   IMT - Institute for Advanced Studies Lucca, Italy
Alexey Tsymbal   Siemens AG, Germany
Aristides Vagelatos   CTI, Greece
Francisco Veredas   Universidad de Málaga, Spain
Justin Wan   University of Waterloo, Canada
Lixia Yao   GlaxoSmithKline, USA

Vera Yashina   Dorodnicyn Computing Centre of the Russian
                                Academy of Sciences, Russian Federation
Xiang Sean Zhou   Siemens Medical Solutions, USA
André Zúquete   IEETA/IT/Universidade de Aveiro, Portugal

HEALTHINF Auxiliary Reviewers

Michael Bales   Columbia University, USA
Eduardo C. Cabrera   UAB, Spain
Alessia Dessì   University of Cagliari, Italy
Atif Khan   University of Waterloo, Canada
Shehroz Khan   University of Waterloo, Canada
Massimiliano Masi   Universita di Firenze, Italy
Manel Taboada   University Autonoma of Barcelona, Spain

Invited Speakers

Pedro Gómez Vilda   Universidad Politécnica de Madrid, Spain
Christian Jutten   GIPSA-lab, France
Adam Kampff   Champalimaud Foundation, Portugal
Richard Reilly   Trinity College Dublin, Ireland
Vladimir Devyatkov   Bauman Moscow State Technical University,
                                Russian Federation
Pietro Liò   University of Cambridge, UK

Contents

Biomedical Electronics and Devices

A 2nd Order CMOS Sigma-Delta Modulator Design Suitable for Low-Power Biosensor Applications

Ryan Selby and Tom Chen[✉]

Department of Electrical and Computer Engineering,
Colorado State University, Fort Collins, CO, USA
chen@engr.colostate.edu

Abstract. Design of biosensor circuits requires low power and low noise. One of the important components in a sensor read channel is the analog-to-digital converter (ADC). In addition to the circuit design techniques to achieve low power consumption, the use of low supply voltage is an effective alternative to reduce the overall power consumption of a CMOS circuit. Sigma-delta modulators are well suited for the low noise requirement in biosensors due to its noise shaping property. This paper presents a low-voltage, low-power, 2nd order Sigma-Delta modulator for use in an electrochemical biosensor system. The modulator was designed using a commercial $0.18\,\mu\text{m}$ CMOS process with a supply voltage of $900\,\text{mV}$. With an input signal bandwidth of $1\,\text{kHz}$ it achieves a SNDR of $61.2\,\text{dB}$ using an over-sampling ratio of 500. Power dissipation is $165\,\mu\text{W}$ and it occupies $0.0225\,\text{mm}^2$ of silicon area.

Keywords: Sigma-Delta modulator · Biosensor

1 Introduction

Understanding cell-cell communication is one of the grand challenges in biological science. Integrated silicon biosensors have been proposed as a method for detecting chemical levels within biological systems [7,10]. One application for these biosensors is the study of cell movement and development within brain tissue. Using electrochemistry for signal detection, an array of micro-electrodes in such a biosensor would be able to measure chemical gradients across small pieces of living tissue. These gradients form a chemical image. Current methods for creating chemical images involve dye and marking compounds which could kill the tissue sample. The traditional methods of observing cell movement and changes are not ideal because temporal changes in chemical concentrations and cell movement cannot be observed reliably. An integrated silicon biosensor array

---

The work reported in this paper was partially funded by the National Science Foundation under the grant DGE-0841259 and by generous support from Texas Instruments.

M. Fernández-Chimeno et al. (Eds.): BIOSTEC 2013, CCIS 452, pp. 3–12, 2014.
DOI: 10.1007/978-3-662-44485-6_1

of hundreds or thousands of electrodes would solve these problems by measuring chemical levels in live tissue, allowing scientists to observe important changes over time.

There are many important chemical compounds that contribute to cell development and cell communication. Dopamine and serotonin are the familiar ones among these compounds. Other important chemical compounds include Nitric oxide (NO) which is believed to influence cell movement and growth in many parts of human body. This chemical is electrochemically active and can be measured using electrochemical methods such as cyclic voltammetry and amperometry. Sensors using electrochemical methods are well suited for silicon CMOS implementation. Traditional sensors based on electrochemistry use single set of three electrodes (working, reference, and counter electrodes) to detect the existance of a given chemical compound. However, understanding cell-cell communication requires an array of electrodes to capture a chemical image of a biosample. In order to create the chemical image in the tissue, many circuit components are needed. First, a potentiostat is used to induce a current in the tissue sample proportional to the concentration of a chemical compound of interest. NO is the chemical compound we study. A low power design of the integrated potentiostat with ultra-low offset voltage for improved accuracy can be found in [1,4]. The induced currents are extremely small, on the order of tens of picoamps, and need to be converted to voltage signals and amplified by way of a transimpedance amplifier. Finally, main amplifiers and an analog to digital converter (ADC) further amplify the signal and convert it to a digital value. A low power main-amplifier design using switch-capacitor architecture can be found in [9,11]. A low power bit-serial decimator design that interfaces with the modulator presented in this paper can be found in [8]. Figure 1 shows the electrode arrangement and potentiostat circuit proposed in [1], and Fig. 2 shows the top-level schematic of the biosensor system.

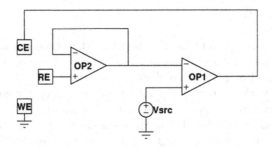

Fig. 1. Potentiostat schematic.

In order to create an accurate chemical image, the electrodes need to be spaced at distances comparable to the size of individual cells within the tissue sample. Electrode pitches of $10\,\mu\text{m}$ to $25\,\mu\text{m}$ are desirable. Ideally, each electrode

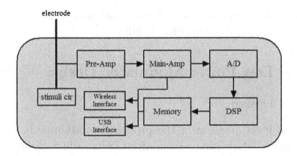

Fig. 2. The top-level schematic of the biosensor system.

should have its own dedicated circuitry; however with the extremely small electrode pitch this is not realistic. Regardless, minimizing the area of the circuitry
is important to reduce the number of shared electrodes per circuit block. Likewise, power consumption should be minimized since the biosensor will contain
many copies of the detection circuitry running in parallel. Power consumption
must also be limited to avoid heat build-up within the tissue, possibly causing
damage.

The chemical signals detected with the potentiostat typically have very low
bandwidths; nitric oxide signals do not normally exceed 1 kHz. While bandwidth
requirements are low, measurement accuracy is much more important. Noise
and signal distortion must be avoided to preserve the integrity of the small
signals inherent in bio-electric systems. Sigma-Delta ADCs are ideal for biosensor
applications because they inherently have higher resolution and lower bandwidth
than other ADC topologies.

Many different designs for low-power Sigma-Delta Modulators have been
presented. The designs in both [3,6] are based on 0.18 μm CMOS processes
with nominal 1.8 V power supplies. These designs use standard architectures
and standard techniques for reducing power. The design in [6] was intended for
audio applications, while [3] presents a design for biomedical applications such
as electo-cardiograms. The designs presented in [2,5] also use 0.18 μm CMOS
processes but they use supply voltages of 0.9 V and 0.8 V respectively to reduce
power. Besides lowering the supply voltage, these designs modify the standard
architecture to save power. Reference [2] shares a single op-amp between multiple
integrator stages and [5] is able to be used at different speeds depending on the
application to maximize efficiency. The design presented in this paper explores
the efficiency of using a reduced power supply to lower power consumption while
still using a standard, easy to implement architecture.

All of these designs are also fully differential architectures while the proposed design is single ended. The use of differential design has the advantage
of expanded output range; however, our application focuses on NO which has a
narrow activation range. Therefore, a single-ended design is chosen for reduced
silicon area. In switched-capacitor circuits where capacitors can occupy a large
percentage of the total area, this is critical. Section 2 of the paper will detail the

design of the modulator and its components, and Sect. 3 will cover the proposed physical layout and present simulation results.

2   Proposed Low Power Modulator Design

2.1   Top Level Design

The overall block level diagram of the proposed modulator is shown in Fig. 3. The design is implemented using a standard $0.18\,\mu\text{m}$ silicon process and a supply voltage of $900\,\text{mV}$. The overall topology is a single ended, second order, Sigma-Delta modulator. This topology uses two integrators, a comparator, and a 1 bit DAC to provide feedback. With accuracy and power consumption prime concerns in this design, the modulator is implemented using switched capacitors which both improve matching in the integration stages and allows for lower power consumption than a continuous time resistive implementation.

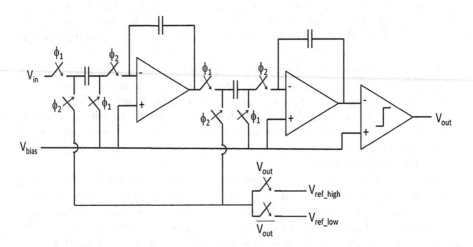

Fig. 3. Top level modulator schematic.

The bio-chemical signals the chip will be measuring have a maximum bandwidth of 1 kHz which makes a Sigma-Delta based ADC with a large oversampling ratio ideal for accurate measurements. The input is sampled at 1 MHz resulting in an oversampling ratio of 500. This clock frequency was chosen based on the performance requirements of the decimation filter presented in [3], and power consumption was reduced as much as possible using this sampling rate.

2.2   Modulator Op-amp Design

The schematic for the proposed op-amp is shown in Fig. 4. The nominal supply voltage used with this process is 1.8 V because the threshold voltages of

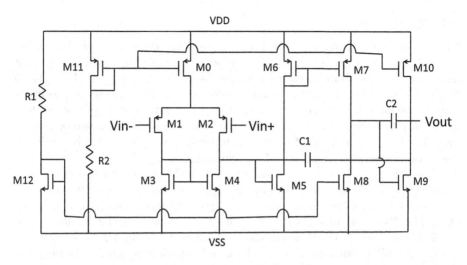

Fig. 4. Op-amp schematic.

the nFETs and pFETs available in this process are approximately 450 mV and −400 mV respectively. When designing with a 900 mV supply, cascoding transistors is not practical in most situations. Thus the op-amp uses multiple, simple, stages to achieve high gain.

The design features three stages comprised of a differential pair input followed by common source amplifiers and current mirrors. Adding extra stages to an op-amp inherently leads to stability problems, and thus two large compensation capacitors are needed to maintain a reasonable phase margin. Capacitors C1 and C2 in the schematic have values of 3.6 pF and 1.3 pF respectively. Because of the relatively large values of the compensation capacitors they are implemented on using poly/n-well capacitors. These capacitors are used because they are much more area efficient than standard metal/poly capacitors. The differential pair uses pFETs for the input to reduce noise, and the input/output common mode voltage is set to 175 mV. With the limited supply voltage, the maximum peak-to-peak output swing is 300 mV. Internal biasing voltages for nMOS and pMOS devices are generated using simple current mirrors with resistors generating the reference currents. The resistors are implemented as thin-film devices which are more area efficient and more accurate than other resistor technologies. Monte Carlo simulations have shown that these bias voltages do not vary widely enough to impact the functionality of the op-amp.

Small bias currents were used throughout the op-amp to reduce power consumption. All branches in the op-amp use 10 μA bias currents except the output stage which uses 30 μA to aid in stability. In total, the op-amp uses 70 μA of current and consumes 63 μW. The proposed design has a DC gain of 112 dB, 8.3 Hz −3 dB bandwidth, and a phase margin of 77 degrees.

2.3   Integrator

The proposed modulator uses two switched capacitor integrators which each use a 100 fF sampling capacitor and 4 pF feedback capacitor. The sampling capacitor uses bottom plate sampling to reduce error due to charge injection. The ratio of feedback capacitance to sampling capacitance is set to 40:1 to prevent the output of the first modulator from saturating. Using this ratio the output of the first modulator uses the entire 300 mV peak-to-peak output swing range of the op-amp. The capacitance ratio of the second integrator was set to 2:1, which allowed the second integrator to use its full output range as well.

While the capacitance ratio is determined by the output swing requirements of the op-amps, the absolute sizes of the capacitors are chosen based on a compromise between noise characteristics and physical size when implemented on chip. The 100 fF input capacitor was chosen because it is the smallest value possible which meets the thermal noise requirements of the ADC system.

The sampling capacitor uses a poly/metal capacitor which is extremely linear but also very large. The feedback capacitor, which is much larger, is implemented with a poly/n-well capacitor which has non-linear properties yet is very area efficient. Obviously, using different types of capacitors makes matching more difficult, but the savings in area is considerable compared to using two poly/metal capacitors. Poly/n-well capacitors are roughly nine times smaller than poly/metal capacitors.

2.4   Comparator Design

The schematic for the comparator is shown in Fig. 5. The comparator uses the same differential pair as the op-amp for an input stage, followed by a high gain common source output stage. After amplifying the input signal the digital output is latched by a simple master-slave D flip-flop. The comparator uses 40 μA of DC current and consumes 36 μW. The output of the comparator settles to its final value within 10 ns of the latch closing.

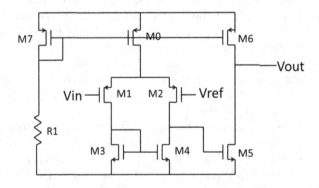

Fig. 5. Comparator schematic.

3 Physical Implementation and Measurement Results

Overall simulation results for the modulator are shown in Table 1 while the power spectrum density plot is shown in Fig. 6. This plot shows regular noise spikes at the 3rd harmonic and above of the input tone which are easily filtered out by the low-pass filter characteristics of the decimation filter. The modulator achieves a SNDR of 61.2 dB resulting in a 10 bit effective resolution while consuming 165 μW.

Table 1. Modulator performance.

Parameter	Results
SNDR	61.2 dB
Effective number of bits	10
Power consumption	165 μW

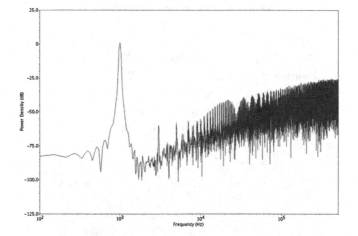

Fig. 6. Modulator output spectrum.

The physical layout of the modulator covers an area of 0.0225 mm². Many measures were taken to ensure that transistor mismatch and signal distortion were minimized throughout the layout. All matched transistors in the op-amps and comparator were implemented using common centroid layout techniques, and all analog transistor blocks are surrounded by guard rings. The 4 phase, non-overlapping clock generation circuit is also isolated via a guard ring. Finally, wires carrying important digital and analog signals throughout the chip are isolated with grounded lines to minimize interference and crosstalk.

The proposed modulator was fabricated and its functionality verified on silicon. An oscilloscope screen capture in Fig. 7 shows the modulator performing as

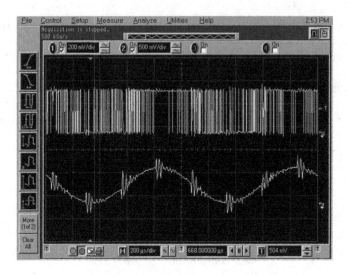

Fig. 7. Silicon test results.

Table 2. Performance comparison.

Design	SNDR	Proposed Auto-zeroed
Lei et al. [6]	68.85 dB	800 μW
Jasutkar et al. [3]	68 dB	400 μW
Goes et al. [2]	80.1 dB	200 μW
Lee and Cheng [5]	50 dB	180 μW
This Work	**61.2 dB**	**165 μW**

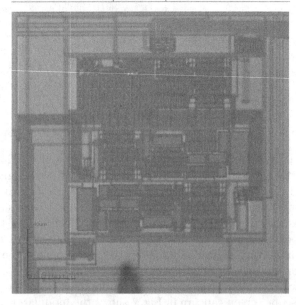

Fig. 8. Die photo.

expected with a 1 kHz sine wave input. A die photo of the test chip is shown in
Fig. 8.

4 Conclusions

A low power, 2nd order, Sigma-Delta modulator for use in an integrated biosensor system was presented in this paper. The proposed modulator uses a 900 mV
supply voltage to reduce power consumption and extend the life of battery
powered biosensors. The modulator achieves a signal-to-noise distortion ratio of
61.2 dB and consumes 165 μW. Compared to existing low-power designs the proposed modulator performs very well balancing SNDR and power consumption.
Table 2 compares the performance of the proposed modulator against existing
designs. The proposed design provides an overall better tradeoff between noise
and power consumption.

References

1. Duwe, M., Chen, T.: Low power integrated potentiostat design for μelectrodes
   with improved accuracy. In: 2011 IEEE 54th International Midwest Symposium
   on Circuits and Systems (MWSCAS), pp. 1–4, 7–10 August 2011
2. Goes, J., Vaz, B., Monteiro, R., Paulino, N.: A 0.9-V Delta-Sigma modulator with
   80 dB SNDR and 83 dB DR using a single-phase technique. In: IEEE International Solid-State Circuits Conference 2006 Digest of Technical Papers, pp. 191–200
   (2006)
3. Jasutkar, R.W., Bajaj, P.R., Deshmukh, A.Y.: GA based low power sigma delta
   modulator for biomedical applications. In: Recent Advances in Intelligent Computational Systems, 2011 IEEE, pp. 772–776, 22–24 September 2011
4. Kern, T., Chen, T.: A low-power, offset-corrected potentiostat for chemical imaging
   applications, In: Proceedings of LASCAS 2013, Cusco, Peru, 27 Febuary–1 March
   2013
5. Lee, S., Cheng, C.: A low-voltage and low-power adaptive switched current sigma-
   delta ADC for bio-acquisition microsystems. Circ. Syst. I 53(12), 2628–2636 (2006)
6. Lei, Z., Xian-li, Z., Xing-hua, W., Ruo-Yuan, Q.: Two-order low-power sigma-
   delta modulator with SC techniques. In: 2010 IEEE International Conference on
   Semiconductor Electronics, pp. 96–99, 28–30 June 2010
7. Pettine, W., Jibson, M., Chen, T., Tobet, S., Henry, C.: Characterization of novel
   microelectrode geometries for detection of neurotransmitters. IEEE Sens. J. 12(5),
   1187–1192 (2012)
8. Scholfield, K., Chen, T.: Low power decimator design using bit-serial architecture
   for biomedical applications. In: Proceedings of DATICS-IMECS 2012, Hong Kong,
   14–16 March 2012
9. Selby, R., Kern, T., Wilson, W., Chen, T.: A 0.18 μm CMOS switched-capacitor
   amplifier using current-starving inverter based op-amp for low-power biosensor
   application. In: Proceedings of LASCAS 2013, Cusco, Peru, 27 February–1 March
   2013

10. Tobet, S., Eitel, C., Dandy, D., Bartels, R., Reynolds, M., Chen, T., Henry, C.: Slices for devices: organotypic slice cultures for in vitro sensor analyses. In: 2013 BMES Annual Meeting, Seattle, WA, 25–28 September 2013
11. Wilson, W., Selby, R., Chen, T.: A current-starved inverter-based differential amplifier design for ultra-low power applications. In: Proceedings of LASCAS 2013, Cusco, Peru, 27 February–1 March 2013

Monitoring Drivers' Ventilation Using an Electrical Bioimpedance System: Tests in a Controlled Environment

Raúl Macías$^{(\boxtimes)}$, Miguel Ángel García, Juan Ramos,
Ramon Bragós, and Mireya Fernández-Chimeno

Department of Electronic Engineering,
Universitat Politècnica de Catalunya, Campus Nord, Edifici C-4,
08034 Barcelona, Spain
{raul.macias,miquel.angel.garcia,juan.jose.ramos,
ramon.bragos,mireia.fernandez}@upc.edu

Abstract. As improving road safety is one of the first aims in the automotive world, several new techniques and methods are being researched in recent years. Some of them consist of monitoring the driver behavior to detect non-appropriate states for driving, *e.g. drowsy driving or drunk driving*. Usually, the appearance of these non-appropriate states is related to changes in several physiological parameters. This work is divided into two main parts. The first one presents an electrical bioimpedance system capable of monitoring the ventilation using textile electrodes. Apart from describing the system, in this part some tests done in a controlled environment are also shown. In the second part of this paper, an enhancement of the system is described and checked using a patient simulator.

Keywords: Bioimpedance · Ventilation · Textile electrodes · Textrodes · Automotive

1 Introduction

According to [1], most of traffic crashes occur during the appearance of non-appropriate states for driving, *e.g. drowsy driving or drunk driving*. Therefore, apart from improving the vehicle performance, automotive companies are also focused on monitoring the driver behavior. To achieve that, several systems are being researched. These systems can be mainly classified on three types. The first type is based on driving performances *i.e. unintended lane departures, steerings and brakes*. The second one is based on camera systems that detect the percentage of eye closure (PERCLOS), head movements and blinkings, as described in [2]. Finally, the third type is based on recording biomedical signals. In [3], signals from electroencephalography (EEG), electrocardiography (ECG) and heart rate variability (HRV) are used. On the other hand, electrooculography (EOG) and ventilation are used in [4] and in [5], respectively.

© Springer-Verlag Berlin Heidelberg 2014
M. Fernández-Chimeno et al. (Eds.): BIOSTEC 2013, CCIS 452, pp. 13–25, 2014.
DOI: 10.1007/978-3-662-44485-6_2

Focusing on the third type, regardless of the physiological parameter to be measured, any system should fulfill three requirements at least. Firstly, the system must be capable of recording signals in a very noisy environment. In a vehicle, there are not only artifacts produced by the car engine but also artifacts caused by other reasons like body motion or the state of the roads. As for the second feature, a long-term monitoring system is required because the appearance of non-appropriate states while driving is a slow process. Moreover, the system should also be non-invasive and non-annoying to allow driving as comfortable as possible. Therefore, the use of hospital devices is not recommended and the design of new biodevices is required.

In the first part of this paper, a new biodevice suitable for automotive applications is shown. This device consists of an electrical bioimpedance (EBI) system capable of monitoring the ventilation, and also the heart rate, using textile electrodes placed on the steering wheel and also on the car seat. In addition, to check the device, two kinds of tests are done. In the first one, the signal obtained by the EBI system is compared to one which has been recorded by a commercial thoracic band. In the second group, the system is checked according to several parameters such as the electrode configuration, the frequency of the injected signal or the clothing thickness.

In addition, there is also a second part in this work where an enhancement of the EBI system is proposed. Finally, this improved system is also checked using a patient simulator.

2   System Description

As mentioned above, the biodevice is based on an EBI instrumentation system. The proposed system and also its enhancement are shown in Fig. 1. It is worth mentioning that both systems are designed following the guidelines described in [6]. Comparing both block diagrams, three main blocks are observed in both systems: Signal Generator (GEN), Analog Front-End (AFE), and Demodulator & Acquisition (DEM). In addition to these main blocks, there are also several textile electrodes placed on the steering wheel and the car seat.

The main difference between both systems is in the AFE block. The enhanced system adds a multiplexing-demultiplexing stage capable of selecting the best electrode configuration to measure in order to allow measuring when the driver is not in contact with all the electrodes, e.g. with only one hand in the steering wheel.

2.1   EBI System

Briefly, the EBI system works as follows. The voltage-driven signal generated by the generator is converted into a current-driven signal. Then, this current-driven signal is injected into the driver's body using a pair of textile electrodes. Using the same, or other, pair of textile electrodes, the voltage drop due to the injected current is measured. Finally, in the demodulation and acquisition

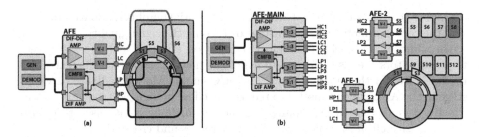

Fig. 1. Block diagrams of the EBI device (left) and the enhanced one (right). In both cases, whereas electrodes connected to (LC) and (HC) are the driving electrodes, the ones connected to (LP) and (HP) are the sensing electrodes.

block, the value of the impedance is obtained. It is worth mentioning that changes in the value of this impedance are related to changes in the gas-fluid ratio during the inhalation-exhalation process, i.e. the ventilation.

Signal Generator. Although it can take several forms, e.g. from a simple linear oscillator or digital clock to a Direct Digital Synthesizer (DDS) able to produce sinusoidal waveforms or arbitrary waveforms in a wide range of frequencies and amplitudes, [6], in the present system, two different kinds of signal generator are used. In the case where the overall system is checked, the excitation signal is an adjustable amplitude single tone waveform of 62.5 kHz generated by a microcontroller, PIC18F1320. In the other case, if only the proper behavior of the AFE is checked, a sine wave with an adjustable frequency is generated using a PXI solution by National Instruments.

Analog Front-End. Once the signal is generated, this is sent to the AFE. Basically, the AFE consists of two main stages: the current injection stage and the voltage sensing one. In the first stage, the excitation signal is sent to a differential-differential amplifier, AD8138, in order to have a differential excitation signal. However, instead of appling directly the voltage of these two outputs to the driving electrodes, each one is used as an input of a voltage-current (V-I) converter based on a second-generation current conveyor (CCII). Thus, the V-I converter acts as a voltage-controlled current source (VCCS), [7]. In this way, using current driving instead of voltage driving, a current limiting mechanism is achieved and the possible nonlinearities are also reduced. So the guidelines of safety risks provided by the IEC-60601-1 standard can be fulfilled. Furthermore, using differential-floating excitation helps reducing the effect of electrode impedance mismatch effect.

In the voltage sensing stage, a differential to single ended voltage conversion is done by a wideband differential amplifier. However, before this conversion, the voltage difference between the pair of sensing electrodes is measured by a pair of high-impedance buffers. These buffers are used because of their input impedances are higher than the input impedance of the differential amplifier, allowing to measure the common-mode voltage without disturbing the signal quality.

In addition, as much in the current injection stage as in the sensing one, filter capacitors are also required to avoid the flow of the Direct-Current (DC) to the body. Furthermore, there is also a Common-Mode Feedback (CMFB) stage in the AFE which is useful to reduce the effect of high electrode mismatch. Finally, it is also worth mentioning that active shielding is also used to reduce the capacitive effect of coaxial cables, interferences and crosstalk.

Textile Electrodes. As the electrodes should be as non-invasive and non-annoying as possible for the driver, using standard metal electrodes seems to be not the better option. In addition, as cited in [8], during long-term monitoring, the hidrogel used with this kind of electrodes can cause irritation and allergy problems.

So, in this system instead of using standard metal electrodes, electrodes made of textiles, also called textrodes, are chosen. In this way, not only irritation and allergy problems are solved but also a higher comfort for the driver is achieved. However, the main drawback of the textrodes is that the electrode impedance shows a strong capacitive behavior, [9]. In addition, as the textrode is not directly in contact the skin, this capacitive behavior depends on the exerted pressure and factors related to the clothing of the driver like material, thickness or number of layers, [10]. Furthermore, in order to cope with the impedance of the electrodes, to use a different pair of electrodes to the driving and the sensing stage is required, i.e. to use the four-wire technique, [11].

Demodulation & Acquisition. Two different demodulation techniques are used. In the case where the overall system is checked, a switching demodulator is used. This switching demodulator is based on a switch controlled by the same square signal generated by the microcontroller. Furthermore, after the switching demodulator, the signal is driven to a third-order Sallen-Key low-pass filter (LPF). Then using this output signal from the LPF, the measured voltage is acquired. In addition, by using a high-pass filter (HPF) and a basic circuitry, the relative variations of the measured voltage are also amplified and acquired. These voltage variations should be amplified before recording because of their low amplitude and also the poor accuracy that the 10-bit Analog-to-Digital Converter (ADC) of the microcontroller can provide. Later the acquired data are sent from the microcontroller to a computer by a mini USB-Serial UART development module. Finally, the impedance value is estimated using a LabVIEW application.

The second demodulation technique is an In-phase & Quadrature (IQ) demodulation. Using the PXI solution, the real and imaginary part of the measured signal are obtained. Later, the magnitude and the phase of the impedance are estimated using a LabVIEW application.

2.2 Enhanced System

As mentioned previously, the enhanced system adds a multiplexing-demultiplexing stage in the AFE block. In this way, instead of fixing the electrode configuration during the monitoring, the system is capable of changing automatically to the best electrode configuration. To achieve this, a pair of 1-to-8 multiplexers (MUXs) and 8-to-1 demultiplexers (DEMUXs) are added in the sensing and the injection stages, respectively. Thus, in the injection stage, previously to the voltage-current conversion, the pair of driving electrodes is chosen. In the same way, in the sensing stage, the two inputs to the wideband differential amplifier are chosen by the pair of 8-to-1 MUXs which are connected to the high-impedance buffers. Note that in the enhanced system, the AFE block is divided into several boards instead of being only one board. In that way, as the current-drivers and voltage buffers are closer to the textile electrodes, i.e. shorter wires, the system is stronger in front of parasitic effects.

It is also worth mentioning that an integrated circuit (IC), AD5933 by Analog Devices, is chosen as solution in the generation and demodulation blocks. Basically, AD5933 is a low cost, system-on-chip high precision impedance converter system solution that combines an on-board frequency generator with a 12-bit, 1 MSPS, analog-to-digital converter (ADC). Furthermore, the signal sampled by the on-board ADC and a discrete Fourier transform (DFT) is processed by an on-board DSP, [12]. Finally, the real and imaginary values of the estimated impedance are sent to a LabVIEW application using the I2C protocol.

3 Methods

A set of tests is done to check the proper functioning of both systems. It should be reminded that any test is done in a controlled environment where no artifacts due to engine vibrations or the state of the road are present.

3.1 EBI System

To check that the EBI system works properly, several measurements are carried out. These measurements can be classified into three groups according to:

- Comparison to a reference signal
- Configuration of electrodes, i.e., the placement of the driving and sensing electrodes in the car seat and steering wheel
- Influence of the thickness of clothing

Comparison to a Reference Signal. In this group, several subjects are monitored by the designed EBI system and also by a commercial device made by BIOPAC Systems. The commercial device acquires the ventilation signal at a sampling frequency of 1000 samples per second using a piezoresistive thoracic band. Then, this signal is used as reference signal to verify the correct operation of the designed device.

About the EBI system, in this case, only the proper behavior of the AFE is checked, i.e. the PXI solution is used in the GEN and DEM blocks. Thus, a single frequency sine wave of 300 kHz is generated as excitation signal and the measured signal is sampled at 25 samples per second in the DEM block. The reasons to apply a frequency of 300 kHz are mainly two. First, using the PXI solution, the hardware limitation is less strong as the one imposed by the PIC solution. Second, the higher the frequency, the better response of the system is achieved because of the capacitive behavior of the textile electrodes.

It is worth mentioning that the system is considered to work properly if the inhalation-exhalation ratio is the same that the obtained by the commercial device.

Configuration of Electrodes. In the second group of measurements, the overall system, i.e. a single tone at 62.5 kHz as excitation signal and switching demodulation, is checked according three differents configurations according to the placement of electrodes. It is worth mentioning that the protocol done in all measurements consists of a two-minute monitoring and, around the last 30 s, five deep breathing are taken.

The first one is the steering wheel-steering wheel configuration. In this case, whereas a driving electrode and a sensing electrode are in contact with the right hand of the driver, the other pair of driving-sensing electrodes are in contact with the left hand. The second configuration is the steering wheel-back seat configuration. In this, whereas a driving and a sensing electrode remain in the steering wheel, the other driving and sensing electrode are moved to the upper half of the back seat. The last configuration tested is the back seat-back seat configuration. Here, whereas a pair of driving-sensing electrodes are in the left electrode of the back seat, the other pair is in the right electrode of the back seat. Note, in the back seat-back seat configuration, instead of using the 4-wire technique, the 2-wire technique is carried out because of both textile electrodes on the back seat act as driving and sensing.

Influence of the Thickness of Clothing. Using the same protocol than before, i.e. two-minute monitoring and at the end five deep breathing, and fixing the electrode configuration to the steering wheel-back seat configuration, the ventilation is monitored under certain situations according to the clothes that the subject is wearing. In the best situation tested, the subject is only wearing a thin T-shirt. On the other hand, in worse situations, the subject is wearing over the same thin T-shirt either a thin jacket or a thick sweater.

It should be pointed that as the capacitive behavior of the textile electrodes is strong in these situations, the PXI solution is chosen to be able to use excitation signals with a higher frequency than 62.5 kHz.

3.2 Enhanced System

Due to the enhanced system is only in the first stages of development, the tests are done using a patient simulator, Fluke PS420. Using this device, a rate of a 15 breathings per minute with 3-ohm impedance variations over a 500-ohm baseline is chosen. Under these conditions, the patient simulator is connected not only to the four electrodes placed on the steering wheel but also to the four electrodes placed on the upper side of the back seat. Therefore, during 4 min and 30 s, the ventilation is recorded and, at any arbitrary time, an electrode is disconnected. Furthermore, two 12-second apneas are also simulated at the middle and at the end of the test.

It is worth mentioning that a priority table is allocated to the system. Thus, the steering wheel-steering wheel configuration and the back seat-back seat configuration are the most and the least priority, respectively.

4   Results

As in previous section, the results are discussed based on the two systems.

4.1   EBI System

Comparison to a Reference Signal. As mentioned previously, the signal acquired by the thoracic band acts as a reference to check the signal from the designed biodevice. Thus, the biodevice works properly if the measured signal fits to the reference, i.e. for the same period of time, the exhalation-inhalation ratio is the same in both signals.

For each volunteer, two different configurations are tested. In the upper graphs of the Figs. 2, 3 and 4, as a driving electrode and a sensing electrode are in contact to the left hand, the second driving and the second sensing electrode are placed in the right and left side of the back seat, respectively. On the other hand, in the bottom graphs whereas the electrodes on the back seat remain at the same point, both electrodes on the steering wheel are moved to the right side.

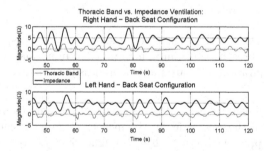

Fig. 2. Comparison between the Thoracic Band and the EBI system for the first volunteer. (Top) Configuration where both driving electrodes are in the right side of the body. (Bottom) Configuration where the driving electrode of the steering wheel is in the left hand and the other driving electrode is in the right side of the back. In both plots, the upper line is related to the bioimpedance device and the bottom one comes from the thoracic band.

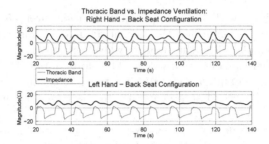

Fig. 3. Comparison between the Thoracic Band and the EBI system for the second volunteer. (Top) Both driving electrodes are in the right side of the body. (Bottom) The driving electrode of the steering wheel is in the left the other driving electrode is in the right side of the back.

Note that for all cases except one (bottom graph in Fig. 4), both signals, from the bioimpedance device and from the thoracic band, match up. The special case can be due to the lack of contact between any textrode and the volunteer.

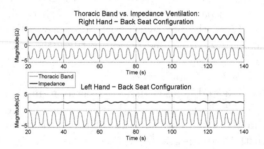

Fig. 4. Comparison between the Thoracic Band and the EBI system for the third volunteer. (Top) Both driving electrodes are in the right side of the body. (Bottom) The driving electrode of the steering wheel is in the left the other driving electrode is in the right side of the back.

Furthermore, whereas the Figs. 2 and 4 show a normal respiration rate, i.e. between 12 and 20 breaths per minute in adults and in normal conditions, in the Fig. 3 a slower respiration rate, around 6–7 breaths per minute, can be observed.

Configuration of Electrodes. In this group of measurements, the influence of the placement of the electrodes in the measured signal is checked. Then, as mentioned in the previous section, three configurations are tested.

As shown in Fig. 5, three signals are plotted. The middle one is the raw signal measured by the EBI system and without processing. Note that this signal is based on two components: a low-frequency component, between 0.1 Hz and 0.3 Hz, and a high-frequency component, over 1 Hz. Furthermore, whereas the lower component is related to ventilation, the higher one is related to the heart rate.

Checking the Figs. 5, 6, 7 and 8, a first issue can be observed. As in Figs. 5 and 7, the high frequency component is noticed easily, in Fig. 8 this component is insignificant. Therefore, it seems to be in contact directly to one hand, at least, is required to obtain the component related to the heart rate.

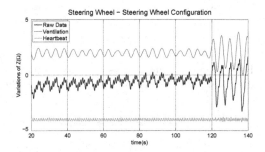

Fig. 5. Steering Wheel - Steering Wheel Configuration. (Middle) The raw data obtained from the bioimpedance device. (Top) Signal related to the ventilation. (Bottom) Signal related to the heart rate.

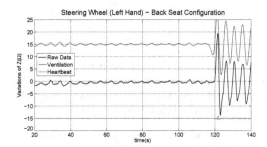

Fig. 6. Steering Wheel - Back Seat Configuration with a driving electrode in the left hand and the other in the right side of the back. (Middle) The raw data. (Top) The ventilation signal. (Bottom) The signal related to the heart rate.

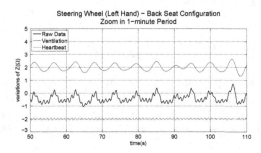

Fig. 7. Zoom in to the Fig. 6 in a one-minute period. (Middle) The raw data. (Top) The ventilation signal. (Bottom) The signal related to the heart rate.

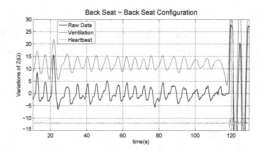

Fig. 8. Back Seat - Back Seat Configuration using a 2-wire measurement technique. (Middle) The raw data. (Top) The ventilation signal. (Bottom) The signal related to the heart rate.

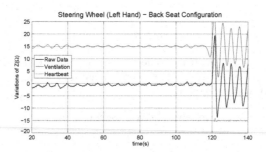

Fig. 9. Steering Wheel - Back Seat Configuration with the driving electrodes in the left hand and in the right side of the back, respectively. The subject is wearing a thin T-shirt made of cotton (100 %). (Middle) The raw data. (Top) The ventilation signal. (Bottom) The signal related to the heart rate.

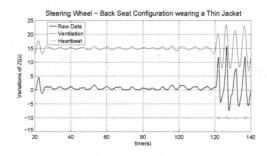

Fig. 10. Steering Wheel - Back Seat Configuration with the driving electrodes in the left hand and in the right side of the back, respectively. The subject is wearing a thin jacket made of polyester (100 %) over a thin T-shirt made of cotton (100 %). (Middle) The raw data. (Top) The ventilation signal. (Bottom) The signal related to the heart rate.

Influence of the Thickness of Clothing. In the last tests, dependencies on clothing are checked. In Fig. 9, applying the previous protocol and the steering

Fig. 11. Steering Wheel - Back Seat Configuration with the driving electrodes in the left hand and in the right side of the back, respectively. The subject is wearing a thick sweater made of woolen (33 %), polyester (27 %), acrylic (27 %) and polyurethane (13 %) over a thin T-shirt made of cotton (100 %). (Middle) The raw data. (Top) The ventilation signal. (Bottom) The signal related to the heart rate.

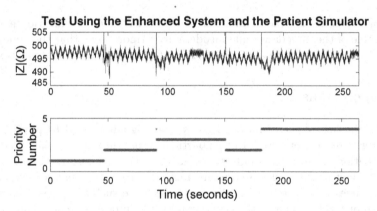

Fig. 12. Test of the enhanced system using a patient simulator. (Top) Ventilation signal acquired without filtering. (Bottom) Priority number of the electrode configuration used during monitoring.

wheel - back seat configuration, a subject is monitored wearing a thin 100 % cotton T-shirt. On the other hand, in Figs. 10 and 11, the subject is wearing over the same T-shirt a thin 100 % polyester jacket and a thick sweater, respectively.

Note that the respiration rate is different in the three figures. Whereas a normal rate of 11 breaths per minute can be observed in Fig. 9, in Figs. 10 and 11, the respiration rate is lower. There seems to be a correlation between clothing and the measured signal. As thicker the clothing, the measured signal is less similar to the reference ventilation signal, i.e. the signal obtained by the thoracic band. Therefore, depending on the clothings, the EBI system could not work properly because of some inhalations or exhalations cannot be monitored.

Table 1. Priority order of the electrode configurations.

Priority	Driving Electrodes	Sensing Electrodes
1	S1–S3	S2–S4
2	S1–S8	S2–S7
3	S3–S5	S4–S6
4	S5–S8	S6–S7

4.2   Enhanced System

Figure 12 shows two plots. Whereas the upper plot is related to the ventilation signal acquired using the enhanced system and the patient simulator, the lower plot shows the priority number according to the electrode configuration described in Table 1.

As observed in both plots, any change of the electrode configuration matches a high value of the impedance magnitude. Furthermore, note that the apneas are done in the second minute and at the end of the test as previously mentioned.

5   Conclusions

As shown in this paper, using an EBI system, signals related to physiological parameters can be monitored. In this particular case, not only signals related to the ventilation are measured but also signals related to the heart rate. However, due to the fact that the use of standard metal electrodes are not recommended, new challenges related to the textile electrodes arise such as the high electrode mismatch or the behavior of the clothing-textrode interface. In addition, testing the bioimpedance system in a real environment is also required. But, in any case, the tests in a controlled environment show a proper operation of the system. This system is capable of monitoring the ventilation signal just like a thoracic band. Furthermore, the system is also able to acquire the signal related to the heart rate.

In addition, although the enhanced system is only in the first stages of development, the automatic selection of the best electrode configuration during monitoring seems to be a solution for the lack of contact between electrodes and the driver. Therefore, this system seems to be a further step to obtain a non-annoying non-invasive biodevice capable of monitoring some physiological parameters in a vehicle environment.

Acknowledgements. The present work has been partially funded by the Spanish Ministerio de Ciencia e Innovación. Proyecto IPT-2011-0833-900000. Healthy Life style and Drowsiness Prevention-HEALING DROP.

References

1. Anund, A., Kecklund, G., Peters, B., Åkerstedt, T.: Driver sleepiness and individual differences in preferences for countermeasures. J. Sleep. Res. **17**, 16–22 (2008)
2. García, I., Bronte, S., Bergasa, L.M., Hernandez, N., Delgado, B., Sevillano, M.: Vision-based drowsiness detector for a realistic driving simulator. In: 13th International IEEE Conference on Intelligent Transportation Systems, Portugal, pp. 887–894 (2010)
3. Michail, E., Kokonozi, A., Chouvarda, I., Maglaveras, N.: EEG and HRV markers of sleepiness and loss of control during car driving. In: Proceedings of the 30th Annual International Conference of the IEEE EMBS, Vancouver, pp. 2566–2569 (2008)
4. Lal, S.K.L., Craig, A.: A critical review of the psychophysiology of driver fatigue. Biol. Psychol. **55**(3), 173–194 (2001)
5. Folke, M., Cernerud, L., Ekström, M., Hök, B.: Critical review of non-invasive respiratory monitoring in medical care. Med. Biol. Eng. Comput. **41**(4), 377–383 (2003)
6. Riu, P., Bragos, R., Rosell, J.: Instrumentation for Bio-Impedance Measurements. In: An Anthology of Developments in Clinical Engineering and Bioimpedance, Norway, pp. 225–250 (2009)
7. Bragós, R., Rosell, J., Riu, P.: A wide-band AC-coupled current source for electrical impedance tomography. Physiol. Meas. **15**, A91–A99 (1994)
8. Wheelwright, C.D.: Physiological Sensors for Use in Project Mercury. In: NASA Technical Note D-1082. Washington, D.C. (1962). http://catalog.hathitrust.org/Record/011447015. Accessed 12th April 2012
9. Beckmann, L., Neuhaus, C., Medrano, G., Jungbecker, N., Walter, M., Gries, T., Leonhardt, S.: Characterization of textile electrodes and conductors using standardized measurement setups. Physiol. Meas. **31**(2), 233–247 (2010)
10. Macías, R., Fernández, M., Bragós, R.: Textile electrode characterization: dependencies in the skin-clothing-electrode interface. J. Phys.: Conf. Ser. **434**, 1–4 (2013)
11. Schwan, H.P., Ferris, C.D.: Four-electrode null techniques for impedance measurement with high resolution. Rev. Sci. Instrum. **39**(4), 481–485 (1968)
12. Analog Devices AD5933. http://www.analog.com/en/rfif-components/direct-digital-synthesis-dds/ad5933/products/product.html

Opened-Ring Electrode Array for Enhanced Non-invasive Monitoring of Bioelectrical Signals: Application to Surface EEnG Recording

J. Garcia-Casado[1(✉)], V. Zena[1], J.J. Perez[1], G. Prats-Boluda[1], Y. Ye-Lin[1], and E. Garcia-Breijo[2]

[1] Grupo de Bioelectrónica, CI2B-UPV, Camino de Vera SN,
46022 Valencia, Spain
{jgarciac,vfzena,jjperez,gprats,yiye}@gbio.i3bh.es
[2] Centro de Reconocimiento Molecular y Desarrollo Tecnológico,
Ud. Mixta UPV-UV, Camino de Vera SN, 46022 Valencia, Spain
egarciab@eln.upv.es

Abstract. The estimation of Laplacian potential on the body surface obtained by means of concentric ring electrodes can provide bioelectrical signals with better spatial resolution and less affected by bioelectrical interferences than monopolar and bipolar recordings with conventional disc electrodes. In this paper an array of flexible concentric opened-ring electrodes for non-invasive bioelectrical activity recordings is presented. A preconditioning circuit module is directly connected to the electrode array to perform a first stage of filtering and amplification. A computer model reveals differences minor than 1 % between the sensitivity of the proposed opened-ring electrode vs a closed-ring electrode. Simultaneous recordings of intestinal myoelectrical activity proved that signals from the developed array of electrodes presented lower electrocardiographic and respiratory interference than conventional bipolar recordings with disc electrodes. The small bowel's slow wave myoelectrical activity can be identified more easily in the recordings with the presented ringed-electrodes.

Keywords: Ring electrode · Laplacian recording · Non-invasive myoelecrical recording · Electroenterogram

1 Introduction

1.1 Bioelectrical Laplacian Recordings

Surface recordings of bioelectrical signals are usually recorded by means of disc electrodes in bipolar or unipolar configuration. In the first method the potential difference between a pair of electrodes is measured. In the latter method the potential of each electrode is compared either to a neutral electrode or to the average of several electrodes. One drawback of using conventional disc electrodes in bioelectrical surface recordings is their poor spatial resolution which is mainly caused by the blurring effect of the different conductivities of the volume conductor [1]. In this respect, Laplacian has been

© Springer-Verlag Berlin Heidelberg 2014
M. Fernández-Chimeno et al. (Eds.): BIOSTEC 2013, CCIS 452, pp. 26–40, 2014.
DOI: 10.1007/978-3-662-44485-6_3

shown to reduce the smoothing effects caused by the volume conductor and to increase the spatial resolution in localizing and differentiating multiple dipole sources [2, 3].

There are different approaches to estimate the Laplacian potential on the body surface. The first ones to be used were discretization techniques like the one introduced by Hjorth as early as in 1975 [4]. In that study, the Laplacian of the EEG signal was estimated as the difference between the average potential of four disc electrodes in the form of a cross and the potential of a fifth disc electrode placed in the center of the cross. In the late 80 s, analytic solutions to estimate the Laplacian of the surface potential were proposed in order to reduce discretization errors [5]. These are complex discrete computational techniques, generally not suitable for real-time applications. Nevertheless, Laplacian potential can also be directly estimated by means of concentric ring electrodes in tripolar, bipolar or tripolar in bipolar configuration (TCB, where the outer ring and the center disc were electrically shorted) [3, 6, 7].

Ring electrodes have already been used to estimate the Laplacian potential of bioelectrical signals such as the electrocardiogram (ECG), electroencephalogram (EEG), electroenterogram (EEnG) and the electrohysterogram (EHG), so as to increase the spatial resolution of and the signal quality of conventional surface potential recordings [6–9]. These electrodes are usually active since the signals sensed by concentric ring electrodes in Laplacian configuration are weaker than the ones obtained by conventional monopolar or bipolar recordings, and the output impedance is bigger. Nevertheless, the ringed-electrodes used in these studies were developed on rigid substrates like printed-circuit boards, what can provoke a poor skin-to-electrode contact and discomfort to the patient.

Therefore, the aim of this study is to develop concentric ring electrodes on a flexible substrate to join the advantages of Laplacian recordings with the comfort and better adaptation to the body surface curvature of conventional disposable disc electrodes.

1.2 Intestinal Myoelectrical Activity

The study of intestinal motility is an outstanding field in gastroenterology due to the fact that abnormal motility patterns are related with several intestinal pathologies [10]. This is the case in irritable bowel syndrome, intestinal obstruction, paralytic ileus, and bowel ischemia. The main problem in monitoring intestinal activity is the difficult anatomic access, hence most methods of studying this activity are considered to be invasive. One possible solution would be the recording of intestinal myoelectrical activity on abdominal surface. This signal is named Electroenterogram (EEnG) and it is composed of two waves: slow waves (SW) and spike burst (SB). The former are periodical, omnipresent electrical potentials that regulate the maximum rate of intestinal muscle contractions. The latter are fast action potentials which are located in the plateau of the SW. They are only present when contractions appear. Whereas SW are related to the frequency and propagation velocity of the contractions [11], SB determine the presence and the intensity of the contractions. The frequency of the SW changes along the small intestine with a stepwise gradient from about 12 cpm at duodenum to 8 cpm at ileum [12].

There are few studies about abdominal surface recordings to identify the EEnG in humans [9, 13–15]. The main reason is that human EEnG is a very weak signal, which is severely attenuated especially in the SB frequency range, because of the insulation effects of the abdominal layers and spatial filtering [1, 16]. Surface EEnG is also very sensitive to physiological interferences such as ECG and respiration, being difficult to identify the SW component of the EEnG by visual inspection of abdominal surface recordings. The ECG spectral frequency range overlaps the SB frequency range, therefore it is necessary to eliminate it from abdominal recordings to identify the SB component of the EEnG [17]. As regards to respiration interference, the typical breathing frequency range (12 cpm to 24 cpm) is very close to the frequency of the SW (8 cpm to 12 cpm), so it is not possible to use conventional filters to remove this interference.

Laplacian recordings of the EEnG by means of active concentric ring electrodes on rigid substrates have proven to enhance signal quality in comparison to conventional monopolar and bipolar recordings with disc electrodes [9]. Therefore, a second objective of this work is to test and study the possible benefits of the flexible concentric ring electrodes to be developed in this study, on the surface recordings of the EEnG.

2   Materials and Methods

2.1   Active Electrode Array Design and Implementation

Sensing Part. In this work it has been decided to design an array of electrodes rather than an individual electrode for surface bioelectrical recordings, since this kind of recordings are usually multichannel and moreover Laplacian recordings are often used for body surface mappings. Specifically an array of three concentric ring TCB active electrodes was developed. The sensor is made out of two parts: a disposable sensing part with three TCB electrodes and a reusable battery-powered signal conditioning circuit. Each of the three sensing electrodes consists of an inner disc and two concentric rings in bipolar configuration i.e. the disc and the outer ring are shorted together.

The outer diameter of the external ring was set to 24 mm which is a compromise between bigger electrodes that would yield signals of higher amplitude and smaller electrodes that would provide better spatial resolution. The rest of dimensions of the electrode are designed considering the following criteria:

- The sum of the areas of the outer ring and the inner disc should be equal to the area of the middle ring so as to provide similar input impedances, improving the common mode rejection ratio.
- The distance between the inner disc and the middle ring should be the same as the distance between the middle and outer rings to reduce common mode interferences.

The dimensions of each ringed-electrode of the array are shown in Fig. 1. As it can also be appreciated in this figure, an opened-ring version has been designed in order to avoid shorts in the single layer layout of the tracks from the electrodes to the connectors. The flexible electrode array was implemented by screen-printing technology on polymer substrates. Specifically, a biocompatible silver paste was printed (Dupont

5064 Silver conductor, thickness 17 μm) on Polyester Melinex ST506 substrate (thickness 175 μm). The serigraphy was made by using an AUREL 900 High precision screen stencil printer. Cured period of inks was 130 °C for 10 min.

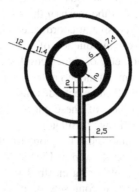

Fig. 1. Dimensions (mm) of the concentric ring electrode to be implemented in the array.

Signal Conditioning Part. As stated before, signals from concentric ring electrodes are of very low amplitude, especially in the cases that the bioelectrical signal to be recorded on the body surface is weak. Therefore it is highly recommended to include an amplification stage as close as possible to the sensing electrode.

In this work a battery-powered conditioning circuit was developed and directly connected to the electrode array. Precisely, a 12 bias (only six are used) flexible-flat-cable-to-flexible-printed-circuit connector (TE Connectivity/AMP-1-84953-2 FFC/FPC) was used for the connection. The circuit is composed of a preamplifier (gain 31.9), followed by a coupling circuit (high pass cut off frequency 0.05 Hz) and an additional differential amplification stage (gain 106.1) for each of the three TCB electrodes of the array. Specifically, the integrated circuits used were 3 OP747 for the operational amplifiers and 3 AD620 for the instrumentation amplifiers. The signal conditioning circuit weights less than 15 g. Its main electrical characteristics were experimentally checked and are shown in the next section.

2.2 Model of the Electrode and Abdomen

In order to compare the sensitivity map of the proposed opened-ring version of the TCB electrode with that of the conventional closed-ring electrode for abdominal recordings, both electrodes and the underlying tissue were modeled and solved by numeric methods.

Similar to the work of Bradshaw et al. [1], the abdomen was modeled as a semi-infinite electric conductor domain composed by five homogeneous and isotropic layers of tissues with its own electrical conductivity (σ) and thickness (th), which represent: skin (th$_1$ = 5 mm, σ_1 = 0.45 S/m), fat (th$_2$ = 12 mm, σ_2 = 0.02 S/m), muscle (th$_3$ = 5 mm, σ_3 = 0.45 S/m), omentum (th$_4$ = 5 mm, σ_4 = 0.02 S/m) and abdominal cavity (th$_5$ = 200 mm, σ_5 = 0.45 S/m). Model width and depth were chosen large enough to ensure a negligible influence of the model's lateral limits on the results. The model of the abdomen can be seen on the bottom of Fig. 2.

So as to accurately compare the results obtained from the geometrical model of the abdomen solved by finite element methods when testing the different types of ring electrodes, it is recommendable that the mesh of the model stays the same. This would permit to get the solutions at fixed spatial coordinates inside the model. However, the mesh and node coordinates are consequence of the geometric characteristic of the model. Therefore, since the geometry of the electrodes to study is different, the mesh will be. To overcome this problem, a unique geometrical model composed by four sets of pieces was used to represent all the electrodes under study in such a way that the electrical conductivity of each one of these pieces was changed between 'High' (conductor, $\sigma_c = 1e5$ S/m) and 'Low' (insulator, $\sigma_i = 1e{-}5$ S/m), depending on the geometry of the electrode to be analyzed. The geometrical model of the electrodes with the four sets of pieces is shown on the top of Fig. 2.

Three different types of concentric ring electrodes were studied; and the 'Low' or 'High' conductivity of each of their four sets of pieces is shown in Table 1. The combination of the first row (CRE) yields a closed-ring TCB electrode that could be considered our reference and gold standard. The combination of the second row (ORE_A) and the third row (ORE_B) yield opened-ring TCB electrodes similar to that shown in Fig. 1. In the case of ORE_A the conductive track that shortcuts the inner disc and the outer ring is in contact with the skin, whereas in the case of ORE_B it is not. It can also be deduced that ORE_B is identical to CRE but with opened rings.

Since results obtained from these electrodes are expected to be very similar, once the geometry and material properties of the model were defined, it was meshed with a very high density of elements (>3 million) to minimize errors in the comparison.

The objective was to compare the sensitivity of each electrode to measure the activity of an electrical dipole located inside the abdominal cavity. For this purpose, it has been shown [18] that the voltage between a pair of electrodes A and B caused by an electrical dipole placed at C, is the scalar product of the electric field on C caused by the injection of a unit current between electrodes A and B, and the dipole vector. From this point of view, the sensitivity of the ring electrodes to measure the electrical activity of a dipole inside the abdominal cavity can be approached to the module of the electric field caused by the injection of a unit current between the ring electrodes.

Electrodes and abdomen were modeled and solved by ANSYS (R).

2.3 Signal Recordings

The study was approved by Committee of Ethics of Universidad Politécnica de Valencia. Five recording sessions, of about three hours, were carried out in healthy human volunteers in fast state (>8 h). Subjects were in a supine position inside a Faraday cage. Firstly the abdominal body surface was exfoliated to remove dead skin cells to reduce contact impedance. The abdominal surface was also shaved in male subjects. Figure 3, shows the location of electrodes for the EEnG recordings. The developed flexible electrode array was placed horizontally 2.5 cm below the umbilicus, providing three Laplacian signals. Three monopolar Ag-AgCl floating gel electrodes of

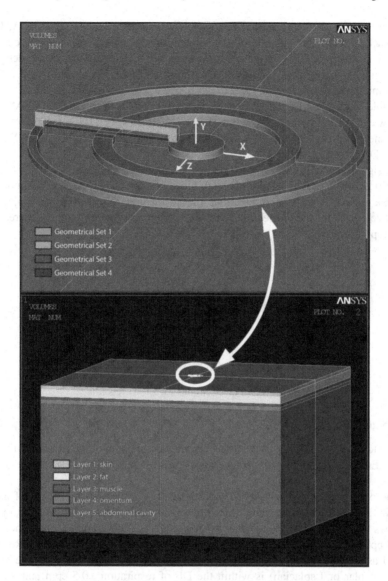

Fig. 2. Model of the electrode (top) and of the abdomen (bottom).

Table 1. Conductivities of the geometrical sets of pieces that conform the electrodes to be tested by the finite elements model. (CRE: closed-ring electrode, ORE: open ring electrode).

Electrode	Geometrical set			
	1	2	3	4
CRE	High	High	Low	High
ORE_A	High	Low	High	Low
ORE_B	High	High	Low	Low

8 mm of sensing diameter (EL258S, Biopac Systems Inc, Santa Barbara, CA, USA) were placed 2.5 cm above the umbilicus. Interelectrode distance was also 2.5 cm. Two bipolar recordings of EEnG were obtained from adjacent monopolar electrodes.

The main sources of physiological interferences usually present on surface EEnG were simultaneously recorded. Specifically, ECG was monitored by Lead 1 using disposable electrodes (EL501, Biopac Systems Inc, Santa Barbara, CA, USA), respiration was recorded by an airflow transducer (1401G, Grass Technologies, Warwick, USA) and body movements were measured by a 3-axis accelerometer (ADXL 335, Analog Devices, LM, Ireland).

All signals, except from acceleration signals, were amplified and band-pass filtered (0.1–100 Hz) by means of commercial bioamplifiers (P511, Grass Technologies, Warwick, USA). A disposable electrode placed on the left ankle of the subject was used as reference for the bioelectrical recordings. Signals were simultaneously recorded at a sampling rate of 1 kHz.

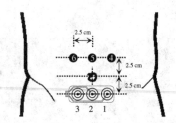

Fig. 3. Location of electrodes for EEnG recordings.

2.4   Signal Analysis

In order to study the activity of the low-frequency component of the EEnG i.e. the slow wave, EEnG and respiratory signals were low-pass filtered (fc = 0.5 Hz) and resampled at 4 Hz.

The power spectral density (PSD) of signals was estimated by means of autoregressive parametrical techniques (AR, order 120). PSD was estimated for moving windows of 120 s every 15 s of the recorded signals. The dominant frequency (DF) over 8 cpm of the PSD of every window was determined. The parameter %Resp was defined as the ratio between the number of windows in which the DF of the surface signal (bipolar or Laplacian) is within the DF of respiration ±0.5 cpm and the total number of windows. Similarly, %TFSW is defined as the ratio of analysed windows whose DF is inside the typical frequency range of intestinal slow wave (8-12 cpm). The rest of cases are included in the parameter %Other.

3   Results

3.1   Sensitivity Map of Opened-Ring Vs Closed-Ring TCB Electrode

Figure 4 displays the normalized spatial distribution on the abdominal cavity of the modulus of the electric field in a radial plane of the CRE. As it could be expected, the sensitivity is maxima at the point closest to the electrode, and presents a decreasing gradient as we move away from the electrode. For this specific electrode, the effect of the connecting track is negligible and the same sensitivity map is obtained for any other radial plane. Analogous sensitivity maps were obtained for ORE_A and ORE_B electrodes which are not shown since no great differences can be found by visual inspection and to reasons of space.

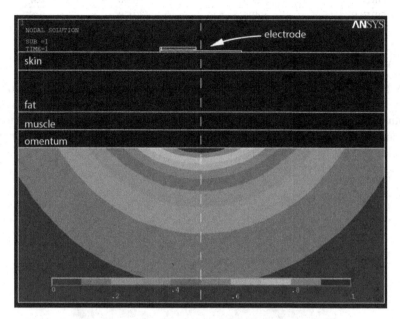

Fig. 4. Normalized spatial distribution of the modulus of the electric field in a radial plane of the closed-ring electrode (CRE).

Table 2 shows the results of the parameters used to describe and compare the sensitivity maps of the three modeled ring electrodes. The results from the model yield that the maximum magnitude of the induced electric field in the abdominal cavity (E_{max}) is almost equal for CRE and ORE_B, but is smaller for ORE_A. This reveals a non-negligible difference in the amplitude of the signals that this kind of electrode would pick-up. Nevertheless, the most important aspect of coaxial ring electrodes is the spatial distribution of its sensitivity which makes them suitable to estimate the Laplacian potencial. Thus, the maximum differences between the normalized spatial distributions of the sensitivity (modulus of the electric field) were obtained and are shown in Table 2. Again, ORE_B presents the most similar spatial distribution of the normalized sensitivity to CRE, providing an error minor than 0.5 %. In the case of ORE_A, in which the connecting track was in contact with the skin, this error is still low but almost reaches a value of 3.8 %.

Table 2. Parameters derived from the sensitivity maps of the modeled electrodes: E_{max} maximum value of the electric field in the abdominal cavity, Δ_{NSD} difference between normalized spatial distribution of the electric field of CRE and ORE.

Electrode	E_{max}(V/m)	Δ_{NSD} (%)	
		Min.	Max.
CRE	4.434	–	–
ORE_A	3.809	−0.46	0.44
ORE_B	4.406	−3.78	3.65

3.2 Active Electrode Array

Figure 5 shows the sensing part of the array of active concentric ring TCB electrodes. It can be appreciated that the substrate is flexible enough to fit the body surface curvature. Moreover the adhesion of the conductor paste to the substrate was checked by means of a sticky tape (8915 Filament APT, 3 M). The paste took off after more than 30 cycles proving the good adherence.

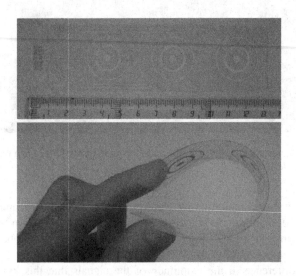

Fig. 5. Implemented flexible array of three TCB concentric ring electrodes.

Both sides of the signal preconditioning circuit of the active electrode array can be seen in Fig. 6. The small size and weight of the circuit and the flexible nature of the array makes it possible to place this part above the electrodes. With the proper fixing strategy, the active electrode array could be used for ambulatory monitoring. Table 3 summarizes the main electrical characteristics of the developed signal conditioning circuit. It can be observed that the battery life is adequate for the recording sessions, and the CMRR and output noise are also appropriate for bioelectrical applications.

Fig. 6. Signal preconditioning circuit: bottom side (left) and top side (right).

Table 3. Main electrical parameters of the signal-preconditioning circuit.

1	0.049 Hz
Differential gain at medium frequency	3386 V/V
CMRR at medium frequencies	116 dB
CMRR at 50 Hz	103 dB
Output noise	0.195 mVrms
Battery life	280 min

3.3   EEnG Monitoring

Figure 7 shows an example of the biosignals simultaneously recorded. It can be appreciated that, as expected, the amplitude of the signals picked up by the ringed-electrodes of the array is smaller than that of the bipolar recordings with disc electrodes. Nevertheless, the conventional bipolar recordings present stronger ECG interference as it can be easily observed in this figure. In the signals from concentric ring electrodes the electrocardiographic interference is almost null. It can only be hardly appreciated in the signal corresponding to electrode 1 (Lp1) which is placed on the left side of the subject. Regarding the intestinal SW activity, approximately five waves can be identified on the Laplacian recordings. In bipolar recordings this is difficult to identify by visual inspection since it seems they are strongly corrupted by the respiration. This can also be observed in the example of PSD of signals shown in Fig. 8. In bipolar recordings the dominant frequency (DF) corresponds to the respiratory frequency; whereas the DF of Laplacian recordings is around 10.5 cpm which fits the normal SW frequency in human jejunum.

The results of the %TFSW, %Resp and %Other parameters, which are presented in Table 4, confirm this behavior. It is shown that in around 25 % of the signal windows studied, the respiratory interference masks the intestinal SW activity. In contrast, this ratio is only around 10 % for the Laplacian recordings. Furthermore, around 75 % of the cases the DF of Laplacians recordings the DFs are in the frequency range associated to the intestinal SW, whereas it is about 65 % for the conventional bipolar recordings.

36   J. Garcia-Casado et al.

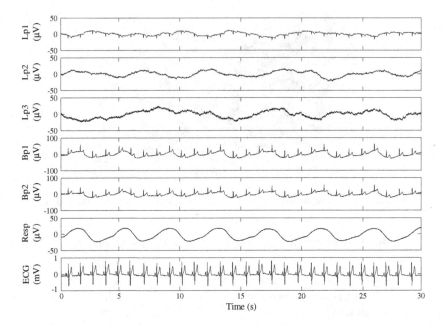

Fig. 7. Simultaneous recording of biosignals: Lp1–3: Laplacian signals from the electrode array; Bp1–2: bipolar signals from disc electrodes; Resp: respiration; ECG: electrocardiogram.

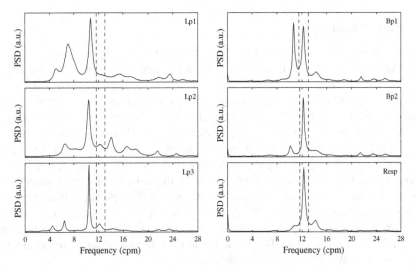

Fig. 8. Example of power spectrum density (AR120) of a 120 s window of recorded biosignals. The bandwidth associated to respiratory components is represented in dashed lines.

Finally to say that the number of cases whose DF is associated to SW harmonics or other components is slightly higher in Laplacian recordings than in conventional bipolar recordings.

Table 4. Percentage of dominant frequency in the bandwidths of the different components (mean ± standard deviation); N = 3385, Lp1–3: Laplacian signals from the electrode array, Bp1–2: bipolar signals from disc electrodes.

Channel	%TFSW	%Resp	%Other
Lp1	76,6 ± 8,6	11,3 ± 5,6	12,0 ± 5,8
Lp2	72,8 ± 8,2	9,5 ± 7,0	17,7 ± 6,2
Lp3	71,2 ± 6,9	11,6 ± 8,0	17,2 ± 3,8
Bp1	63,0 ± 12,1	27,3 ± 12,0	9,7 ± 4,8
Bp2	65,6 ± 13,8	25,2 ± 13,8	9,2 ± 6,0

4 Discussion

To the authors' knowledge, the flexible array of active concentric ring electrodes presented in this paper is the first one of these characteristics. Other authors have developed active concentric ring electrodes but on rigid substrates [3, 6–9]. This new sensor is more comfortable for the subject under study and provides a better contact since it adapts to the body surface curvature. Our group has recently developed other flexible concentric ring electrodes [19]. However such electrodes require the screen-printing of three layers, alternating conductor and dielectric pastes. In flexible substrates it is very complicated to use bias between layers, and the solution proposed in the present work favors an easier manufacturing. The comparison of the sensitivity of TCB opened-ring electrodes and closed-ring electrodes shown in this document reveals that signals can be picked up more attenuated with the opened-ring version, but with differences smaller than 3.8 % in the spatial distribution of the sensitivity. What is more, if a dielectric is used to prevent the contact between the skin and the conductive line that shortcuts the inner disc with the outer ring, these differences are smaller than 0.5 %. Regarding the model we used, it is more realistic than others [20] since it considers the abdominal layers with their different conductivities and thicknesses. It should also be pointed out that, in contrast to individual electrodes used in other studies [6–9, 19], the electrode array developed in this work is a more compact solution that reduces the signal preconditioning cost and space, and it is more suitable for bioelectrical mapping of the body surface. Furthermore, the modularity of the developed sensor permits to reuse the signal conditioning circuit while the sensing part can be disposed for hygienic reasons.

Signal recording experiences of this work show that active concentric electrodes of the flexible array enhance the quality of non-invasive EEnG signals in terms of electrocardiographic and respiratory interferences, in comparison to bipolar recordings with conventional disc electrodes. This is in agreement with previous studies that used this kind of electrodes implemented on rigid substrates [9]. The presence of these interferences is very common in non-invasive recordings of bioelectric signals and limits the bandwidth of analysis of such signals [21] with the corresponding loss of information. In order to cancel these physiological interferences many different signal-processing techniques have been developed [17, 22–24]. In contrast to these techniques, the electrode presented in this work permits to reduce such interferences in the raw signals, with no need of post-processing and no computational cost which makes it ideal for real time monitoring applications. In the specific case of EEnG, on one hand, the reduction of respiratory interference permits to identify more easily the activity of intestinal slow wave [15]. On the other hand, the reduction of ECG interference could help the identification of spike bursts activity [17]. This could provide more robust systems to non-invasively monitor intestinal myoelectrical activity which could bring close the clinical application of this technique. Nevertheless, this should be confirmed in future studies.

Moreover, although it has not been tested in this work, according to other authors [25, 26], the Laplacian potential mapping can enhance spatial sensibility for surface bioelectrical activity. This can be of great importance for the studies of propagation maps of cardiac [27], electroencephalographic [28], intestinal [29] or uterine [30] activity which can provide electrophysiological information of clinical relevance. The developed flexible array of active concentric ring electrodes, with different electrode sizes, would be very suitable for these applications.

5 Conclusions

The flexible array of active concentric ring electrodes developed in this paper joins the benefits of Laplacian techniques in terms of enhancing spatial resolution, with the comfort and adaptation to body surface curvature of conventional disposable electrodes.

The non-invasive recordings of intestinal myolectrical signals with this new kind of electrodes provide enhanced bioelectric signals in terms of robustness to physiological interferences such as ECG and respiration, and permit to identify more easily the intestinal slow wave activity.

Acknowledgements. Research supported in part by the Ministerio de Ciencia y Tecnología de España (TEC 2010-16945) and by and by the Universidad Politècnica de València (SP20120469).

References

1. Bradshaw, L.A., Richards, W.O., Wikswo Jr., J.P.: Volume conductor effects on the spatial resolution of magnetic fields and electric potentials from gastrointestinal electrical activity. Med. Biol. Eng. Comput. **39**, 35–43 (2001)
2. Wu, D., Tsai, H.C., He, B.: On the estimation of the Laplacian electrocardiogram during ventricular activation. Ann. Biomed. Eng. **27**, 731–745 (1999)
3. Besio, W.G., Aakula, R., Koka, K., Dai, W.: Development of a tri-polar concentric ring electrode for acquiring accurate Laplacian body surface potentials. Ann. Biomed. Eng. **34**, 426–435 (2006)
4. Hjorth, B.: An on-line transformation of EEG scalp potentials into orthogonal source derivations. Electroencephalogr. Clin. Neurophysiol. **39**, 526–530 (1975)
5. Perrin, F., Pernier, J., Bertrand, O., Giard, M.H., Echallier, J.F.: Mapping of scalp potentials by surface spline interpolation. Electroencephalogr. Clin. Neurophysiol. **66**, 75–81 (1987)

6. Lu, C.C., Tarjan, P.P.: An ultra-high common-mode rejection ratio (CMRR) AC instrumentation amplifier for laplacian electrocardiographic measurement. Biomed. Instrum. Technol. **33**, 76–83 (1999)
7. Koka, K., Besio, W.G.: Improvement of spatial selectivity and decrease of mutual information of tri-polar concentric ring electrodes. J. Neurosci. Methods **165**, 216–222 (2007)
8. Li, G., Wang, Y., Jiang, W., Wang, L.L., Lu, C.-Y.S., Besio, W.G.: Active laplacian electrode for the data-acquisition system of EHG. J. Phys: Conf. Ser. **13**, 330–335 (2005)
9. Prats-Boluda, G., Garcia-Casado, J., Martinez-de-Juan, J.L., Ye-Lin, Y.: Active concentric ring electrode for non-invasive detection of intestinal myoelectric signals. Med. Eng. Phys. **33**, 446–455 (2011)
10. Quigley, E.M.: Gastric and small intestinal motility in health and disease. Gastroenterol. Clin. North Am. **25**, 113–145 (1996)
11. Weisbrodt, N.W.: Motility of the Small Intestine, 2nd edn, pp. 631–663. J. LR, New York (1987)
12. Fleckenstein, P., Oigaard, A.: Electrical spike activity in the human small intestine. A multiple electrode study of fasting diurnal variations. Am. J. Dig. Dis. **23**, 776–780 (1978)
13. Chen, J.D., Schirmer, B.D., McCallum, R.W.: Measurement of electrical activity of the human small intestine using surface electrodes. IEEE Trans. Biomed. Eng. **40**, 598–602 (1993)
14. Chang, F.Y., Lu, C.L., Chen, C.Y., Luo, J.C., Lee, S.D., Wu, H.C., Chen, J.Z.: Fasting and postprandial small intestinal slow waves non-invasively measured in subjects with total gastrectomy. J. Gastroenterol. Hepatol. **22**, 247–252 (2007)
15. Prats-Boluda, G., Garcia-Casado, J., Martinez-de-Juan, J.L., Ponce, J.L.: Identification of the slow wave component of the electroenterogram from Laplacian abdominal surface recordings in humans. Physiol. Meas. **28**, 1115–1133 (2007)
16. García-Casado, J., Martínez-de-Juan, J.L., Ponce, J.: Effect of abdominal layers on surface electroenterogram spectrum. In: Proceedings of the 25th Annual International Conference of the IEEE, vol. 3, pp. 2543–2546 (2003)
17. Garcia-Casado, J., Martinez-de-Juan, J.L., Ponce, J.L.: Adaptive filtering of ECG interference on surface EEnGs based on signal averaging. Physiol. Meas. **27**, 509–527 (2006)
18. Rush, S., Driscoll, D.A.: EEG electrode sensitivity–an application of reciprocity. IEEE Trans. Biomed. Eng. **16**, 15–22 (1969)
19. Prats-Boluda, G., Ye-Lin, Y., Ibañez, J., García-Breijo, E., García-Casado, J.: Active flexible concentric ring electrode for non invasive surface bioelectrical recordings. Meas. Sci. Technol. **23**, 10 (2012)
20. Besio, W.G., Koka, K., Aakula, R., Dai, W.: Tri-polar concentric ring electrode development for Laplacian electroencephalography. IEEE Trans. Biomed. Eng. **53**, 926 (2006)
21. Rabotti, C., Mischi, M., van Laar, J.O., Oei, G.S., Bergmans, J.W.: Inter-electrode delay estimators for electrohysterographic propagation analysis. Physiol. Meas. **30**, 745–761 (2009)
22. Irimia, A., Bradshaw, L.A.: Artifact reduction in magnetogastrography using fast independent component analysis. Physiol. Meas. **26**, 1059–1073 (2005)
23. Hassan, M., Boudaoud, S., Terrien, J., Karlsson, B., Marque, C.: Combination of canonical correlation analysis and empirical mode decomposition applied to denoising the labor electrohysterogram. IEEE Trans. Biomed. Eng. **58**, 2441–2447 (2011)
24. Vullings, R., Peters, C.H., Sluijter, R.J., Mischi, M., Oei, S.G., Bergmans, J.W.: Dynamic segmentation and linear prediction for maternal ECG removal in antenatal abdominal recordings. Physiol. Meas. **30**, 291–307 (2009)

25. Besio, W., Chen, T.: Tripolar Laplacian electrocardiogram and moment of activation isochronal mapping. Physiol. Meas. **28**, 515 (2007)
26. Soundararajan, V., Besio, W.: Simulated comparison of disc and concentric electrode maps during atrial arrhythmias. Int. J. Bioelectromagn. **7**, 217–220 (2005)
27. Haissaguerre, M., Hocini, M., Shah, A.J., Derval, N., Sacher, F., Jais, P., Dubois, R.: Noninvasive panoramic mapping of human atrial fibrillation mechanisms: a feasibility report. J. Cardiovasc. Electrophysiol. **24**, 711–717 (2013)
28. Hangya, B., Tihanyi, B.T., Entz, L., Fabo, D., Eross, L., Wittner, L., Jakus, R., Varga, V., Freund, T.F., Ulbert, I.: Complex propagation patterns characterize human cortical activity during slow-wave sleep. J. Neurosci. **31**, 8770–8779 (2011)
29. Lammers, W.J., Al-Bloushi, H.M., Al-Eisaei, S.A., Al-Dhaheri, F.A., Stephen, B., John, R., Dhanasekaran, S., Karam, S.M.: Slow wave propagation and plasticity of interstitial cells of Cajal in the small intestine of diabetic rats. Exp. Physiol. **96**, 1039–1048 (2011)
30. Rabotti, C., Mischi, M., Oei, S.G., Bergmans, J.W.: Noninvasive estimation of the electrohysterographic action-potential conduction velocity. IEEE Trans. Biomed. Eng. **57**, 2178–2187 (2010)

Development and Application of a Portable Device for Cutaneous Thermal Sensitivity Assessment: Characterizing the Neuropathic Pain Following Spinal Cord Injury

Renato Varoto[1(✉)], Fábio Casagrande Hirono[1],
Fernando Ometto Zorzenoni[2], Ricardo Yoshio Zanetti Kido[2],
William Barcellos[1], and Alberto Cliquet Jr.[1,2]

[1] Department of Electrical Engineering, University of São Paulo (USP),
São Carlos, Brazil
rvaroto@yahoo.com.br,
{fchirono,william.barcellos}@gmail.com,
cliquet@usp.br
[2] Department of Orthopedics and Traumatology,
University of Campinas (UNICAMP), Campinas, Brazil
{fernandozorzenoni,ricardo.kido}@gmail.com

Abstract. Neuropathic pain is characterized to arise due to injury or dysfunction of Peripheral and Central Nervous Systems, without stimulation of nociceptors, being one of the major problems following spinal cord injury (SCI). It involves altered mechanisms of impulse transmission in somatosensory pathways, causing abnormal sensations. Quantitative and qualitative sensory tests, by the detection of tactile and thermal stimuli and McGill Pain Questionnaire, are methods used to characterize and study the neuropathic pain. Therefore, this work describes the development and application of a portable device for cutaneous thermal sensitivity assessment in spinal cord injured subjects (SCIS). Using method of levels, the assessment was applied in healthy subjects and SCIS with and without neuropathic pain. The thresholds determined for healthy subjects during thermal sensitivity assessment are consistent and other results provided by clinical trials are according to previous works, demonstrating the device feasibility as an auxiliary tool for neuropathic pain study.

Keywords: Thermal sensitivity assessment · Neuropathic pain · Spinal cord injury

1 Introduction

Spinal cord injury (SCI) causes disruption of nerve fibres that transmit ascending sensory and descending motor information. This disruption causes losses in the transmission of sensory-motor information across the site of the lesion, resulting in considerable physical and emotional consequences for individual [1, 2]. Sensory-motor dysfunctions occur in the parts of the body innervated by areas below the site of the lesion, being characterized by paralysis, altered sensation and weakness [3].

© Springer-Verlag Berlin Heidelberg 2014
M. Fernández-Chimeno et al. (Eds.): BIOSTEC 2013, CCIS 452, pp. 41–53, 2014.
DOI: 10.1007/978-3-662-44485-6_4

Spinal cord injured subjects (SCIS) also suffer other disorders and numerous secondary pathologies such as losses of bowel and bladder functions, pressure ulcers, spasticity, gastrointestinal and sexual dysfunctions and heterotopic ossification [2, 4, 5]. However, one of the major problems following SCI is the neuropathic pain [6].

Neuropathic pain is characterized to arise without stimulation of nociceptors (sensory pain fibres that detect tissue damage by physical, chemical or thermal phenomena), but due to injury or dysfunction of Peripheral and Central Nervous Systems. Thus, neuropathic pain is an aggravating for the already weakened patient, imposing severe limitations in performing the activities of daily living [7, 8].

The pathophysiology of neuropathic pain involves altered mechanisms of impulse transmission in somatosensory pathways, so that axonal injury leads to a gain in excitatory transmission, in other words, there is a massive axonal input. It results from an axonal hyperexcitability, with the generation of ectopic electrical impulses, causing abnormal sensations [9].

In SCIS, partially preserved pathways spinothalamic tract may be the local generator of pain [10]. Fibres Aδ and C present little myelin and follow the column via anterolateral spinothalamic tract. These fibres are the main components of the fibres that lead thermal sensitivity [11]. Thus, the thermal sensitivity follows the same neurological path of the pain.

Some methods are applied to characterize and study the neuropathic pain, for example, McGill Pain Questionnaire, quantitative sensory testing (QST) and somatosensory evoked potential [12, 13].

The McGill Pain Questionnaire is an instrument that evaluates qualitatively and quantitatively pain, providing quantitative measures of clinical pain that can be treated statistically [14]. Reference [15] adapted (translation and validation) the questionnaire to Portuguese. The present pain intensity (PPI) is the number chosen by the subject at the time of administration of the questionnaire, ranging from 0 (no pain) to 5 (excruciating). The pain rating index based on the subjects' mean scale values (PRI(S)) obtained by [16] is described as the sum of all values of words chosen by subject for all categories (sensory, affective, evaluative, motor and miscellaneous). And other important value is the number of words chosen (NWC) that is the sum of all words chosen by the subject [14].

QST assess and quantify sensory function in subjects with losses in the neurological system, measuring the detection threshold of tactile, vibratory, thermal or painful stimuli [12, 17]. Especially for thermal stimuli, some equipments utilize the Peltier effect, in which the intensity and direction of electrical current controls the surface temperature of a test electrode [18, 19]. The skin was touched by the electrode and the subject reports the sensation in relation to the temperature [12, 17, 20].

This paper describes the development and application of a portable device for cutaneous thermal sensitivity assessment based on Peltier effect. This device is designed for quick and practical assessment of thermal sensitivity in SCIS, representing an auxiliary tool for neuropathic pain study. Using method of levels, the device was used in healthy subjects and SCIS with and without neuropathic pain. In general, the obtained results were compared with previous works to verify the device feasibility.

2 Materials and Methods

About this work, instrumentation development was done at Laboratory of Biocybernetics and Rehabilitation Engineering - USP, and clinical application was performed at Laboratory of Biomechanics and Rehabilitation of the Locomotor System – UNICAMP.

Basically, the portable device is composed by microcontroller, thermoelectric module and temperature transducer. The microcontroller associated with amplifier circuits offers electrical energy to supply the thermoelectric module and provides information about device operation condition. Furthermore, it allows setting the probe operating temperature (Fig. 1).

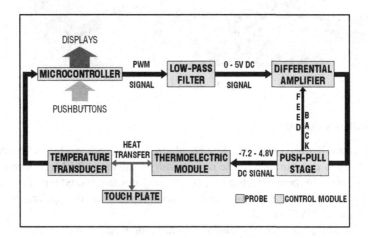

Fig. 1. Block diagram of the device.

2.1 Thermoelectric Module, Temperature Transducer and Probe Assembly

The thermoelectric module used was a solid state heat pump (Melcor Corporation, Trenton, NJ, USA), based on Peltier effect. This heat pump contains 66 thermocouples, being able to transfer until 3.56 W of heat from cold to hot faces (Q_{max}); it results in the maximum temperature difference of the 67 °C between two faces (ΔT_{max}) with low power consumption - input electrical current (I_{max}) of 0.8 A and dc voltage (V_{max}) of 7.98 V – to achieve ΔT_{max}.

Temperature transducer with analog output based on semiconductor junctions was used to monitor the temperature of thermoelectric module. The transducer used was LM35 (National Semiconductor Corporation, Santa Clara, CA, USA) that is a precision integrated-circuit temperature transducer, whose output voltage is linearly proportional to the Celsius temperature. The LM35 does not require external calibration to provide readouts with accuracies of ±0.75 °C over a range −55 to +150 °C. Other important features of LM35 make it suitable for control circuits as low output impedance, very low self-heating and sensitivity of 0.01 V/°C.

The probe is composed by aluminium touch plate (16 × 16 mm), thermoelectric module, LM35 transducer, heat sink and auxiliary fan. In the first stage of probe assembly, LM35 was coupled to the touch plate by aluminium clamp, and the thermoelectric module was fixed on heat sink. Touch plate with LM35 was fixed on thermoelectric module by thermal compound, in the second stage. Moreover, the assembly was placed in a suitable case (Fig. 2).

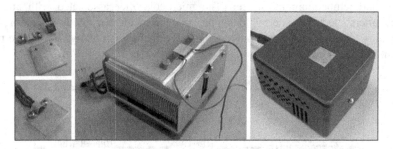

Fig. 2. Probe assembly.

2.2 Electronic Apparatus

The microcontroller used was PIC18F252 (Microchip Technology Inc., Chandler, AZ, USA) and it was programmed to use pulse-width modulation (PWM) as technique for controlling power to thermoelectric module. It compares the desired temperature of the probe with the actual temperature and, in accordance with this difference, adjusts the duty cycle of PWM signal.

An active low-pass filter was used to generate a dc signal from 0–5 V PWM signal. This filter was configured in Sallen-Key topology of second order with cut-off frequency of approximately 35 Hz and quality factor of 0.5.

During bench tests with the probe, dc voltages of thermoelectric module to ensure that the temperature probe reached 0 and 60 °C were determined. The voltage polarity was defined so that a negative voltage decreases the probe temperature and a positive one would increase the temperature. These voltages were −7.2 V and 4.8 V, respectively.

Therefore, it was necessary to convert 0–5 V dc signal (Vin) obtained from PWM into asymmetric bipolar dc signal (−7.2 V–4.8 V) (Vout) required by thermoelectric module. This conversion was done by differential amplifier using an operational amplifier as active component, whose inverting input was held at 5 V. Thus, an amplifier with linear transfer characteristic was developed (1), resulting in proper range of signal.

$$\text{Vout} = 2.435 * \text{Vin} - 7.353 \tag{1}$$

Although the range of signal was appropriate, the differential amplifier was not able to provide the required electrical power (up to 5.76 W). Thus, a push-pull stage with MOSFETs was applied. These components exhibit a negative thermal coefficient, in other words, its electrical conductivity decreases with increasing temperature, protecting the electronic circuit.

The output of push-pull stage with unitary gain was also used as feedback signal to differential amplifier. This strategy reduces signal distortion, providing a linear signal to thermoelectric module. Whole electronic circuit including the thermoelectric module is powered by two batteries (12 V, 5 Ah).

2.3   Clinical Trials

Twenty SCIS were recruited to participate in this work, and they were divided into two groups, pain SCIS (P) and non-pain SCIS (NP). Subjects were classified according to the American Spinal Cord Injury Association (ASIA) Impairment Scale (AIS) [1, 21–23]. Control group (CT) was formed by 10 healthy subjects (Table 1). This project was approved by the Ethics Committee (local/national). In accordance with institutional policies and the Declaration of Helsinki, informed consent was obtained from each subject prior to participation.

Table 1. Subjects characteristics.

	P	NP	CT
Age (year)*	39.3(12.1)	35.4(12.9)	27.2(10.6)
Gender (Male/Female)	9/1	7/3	9/1
Body mass (kg)*	69.2(10.1)	71.5(14.7)	77.6(13.6)
Height (m)*	1.74(0.07)	1.72(0.09)	177.3(4.8)
Neurological lesion level (Cervical/Thoracic)	7/3	3/7	–
AIS (A/B/C/D)	8/2/0/0	7/2/1/0	–

*Values in mean(SD)

Inclusion criteria for P group were lesion level above T12 with central neuropathic pain after traumatic SCI. Exclusion criteria were based on the presence of any other pain different from central neuropathic pain such as nociceptive or peripheral neuropathic pain; or subjects that were under analgesic treatment.

In relation to NP group, inclusion criteria were lesion level above T12 without central neuropathic pain and spontaneous dysesthesia.

To study the sensitivity of pain and losses of sensory pathways following SCI, McGill Pain Questionnaire (Portuguese version), tactile and thermal sensitivity assessments were applied.

For McGill Pain Questionnaire, the values used for data analysis were PRI, NWC and PPI. The subjects were also asked to point out the site of pain.

Semmes-Weinstein monofilaments (aesthesiometer) (Sorri-Bauru, Bauru, SP, Brazil) were used to perform tactile sensitivity test (Fig. 3). Tactile and pain thresholds were measured as the force required for bending the monofilament during 3 s. The tactile stimuli were applied to the dominant leg at a point 100 mm distal from the patella, in the anterolateral side of the leg, corresponding to the L5 dermatome.

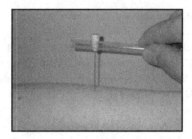

Fig. 3. Tactile sensitivity test with Semmes-Weinstein monofilament.

The thermal stimuli were applied to the same site of the tactile stimuli (Fig. 4). For temperature range from 30 °C to 60 °C, with increment of 5 °C, the skin was stimulated by probe (aluminium touch plate, specifically) by over 3 s. Subsequently, the temperature range was from 30 °C to 0 °C, with decrement of 5 °C. Warm and cold thresholds (temperature at which the patient feels the stimulus) and pain thresholds were recorded using the method of levels [17]. For each subject, three measurements with interstimulus interval of 3–6 s were used to calculate the thresholds.

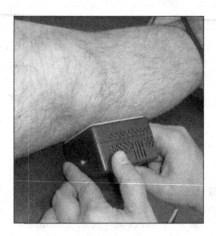

Fig. 4. Application of the thermal stimuli at the right leg.

3   Results

Figure 5 shows portable device for thermal sensitivity assessment; probe and control module.

On the front panel, the control module has two pushbuttons that set the desired probe temperature; the red pushbutton (+) increases probe temperature of 1 °C while the black one (−) decreases it of 1 °C, at range of 0 °C to 60 °C. This desired probe temperature and the instantaneous one are shown on the smaller green display and larger red display, respectively. When temperatures become equal, LED turns on,

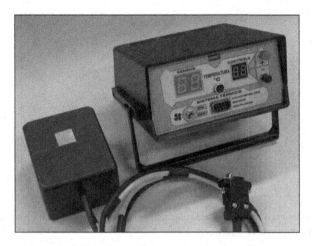

Fig. 5. Portable device for thermal sensitivity assessment.

indicating that probe is ready to use. Besides, the front panel has a toggle switch for the auxiliary fan and a DB9 connector for the probe cable.

According to the McGill Pain Questionnaire applied to the P group, half of subjects feel pain at injury level and half of them below the injury level (Fig. 6). Figure 7 shows the relation between reported words and number of subjects for each group of questionnaire; and Table 2 presents the scores for each variable.

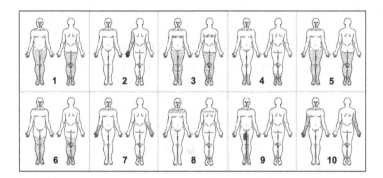

Fig. 6. Site of pain pointed out by subjects of the P group.

In relation to the tactile sensitivity test, two subjects felt the tactile stimuli and reported pain, in the NP group (Table 3). One subject (3) reported pain, which was progressively increasing according to the thickness of the monofilament (up to 8 score in a visual scale of 0–10 in the 300 g monofilament). Furthermore, this subject presents a delay of 2–3 s between the stimulus and the sensation. Another subject (4) experienced increased paresthesia throughout the right leg, from the 2 g to the 300 g monofilament. One subject (8) reported paresthesia with the 300 g monofilament.

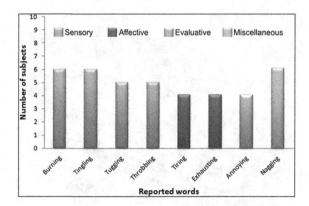

Fig. 7. Relation between reported words and number of subjects for each group of McGill Pain Questionnaire (Portuguese version).

Table 2. McGill Pain Questionnaire scores.

Variables	McGill Pain Questionnaire scores
PRI	28.5(13.7)
NWC	12.7(3.0)
PPI	3.2(1.4)

*Values in mean(SD)

Table 3. Tactile sensitivity in the NP group.

Subjects (AIS)	Detected threshold ■/Force (grams)	
	Tactile	Pain
1(A)	–	–
2(B)	–	–
3(C)	■/0.05	■/0.2
4(A)	–	–
5(A)	–	–
6(A)	–	–
7(A)	–	–
8(A)	–	–
9(B)	■/2.0	■/300.0
10(A)	–	–

Table 4. Tactile sensitivity in the P group.

Subjects (AIS)	Detected threshold ■/Force (grams)	
	Tactile	Pain
1(A)	–	–
2(B)	–	–
3(B)	–	–
4(A)	–	–
5(A)	–	–
6(A)	■/300.0	–
7(A)	–	–
8(A)	–	–
9(A)	–	–
10(A)	■/300.0	–

In the P group, two subjects felt the tactile stimuli (Table 4). One subject (6) felt stimulus of the 300 g monofilament at the contralateral leg (aloestesia). One subject (8) reported muscular spasticity with the 300 g monofilament.

During thermal sensitivity assessment, three subjects detected cold, two detected warm and one detected pain due to heat, in the NP group (Table 5). Furthermore, none reported pain due to cold and one subject (3) presented muscle spasms with stimulus of 50 °C.

In the P group, four subjects detected cold, one detected pain by cold at 5 °C and seven subjects detected warm. Four subjects detected pain due to heat and heat pain tolerance, with three subjects (1, 7 and 8) presenting muscular spasms at 55 °C (Table 6). One subject (3) felt dysesthesia at the stimulation site with stimuli of 0 °C and 45 °C. Another subject (7) detected warm all around the knee (not only at the stimulation site) detecting warm at 0 °C. And one subject (9) felt a non specific vibration in the L5 dermatome with 45 °C in the right leg.

Table 5. Thermal sensitivity in the NP group.

Subjects (AIS)	Detected threshold ▲/Temperature (°C)			
	Cold	Pain due to cold	Warm	Pain due to heat
1(A)	–	–	–	–
2(B)	–	–	–	–
3(C)	▲/23.3	–	▲/38.3	▲/50.0
4(A)	–	–	–	–
5(A)	▲/25.0	–	–	–
6(A)	–	–	–	–
7(A)	–	–	–	–
8(A)	–	–	–	–
9(B)	▲/13.3	–	▲/38.3	–
10(A)	–	–	–	–

Table 6. Thermal sensitivity in the P group.

Subjects (AIS)	Detected threshold ▲/Temperature (°C)			
	Cold	Pain due to cold	Warm	Pain due to heat
1(A)	–	–	▲/50.0	▲/55.0
2(B)	–	–	▲/45.0	–
3(B)	▲/0.0	–	▲/35.0	▲/45.0
4(A)	–	–	–	–
5(A)	–	–	▲/48.3	–
6(A)	▲/20.0	–	▲/26.7	▲/41.7
7(A)	▲/25.0	–	▲/23.3	–
8(A)	–	–	–	–
9(A)	–	–	–	–
10(A)	▲/15.0	▲/5.0	▲/40.0	▲/50.0

Table 7 indicates cold (C), warm (W), pain due to heat (HP) and heat pain tolerance (HPT) thresholds for each group.

Table 7. Temperature thresholds for each group.

Threshold (°C)	Group		
	P	NP	CT
C	15(10.8)	20.6(6.3)	19.2(5.2)
W	38.3(10.4)	38.3(0)	36.4(3.3)
HP	47.9(5.8)	50(0)	50.8(2.6)
HPT	48.3(4.9)	50(0)	52.8(3.1)

*Values in mean(SD)

4 Discussion

For touch plate construction, aluminium, copper and stainless steel were available. The choice was based on coefficient of thermal conductivity and oxidation resistance of metals. According to the coefficient of thermal conductivity, copper (398 W/mK) is a better conductor than aluminium (247 W/mK) and stainless steel (15.9 W/mK), but stainless steel presents high oxidation resistance. Therefore, aluminium was chosen because presents high coefficient of thermal conductivity and intermediate oxidation resistance [24].

These properties associated to the low mass (0.7 g) and small dimension of touch plate enabled a fast thermal equilibrium between both plate surfaces. Thus, the surface temperature acquired by the transducer is the same of surface dedicated to the touch.

Reach and stabilization of desired probe temperature can be attributed to the technique for controlling power to thermoelectric module and the use of heat sink and auxiliary fan. For each increment or decrement of 5 °C, this strategy allowed temperature stabilization in around 5 s during clinical trials.

In relation to the technique for controlling power to thermoelectric module, another alternative based on the use of PWM signal and an H-bridge can be applied, replacing low-pass filter and differential amplifier. However, this configuration provides a dc voltage range from -12 V to 12 V for the PWM duty cycle of 0 and 100 %, respectively; values which are not in accordance with the asymmetric bipolar dc signal (-7.2 V–4.8 V) required by the thermoelectric module to operate in proper temperature range (0–60 °C). Thus, the PWM duty cycle should be limited between values of 20 % and 70 %, also avoiding damage to the module since this operates at a maximum dc voltage of 7.98 V. Therefore, the use of the low-pass filter and the differential amplifier is more appropriate to the objectives of this work.

Generally, in healthy subjects, the activity of cold-sensitive neurons increases below 35 °C, and maximum cutaneous cold sensitivity is around 25 °C, while cold fibre activity is ceased at temperatures below 12 °C. The firing rates of warm-sensitive neurons increase above 25 °C, and their range of thermosensitivity extends from 35 °C to 43 °C. Temperatures above 43 °C and below 12 °C cause pain, whose stimulus is transmitted by Aδ and C fibres. Furthermore, nociceptive heat activates Aδ fibres around 43 °C, while temperatures above 52 °C activate C fibers. Cold stimuli below 12 °C also cause pain and, in addition, nociceptives heat and cold are transmitted by polymodal C fibers [25, 26]. Therefore, the thresholds determined for CT group during thermal sensitivity assessment using portable device are consistent (Table 3).

In relation to SCIS, P group was more sensitive to thermal stimuli than NP group, where 70 % of subjects in P group detected some kind of thermal sensitivity against 30 % of NP group (Tables 3 and 4). This difference between P and NP groups is in agreement with the study of [10]. In this study, it was reported that there is some preservation of the spinothalamic tract in pain SCIS greater than in non pain SCIS, which may be involved in the development of neuropathic pain. From the three subjects who experienced thermal stimuli in NP group, only one presents complete injury. For the P group, two subjects are AIS B and five are AIS A.

This finding can be justified through theories that explain why subjects with complete SCI have some sensibility, characterizing the discomplete injury. References [27] and [28] found motor remnants in complete SCIS due to a neural control. Thus, discomplete injury is an incomplete injury that fits the AIS criteria for grade A. Moreover, some subjects with complete SCI can present some semblance of sensibility, which can be evoked below the level of injury due to incomplete injuries in the spinothalamic tract. These subjects have subclinical functions of ascending and descending tracts [29].

5 Conclusions

Due to feedback control, the portable device provides easy temperature control with resolution of 1 °C. The device is simple to build and can stabilize its temperature in about 5 s for a 5 °C temperature change, therefore representing a simple alternative for quick and practical assessment. It can provide quantified information about sensory performance of subjects, and the results obtained from clinical trials are in accordance to previous works, thus demonstrating the device feasibility for thermal sensitivity assessment. Spinal cord injured subjects that refer neuropathic pain are more sensitive to thermal stimuli and less sensitive to tactile stimuli than patients do not present neuropathic pain.

Acknowledgements. We thank the support by grants from São Paulo Research Foundation (FAPESP) and National Council for Scientific and Technological Development (CNPq).

References

1. Maynard Jr., F.M., Bracken, M.B., Creasey, G., Ditunno Jr., J.F., Donovan, W.H., Ducker, T.B., Garber, S.L., Marino, R.J., Stover, S.L., Tator, C.H., Waters, R.L., Wilberger, J.E., Young, W.: International standards for neurological and functional classification of spinal cord injury. Spin. Cord **35**, 266–274 (1997)
2. Eng, J.J., Miller, W.C.: Rehabilitation: From bedside to community following spinal cord injury (SCI). In: Eng, J.J., Teasell, R.W., Miller, W.C., Wolfe, D.L., Townson, A.F., Aubut, J., Abramson, C., Hsieh, J.T.C., Connolly, S. (eds.) Spinal Cord Injury Rehabilitation Evidence, pp. 16–29. Vancouver, Canada (2006)
3. Raineteau, O., Schwab, M.E.: Plasticity of motor systems after incomplete spinal cord injury. Nat. Rev. Neurosci. **2**, 263–273 (2001)
4. Kaplan, S.A., Chancellor, M.B., Blaivas, J.G.: Bladder and sphincter behavior in patients with spinal cord lesions. J. Urol. **146**, 113–117 (1991)
5. Verschueren, J.H.M., Post, M.W.M., de Groot, S., van der Woude, L.H.V., van Asbeck, F.W.A., Rol, M.: Occurrence and predictors of pressure ulcers during primary in-patient spinal cord injury rehabilitation. Spin. Cord **49**, 106–112 (2011)
6. Bonica, J.J.: Introduction: Semantic, epidemiologic, and educational issues. In: Casey, K.L. (ed.) Pain and Central Nervous System Disease: the Central Pain Syndromes, pp. 13–29. Raven Pr, New York (1991)
7. Richards, J.S., Meredith, R.L., Nepomuceno, C., Fine, P.R., Bennett, G.: Psychosocial aspects of chronic pain in spinal cord injury. Pain **8**, 355–408 (1980)
8. Summers, J.D., Rapoff, M.A., Varghese, G., Porter, K., Palmer, R.E.: Psychosocial factors in chronic spinal cord injury pain. Pain **47**, 183–189 (1991)
9. Catafau, S., Bosque, Q.: Mecanismos Fisiopatológicos da Dor Neuropática. Medica Panamericana, Madrid (2003)
10. Wasner, G., Lee, B.B., Engel, S., Mclachlan, E.: Residual spinothalamic tract pathways predict development of central pain after spinal cord injury. Brain **131**, 2387–2400 (2008)
11. Kirillova, I., Rausch, V.H., Baron, R., Jänig, W.: Mechano- and thermosensitivity of injured muscle afferents. J. Neurophysiol. **105**, 2058–2073 (2011)
12. Finnerup, N.B., Johannesen, I.L., Fuglsang-Frederiksen, A., Bach, F.W., Jensen, T.S.: Sensory function in spinal cord injury patients with and without central pain. Brain **126**, 57–70 (2003)
13. Kido, R.Y.Z., Zorzenoni, F.O., Tanhoffer, R., Varoto, R., Lima, V.M.F., Beinotti, F., Gaspar, M.I.F.A.S., Cliquet Jr., A.: Neuropathic pain in spinal cord injury subjects: somatosensory differences. In: 16th International Functional Electrical Stimulation Society Annual Conference, CD-ROM. Atha Comunicação e Editora, São Paulo (2011)

14. Melzack, R.: The McGill pain questionnaire: Major properties and scoring methods. Pain **1**, 277–299 (1975)
15. de Mattos Pimenta, C.A., Teixeira, M.J.: Questionário de Dor McGill: Proposta de Adaptação para a Língua Portuguesa. Rev. Esc. Enferm. USP **30**, 473–483 (1996)
16. Melzack, R., Torgerson, W.S.: On the language of pain. Anesthesiology **422**, 50–59 (1971)
17. Shy, M.E., Frohman, E.M., So, Y.T., Arezzo, J.C., Cornblath, D.R., Giuliani, M.J., Kincaid, J.C., Ochoa, J.L., Parry, G.J., Weimer, L.H.: Quantitative sensory testing. Neurology **60**, 898–904 (2003)
18. Arezzo, J.C., Schaumburg, H.H., Laudadio, C.: Thermal sensitivity tester. Diabetes **35**, 590–592 (1986)
19. Yang, G.H., Kyung, K.U., Srinivasan, M.A., Kwon, D.S.: Qualitative tactile display device with pin-array type tactile feedback and thermal feedback. In: 2006 IEEE International Conference on Robotics and Automation, pp. 3917–3922, Orlando (2006)
20. Kenshalo, D.R., Bergen, D.C.: A device to measure cutaneous temperature sensitivity in humans and subhuman species. J. Appl. Physiol. **39**, 1038–1040 (1975)
21. Dahlberg, A., Alaranta, H., Sintonen, H.: Health-related quality of life in persons with traumatic spinal cord lesion in helsinki. J. Rehabil. Med. **37**, 312–316 (2005)
22. Wolfe, D.L., Hsieh, J.T.C.: Rehabilitation practice and associated outcomes following spinal cord injury. In: Eng, J.J., Teasell, R.W., Miller, W.C., Wolfe, D.L., Townson, A.F., Aubut, J., Abramson, C., Hsieh, J.T.C., Connolly, S. (eds.) Spinal Cord Injury Rehabilitation Evidence, pp. 44–90. Vancouver, Canada (2006)
23. Kirshblum, S.C., Burns, S.P., Biering-Sorensen, F., Donovan, W., Graves, D.E., Jha, A., Johansen, M., Jones, L., Krassioukov, A., Mulcahey, M.J., Schmidt-Read, M., Waring, W.: International standards for neurological classification of spinal cord injury (Revised 2011). J. Spinal Cord Med. **34**, 535–546 (2011)
24. Callister Jr., W.D.: Fundamentals of Material Science and Engineering. Wiley, New York (2001)
25. Nomoto, S., Shibata, M., Iriki, M., Riedel, W.: Role of afferent pathways of heat and cold in body temperature regulation. Int. J. Biometeorol. **49**, 67–85 (2004)
26. Schepers, R.J., Ringkamp, M.: Thermoreceptors and thermosensitive afferents. Neurosci. Biobehav. Rev. **34**, 177–184 (2010)
27. Dimitrijevic, M.R.: Residual motor functions in spinal cord injury. Adv. Neurol. **47**, 138–155 (1988)
28. Sherwood, A.M., Dimitrijevic, M.R., Mckay, W.B.: Evidence of subclinical brain influence in clinically complete spinal cord injury: Discomplete SCI. J. Neurol. Sci. **110**, 90–98 (1992)
29. Finnerup, N.B., Gyldensted, C., Fuglsang-Frederiksen, A., Bach, F.W., Jensen, T.S.: Sensory perception in complete spinal cord injury. Acta Neurol. Scand. **109**, 194–199 (2004)

Pine Decay Assessment by Means of Electrical Impedance Spectroscopy

Elisabeth Borges[1(✉)], Mariana Sequeira[1], André Cortez[1],
Helena Catarina Pereira[1], Tânia Pereira[1], Vânia Almeida[1],
Teresa Vasconcelos[2], Isabel Duarte[2], Neusa Nazaré[2], João Cardoso[1],
and Carlos Correia[1]

[1] Instrumentation Center, Physics Department of the University of Coimbra,
Rua Larga, Coimbra, Portugal
eborgesf@gmail.com
[2] Centro de Estudos de Recursos Naturais Ambiente e Sociedade, Escola
Superior Agrária de Coimbra of the Instituto Politécnico de Coimbra,
Bencanta, Coimbra, Portugal
tvasconcelos@esac.pt

Abstract. Plant diseases, such as the pinewood disease, PWD, have become a problem of economical and forestall huge proportions. These diseases, that are asymptomatic and characterized by a fast spread, have no cure developed to date. Besides, there are no technical means to diagnose the disease in situ, without causing tree damage, and help to assist the forest management. Herein is proposed a portable and non-damage system, based on electrical impedance spectroscopy, EIS, for biological applications. In fact, EIS has been proving efficacy and utility in wide range of areas. However, although commercial equipment is available, it is expensive and unfeasible for in vivo and in field applications. The developed EIS system is able to perform AC current or voltage scans, within a selectable frequency range, and its effectiveness in assessing pine decay was proven. The procedure and the results obtained for a population of 24 young pine trees (*Pinus pinaster* Aiton) are presented. Pine trees were kept in a controlled environment and were inoculated with the nematode (*Bursaphelenchus xylophilus* Nickle), that causes the PWD, and also with bark beetles (*Tomicus destruens* Wollaston). The obtained results may constitute a first innovative approach to the diagnosis of such types of diseases.

Keywords: Electrical impedance spectroscopy · Bioimpedance · Early detection · Physiological states · Pinewood disease · Pinewood nematode · Plant diseases · Hydric stress · *Pinus pinaster* aiton · *Bursaphelenchus xylophilus* nickle

1 Introduction

Electrical impedance measurements performed in a wide frequency range give rise to a great number of techniques able to characterize solids, liquids and suspensions [1]. Lately, the method has proved its value also in the characterization of biological tissues and fluids, either in vitro or in vivo [1], and also to living plants [2–5], animals [6, 7]

© Springer-Verlag Berlin Heidelberg 2014
M. Fernández-Chimeno et al. (Eds.): BIOSTEC 2013, CCIS 452, pp. 54–73, 2014.
DOI: 10.1007/978-3-662-44485-6_5

and humans [8, 9]. Concerning the vegetal field, the applications of electrical impedance spectroscopy, EIS, techniques have been claiming significant and growing
acceptance, especially as a measure of the water content in the quality control processes
[2, 10, 11] of fruits [12] and vegetables [13, 14], as well as in the monitoring of the
maturation process of fruits [5, 15] and the physiological state of living plants under
adverse environmental conditions [3, 4].

The electrical impedance of a biological material, or simply bioimpedance, is a
passive electrical property that measures the opposition relatively to an alternating
current flow applied by an external electric field. The current I, as it passes across a
section of a material of impedance Z, drops the voltage V, established between two
given points of the same section, yielding the well-known generalized Ohm's law:
$V = IZ$, where V and I are complex scalars and Z is the complex impedance. The
law can be rewritten as $V = I|Z|e j^{\Theta}$ since, at a given frequency, the current flow I lag
the voltage V by a phase of Θ (i.e. the current signal is shifted $(\Theta/2\pi)T$ s to the right
with respect to the voltage signal, in the time domain). Hence, the result of the EIS
measurements is a set of complex (magnitude and phase) of impedance versus
frequency.

Cell membranes, intracellular fluid (cytosol) and extracellular fluid are the major
contributors of the impedance of biological tissues [8, 11]. The cytosol and the extracellular fluid are mostly constituted by water, consist in electrolytes, and act like ohmic
resistors, while the insulating membranes behave like capacitors [8, 11]. It is, therefore,
possible to depict the behaviour of a biological tissue by the representation of capacitive
and resistive elements of a respective equivalent electrical circuit. A commonly used
circuit to represent biological tissues consists of a parallel arrangement between
a resistor, simulating the extracellular fluid, and a second serial arrangement connecting
a resistor, this one of the cytosol, and a capacitor, of the membrane [8] - see Fig. 1. Since
the time constant for loading cell membranes is typically of the order of the microsecond
[11], tissue impedance can be measured in a frequency range up to tens of MHz [1]. In
this range of frequencies the membrane performs like an almost perfect capacitor,
allowing an estimation of the combined ohmic value of the cytosol and the extracellular
fluid. On the other hand, using direct current level, DC, (low frequency), the current
does not cross the membrane due to its insulator behaviour. This short circuit-like
actuation forces the current to flow exclusively through the extracellular fluid providing,
thus, a measure of its ohmic value. However, due to technical limitations and multiple
dispersions (α dispersions at low frequencies – tissues' electrolyte behaviour – and γ
dispersions at very high frequencies – tissues' aqueous behaviour [16]), the usage of DC
and very high frequency AC currents is restricted [8]. Therefore, it becomes quite more
convenient to determine the ohmic values by prediction. The model commonly used to
predict such values is the Cole bioimpedance model, in which the bioimpedance spectra
is represented by means of a Cole-Cole plot (see Fig. 1), that explores resistance versus
reactance, allowing the determination of the ohmic values of the cytosol and the
extracellular fluid. The mathematical expression descriptive of the Cole-Cole plots is the
Cole equation (here expressed has in [17]):

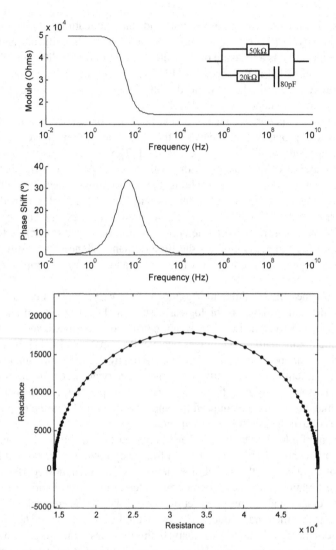

Fig. 1. Bode and Cole-Cole diagrams obtained by simulation with Matlab® for an electrical circuit representing a hypothetical biological tissue (right top of the figure).

$$Z = Z_\infty + \frac{\Delta R}{1 + (j\omega\tau)^\alpha}, \Delta R = R_0 - R_\infty \qquad (1)$$

Where Z is the impedance value at frequency ω (with $\omega = 2\pi f$), Z_∞ is the impedance at infinite frequency (high frequencies) (note: this term is misleading and is replaced by an ideal resistor R_∞), j is the complex number, R_0 is the impedance at DC frequency, τ is the characteristic time constant and α is a dimensionless parameter with a value between 0 and 1.

The relationship between reactance and resistance, perceived in a Cole-Cole plot, expresses the electrical properties of tissues. Diseases and nutritional or hydration levels may change their physiological state. These changes have direct influence in the impedance spectra. The phase angle and other interrelated indices, such as Z_0/Z_∞ [8] and Z_0/Z_{50} [13], have been used to extract information about the physiological condition of biological materials. The index Z_0/Z_{50} gains some significance since it is at the 50 kHz that the current starts passing through both cytosol/membranes and extracellular fluid, although the proportion varies from tissue to tissue [8].

The nature of the impedance excitation signal varies depending on the application. It is possible to excite the sample with a current and measure a voltage or to do the exact opposite. The discussion on what source, voltage or current, is the most convenient remains. Current sources, CS, provide suitably controlled means of current injection [18] and present reduced noise due to spatial variation when compared with voltage sources, VS [19]. However, CS accuracy decreases with high frequency [20], especially due to their output impedance degradation [19]. Since the impedance measurements are limited to field strength where the current is linear with respect to the voltage applied [11], or vice versa, CS need high-precision components [21] and a limited bandwidth operation range [20, 21] to overcome the stated limitation. On the other hand, VS, although producing less optimal EIS systems [21], can operate over a sufficient broad frequency range [20, 21] and are built with less expensive components [21].

Nowadays, instruments with high precision, high resolution and frequency ranges extending from some Hz to tens of MHz are commercially available [1]. However, in what concerns to the range of low or high frequencies (already above 100 kHz), the degradation of the excitation signal affects the accuracy of the measurements [1]. Besides, the typical solutions consist in impedance analyzers and LCR meters which are desktop instruments [1], unfeasible for in vivo [1] and in field applications.

Those EIS features, together with the lately demand in the vegetal applications, fundament this work. In fact, there are several plant pest and diseases affecting different cultures of huge economic and forestall importance, not only in our country but also around the world. This is the case of esca disease in vineyard, ink disease in chestnuts or pinewood disease, PWD, and bark beetles in pinus stands, among others. It is known that PWD, the case study presented in this paper, is caused by the nematode *Bursaphelenchus xylophilus* Nickle, that is housed in the tracheas of pine sawyer *Monochamus galloprovincialis* Olivier. Bark beetles in general and pine shoot beetle (*Tomicus destruens* Wollaston), in particular, play an important role in nematode establishment since they are responsible of pine decay, condition required for *M. galloprovincialis* oviposition. The PWD disease leads to a rotting process from within the plant (therefore, inside the stem) so that symptoms are difficult to see from the outside. Furthermore, there is still no cure available and the only solution to discontinue the progress of the disease throughout the culture is to identify and isolate the specimens that seem to have contracted the disease.

The authors propose a portable EIS system able to perform AC scans within a selectable frequency range. The system implements the phase sensitive detection, PSD, method and can drive either a current or a voltage signal to excite a biological sample in field or in vivo. The instrumentation was designed to be cost-effective and usable in several applications.

The design specifications are listed in Table 1.

A first interesting case study is presented for a population of 24 young pine trees (*Pinus pinaster* Aiton), from a controlled environment. Pine trees were inoculated with the nematode that causes the PWD and also with bark beetles (*T. destruens* Wollaston).

Table 1. Summary of specifications of the EIS system.

	Range	
Parameter	Current mode	Voltage mode
Measuring method	2 electrodes	
Frequency	1 kHz to 1 MHz	
Signal amplitude	25 µA	4.6 V
Impedance magnitude	2.5 kΩ to 100 kΩ	1.5 kΩ to 2.2 MΩ
Impedance phase	$-\pi$ rad to π rad	$-\pi$ rad to π rad
Mean absolute magnitude error	1675.45 Ω	709.37 Ω
Mean absolute phase error	2.45°	2.06°
Mean harmonic distortion	0.29 %	0.48 %
Mean SNR	117.0 dB	118.8 dB
Calibration	Automatically calibrated by software	

2 System Design

2.1 General Description

The developed EIS system employs two electrodes and consists of three main modules: signal conditioning unit, acquisition system (PicoScope® 3205A) and a laptop for data processing (Matlab® based software), as Fig. 2 depicts. There were built two different versions: one OEM for lab studies and another miniaturized version for field acquisitions.

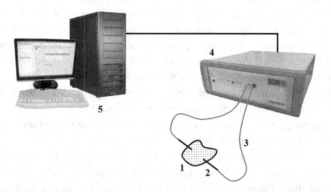

Fig. 2. Schematics of the EIS OEM system – (1) Biologic sample; (2) electrodes; (3) short coaxial cables; (4) EIS system conditioning unit and acquisition system, with the Picoscope® 3205A incorporated; (5) laptop/PC.

The electrodes being used are beryllium cooper gold platted needles with around 1.02 mm in diameter. The bioimpedance measurement requires the most superficial possible penetration of the electrodes in order to reduce the dispersion of the needles surface current density [18], and also to reduce damage on the biologic sample.

The digital oscilloscope PicoScope® 3205A has dual functionality: (1) synthesizes and provides the excitation AC signal to the conditioning unit (ADC function); (2) digitizes both excitation and induction signals at high sampling rates (12.5 MSps) and transfers data to the computer via USB where it is stored. The signal conditioning unit receives the exciting AC signal, coming from the PicoScope®, and amplifies it to be applied, through an electrode, to the specimen under study. The induced AC signal is collected by a second electrode and is redirected to the conditioning unit where it is also amplified. Both excitation and induced signals are conduced to the PicoScope® to be digitized.

It is also important to remark that the conditioning unit has an external switch that allows the user to select the mode type of excitation: by AC current or AC voltage. As previously mentioned, it is more advantageous to choose a mode of excitation over another, depending on the type of application.

The features of both excitation modes are described below.

2.2   Design Specifications

The current mode circuit employs the current-feedback amplifier AD844 in a non-inverting ac-coupled CS configuration (see Fig. 3), already studied by Seoane, Bragós and Lindecrantz, 2006 [22].

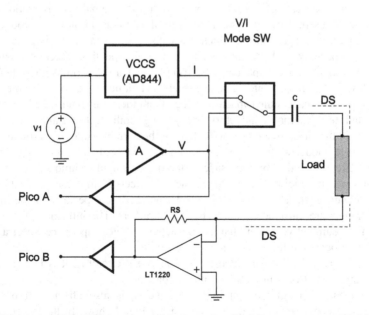

Fig. 3. Schematics of the EIS system conditioning unit – (1) AC current source; (2) AC voltage source; (3) current/voltage sense.

A common problem inherent to bioimpedance measurements is the charging of the dc-blocking capacitor between the source and the electrode due to residual DC currents [22]. This effect lead to saturation of the transimpedance output of the AD844. The DC feedback of the implemented configuration maintains dc voltage at the output close to 0 V without reducing the output impedance of the source. Subsequently, the output current, is maintained almost constant over a wide range of frequencies.

The high speed voltage-feedback amplifier LM7171 is employed in the voltage mode circuit (see Fig. 3). This behaves like a current-feedback amplifier due to its high slew rate, wide unit-gain bandwidth and low current consumption. Nevertheless it can be applied in all traditional voltage-feedback amplifier configurations, as the one used. These characteristics allow the maintenance of an almost constant voltage output over a wide range of frequencies.

Current or voltage signals resulting from voltage or current excitation modes, respectively, are sensed by a high speed operational amplifier, LT1220 (see Fig. 3), which performs reduced input offset voltage and is able of driving large capacitive loads.

Gain values of both current excitation source and voltage excitation source can be changed in order to extend the range of impedance magnitude. The transductance gain of the LT1220 is currently set to 5.1 kΩ and defines the gain of the system. Since the gain values are known and also the amplitude of the AC excitation signal, Vsin, from the PicoScope®, the EIS system is calibrated automatically by software.

2.3   Cables Capacitance

The characteristics of the cables that connect between the conditioning unit and the sample under study are also crucial. For an optimized signal-to-noise ratio, coaxial cable must be used. Nevertheless, this type of cable is prone to introduce high equivalent parasitic capacitances, which translate in errors in the bioimpedance measurements, especially at high frequencies. To overcome this effect, the employed RG174 RF coaxial cables (capacitance of 100 pF.m^{-1}) are as short as possible (around 15 cm). It was also implemented a driven shield technique to the coaxial cables, which permits to partially cancel the capacitive effect, that otherwise is generated between the internal and the external conductors, by putting both at the same voltage [23]. Reductions in the capacitive effect of 20.4 %, in the current mode, and around 35.8 %, in the voltage mode, at the highest frequencies are verified. Figure 4 depicts the capacitive effect reduction by the usage of the driven shield technique.

When assessing bioimpedance, the capacitive effects from cables are not the only exerting influence. In fact, phase shift effects, perceptible especially in the high frequencies range, are introduced mainly by the amplifiers. The influence of phase shift errors has a cumulative effect that is translated, in the impedance spectra, as an inflexion that occurs at high frequencies (see Fig. 4).

This behaviour can be simulated by an equivalent circuit as it is like the system analyzes any load always in parallel with a capacitor.

The impedance magnitude, at high frequencies, is also affected. It presents a characteristic decline as the bode diagrams of the Fig. 4 show. In the developed EIS system, the slight decline of the impedance magnitude is due to the loss of the product gain-bandwidth of the LT1220 for high frequencies.

Since stray capacitances are considered systematic errors of the system, thus affecting all the measurements, theirs influence doesn't directly affect the results. Although, it is convenient to have an approached sense of the real equivalent circuit (see Fig. 5), in such a way that the effect of all the parasitic elements can be considered and/or discounted where justified.

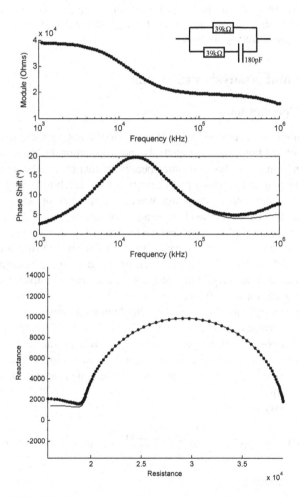

Fig. 4. Bode and Cole-Cole diagram showing the reduction of cables capacitive effect by the application of the driven shield technique. The voltage mode excitation was used to analyze the circuit at the right top. The reduction is more noticeable at high frequencies where the capacitive effects have more influence.

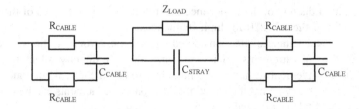

Fig. 5. Equivalent electric circuit of all parasitic elements affecting impedance measurements of a load, Z_{LOAD}. The effect of the stray capacitances from cables, C_{CABLE}, is minimized by the driven shield. Other stray capacitance effect, C_{STRAY}, due primarily to the phase shift of amplifiers, can be minimized by software.

3   Software and Analysis Processing

3.1   General Specifications

The software interface, developed with Matlab® tools, allows the operator to choose the parameters of the bioimpedance analysis and to monitor the data acquisition. The operator can perform an analysis for one specific frequency or alternatively can carry out a true bioimpedance spectroscopy. These two software functioning modes can be programmed for continuous monitoring, where the number of acquisitions and the intervals between them are specified by the user. The interface includes a basic function that allows a preview of the Bode and Cole-Cole diagrams of the acquired data. Bioimpedance *txt or *mat files are saved in a pre-determined directory with a file-name, previously chose by the user, to which date and time are associated. Each file contains information about magnitude, phase shift and real and imaginary parts of the measured impedance, for each frequency.

The type of bioimpedance spectroscopy implemented consists in a frequency AC sweep, whose limit values are 1 kHz and 1 MHz. Notwithstanding, the software allows the operator to choose other frequency limits, as well as the number of intervals between them. In addition, it can be choose a linear or logarithmic analysis. Therefore, the frequencies, f(i), over which the impedance of a sample is analyzed, are determined by the following equations:

For a linear analysis:

$$f(i) = f\_start + i * \frac{f\_stop - f\_start}{n-1}, \forall \subset [0, n-1] \wedge n \in N \qquad (2)$$

For a logarithmic analysis:

$$f(i) = f\_start^* \left[10^{\frac{\log_{10}\left(\frac{f\_stop}{f\_start}\right)}{(-1)}} \right]^i, \forall i \subset [0, n-1] \wedge n \in N \qquad (3)$$

Where f_star and f_stop are, respectively, the first and final frequencies of the AC sweep, and n the number of intervals between them.

3.2 PSD Method

To assess the impedance phase shift it is implemented a digital Phase Sensitive Detection, PSD, method with a novel implication. As stated in the literature, the PSD method is a quadrature demodulation technique that implements a coherent phase demodulation of two reference (matched in phase and quadrature) signals [24]. It is also known that this method is preferable over others especially when signals are affected by noise [24].

The signal from the Picoscope® that corresponds to the current, $V_I = B\sin(\omega t + \varphi_2)$, is set as the reference signal. Since the phase of the signal V_I is not controlled, it is easily understandable that it doesn't necessarily contain a null phase. This statement remains valid whether V_I is used to excite the sample, in the current mode, or whether it corresponds to the current passing through the sample, in the voltage mode. The signal from the Picoscope® that corresponds to the voltage, $V_V = A \sin(\omega t + \varphi_1)$, also contains a non-null phase. Both amplitudes, A and B, are also different from each other and none equals to 1.

The developed PSD algorithm was tested with Matlab® for several phases and amplitudes without the theoretical requirements (i.e., ensure that the reference signal has null phase at the origin and that its amplitude equals to 1 [24]). For all of them it was showed an always corrected phase shift assessment, when compared to the results obtained for a reference signal with the theoretical characteristics.

In addition, the mathematical resolution for the demodulation of two signals with non-null phases and amplitudes not equal to 1, corresponds to the phase difference between both signals. The following mathematical demonstration and the schematic block diagram (shown in Fig. 6) support the results obtained with the simulation.

Assuming that the analog input signals $V_V(t)$ and $V_I(t)$ are sine waves of frequency f, amplitude A and B, respectively, and initial phase φ_1 and φ_2, respectively:

$$V_V(t) = A \sin(2\pi f + \varphi_1) \tag{4}$$

$$V_I(t) = B \sin(2\pi f + \varphi_2) \tag{5}$$

The digitized input signals $V_V(n)$ and $V_I(n)$ are obtained from $V_V(t)$ and $V_I(t)$, respectively, by sampling at a frequency f_s, where f_s is a multiple of the f:

$$V_V(n) = A \sin\left(\frac{2\pi f n}{f_s} + \varphi_1\right), n \in [0, N-1] \tag{6}$$

$$V_I(n) = B \sin(\frac{2\pi f n}{f_s} + \varphi_2), n \in [0, N-1] \tag{7}$$

Where N is the number of samples. N/f_s is the measurement time and must be an exact multiple of $1/f$, so that there is whole number of cycles of the sine wave.

The signal $V_I(n)$ is set as reference. The quadrature reference signal, $V_{Iq}(n)$, results from the reference signal shifted by a phase of 90°. Consequently, $V_{Iq}(n)$ is cosine with the same frequency, amplitude and initial phase as $V_V(n)$:

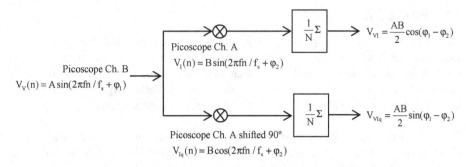

Fig. 6. Schematic of phase-sensitive demodulator implemented in the EIS system.

$$V_I(n) = B \sin\left(\frac{2\pi fn}{f_s} + \varphi_2\right), n \in [0, N-1] \tag{8}$$

$$V_{Iq}(n) = B \cos\left(\frac{2\pi fn}{f_s} + \varphi_2\right), n \in [0, N-1] \tag{9}$$

The output voltages of the system shown at Fig. 6 are:

$$V_{VI} = \frac{1}{N} \sum_{n=0}^{N} V_V(n) V_I(n) \tag{10}$$

$$V_{VIq} = \frac{1}{N} \sum_{n=0}^{N} V_V(n) V_{Iq}(n) \tag{11}$$

The multiplication between two sine signals, with the same frequency, results in a sum of a DC signal and a sine signal with a frequency that is the double of the original. The double frequency component can be suppressed since the time is a multiple of the period of the input sine signal. Therefore, it remains only the DC component which amplitude is dependent on the amplitude of the individual sine signals and their relative phase:

$$V_{VI} = AB \cos(\varphi_1 - \varphi_2) \tag{12}$$

$$V_{VIq} = AB \sin(\varphi_1 - \varphi_2) \tag{13}$$

From the expressions above, the resulting amplitude and phase can be determined:

$$\varphi_1 - \varphi_2 = \arctan\left(\frac{V_{VIq}}{V_{VI}}\right) \tag{14}$$

$$A = \frac{2}{B} \sqrt{(V_{VI})^2 + (V_{VIq})^2} \tag{15}$$

The determined phase is actually a phase difference between the demodulated signal, V_V and the reference signal, V_I, i.e., it corresponds to the phase difference between voltage and current signals. Figure 7 shows the consistence of the algorithm when the impedance phase of a real data is compared with a spice simulation in Cadence®.

The determination of impedance magnitude cannot be achieved by the PSD method, since the amplitude equation (Eq. 15) shows a dependence on the amplitude of the reference signal, which, in this case, is not equal to 1. Hence, to assess amplitude, the EIS system algorithm processes the root mean square, RMS, of both signals $V_V(t)$ and $V_I(t)$ from de channel B and A, respectively, of the Picoscope®. In this manner, the impedance magnitude is given by the ratio between the RMS value of the signal $V_V(t)$ and the rms value of the signal $V_I(t)$:

$$|Z| = \sqrt{\frac{\sum_{i=0}^{N}(V_v^2)_i}{\sum_{i=0}^{N}(V_I^2)_1}}\text{Gain}, \forall i \in [0, N-1] \qquad (16)$$

Where *Gain* is the EIS system gain defined by the transconductance gain of the LT1220 (see Sect. 2.2).

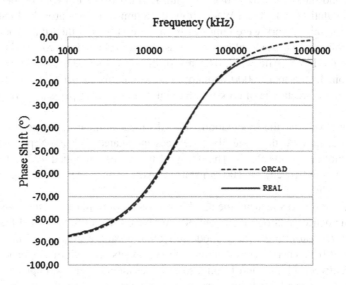

Fig. 7. Comparison between impedance phase of a real data and Cadence® simulated data for a RC circuit. The deviation that occurs between the graphics, at high frequencies, is due to the influence of stray capacitances (see Sect. 2).

4 Biology Application Study

4.1 Materials and Methods

Twenty four young healthy pine trees (Pinus pinaster Aiton), with about 2, 5 m tall and 2 to 3 cm in diameter, constituted the population for the conducted study. The pine trees were placed in vases in a controlled water environment at a greenhouse. Half of the tree population was watered during 5 min. per day ($\sim 133,37$ mL/day), while the other half were watered during only 2 min. per day ($\sim 66,67$ mL/day). This second half was less watered to maintain a relevant level of hydric stress.

After one month elapsed since the pine trees were placed in the greenhouse, the inoculations with pinewood nematode, PWN, (*Bursaphelenchus xylophilus* Nickle) and with the bark beetle (*T. destruens* Wollaston) were performed. Six pines were inoculated with PWN, other 6 pines were inoculated with bark beetles, other 6 pines were inoculated simultaneously with PWN and bark beetles, while the remaining 6 were kept under normal conditions, i.e., healthy. The position of the pines in the greenhouse was made so that each sub-group had the same number of pines with normal watering (5 min/day) and with reduced watering (2 min/day).

To perform the inoculations with bark beetles, callow adults were collected immediately after emergence. In each tree, a box containing 15 beetles were placed in the middle and the device was covered using Lutrasil tissue to avoid beetles escape.

The inoculation with the PWN followed an innovative approach. Firstly, three 2×2 cm rectangle of cork were removed from the first tiers of the trunk (about 1,80 m above the soil) and exposed phloem was erased with a scalpel in order to increase the adhesion of the PWN. Afterward, 0,05 mL of of a PWN suspension was placed on in each incision. In the total, 6000 nematodes were inoculated per tree. To finalize the task, the removed rectangle of cork was fixed in the respective place and wrapped with plastic tape.

Seventy days after the inoculations, the EIS measurements were performed in all the tree population. At this time, the pine trees inoculated with PWN presented some visually symptoms of the PWD. The decay of those trees, rounded 40 %. Two of the healthy pines died (decay of 100 %) due to hydric stress. All remaining individual appeared healthy.

To perform the EIS measurements, the electrodes were placed in the trunk of each tree, in a diametric position, and about 30 cm above the soil. It was used the portable EIS system version in the voltage mode of excitation and a frequency range between 1 kHz and 1 MHz. There were taken two measurements for each tree. The acquisitions took place between 11 a.m. and 13 p.m. since it was already verified in previous studies that at this time period the trees impedance is higher and presents few variation (see Fig. 8 – Sect. 4.2).

In order to relate the EIS data with the PWD and the stage of the disease, the trunk of the pine trees inoculated with PWN were cut in three distinct regions to perform a count of nematodes. The cuts were executed: (a) immediately below the inoculation incision (180 cm above the soil); (b) 30 cm above the soil (where EIS measurements took place); and (c) in the middle of the previous two cuts (approximately 80 cm above the soil).

After the EIS measurements, two healthy pines were monitored by two independent portable EIS systems. After a week of monitoring, the same pines were inoculated with PWN, and the measurements continued during 7 more weeks. The main purpose of this last experiment was to study the variation of the pine EIS profiles during the decay due to the PWD.

4.2 Results

For each obtained impedance spectra there were assessed several impedance parameters. Due to paper space limitation and also because it is a well-known impedance parameter, it will only be presented the results obtained for the ratio Z_1/Z_{50}. Note that it is used the index 1, that corresponds to the lowest analyzed frequency (1 kHz), instead of the index 0, as explained in Sect. 1.

4.2.1 Impedance Daily Oscillations

The EIS measurements revealed that EIS Cole profiles have a daily oscillation. To analyze this behavior it was calculated the R_1/R_{50} ratio (R represents module) for a period of 4 days.

To confirm the daily oscillation it was calculated the fast fourier transform of this ratio. A frequency of 11,57 µHz was clearly founded, which corresponds to a frequency of 24 h.

The higher values of the ratio R1/R50 correspond to the night period, while the lower values correspond to the day period where the temperature and luminance are higher (between 11 a.m. and 3 p.m.). Previous studies on plants also shown that, during the day period, the variation of impedance values is lower than the one observed at the night period. This was the main reason that lead to performing the EIS measurements between the 11 a.m. and 3 p.m.

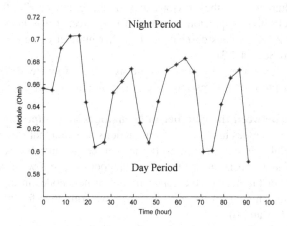

Fig. 8. Variation of the R_1/R_{50} ratio during the monitoring of a healthy pine tree. The impedance values show a daily oscillation that is characteristic of the studied trees.

4.2.2   Discrimination Between Physiological States

In order to compare results between the different physiological states of the trees, there were assessed several impedance parameters. The impedance parameter that showed better results was the Z_1/Z_{50} ratio – see Fig. 9.

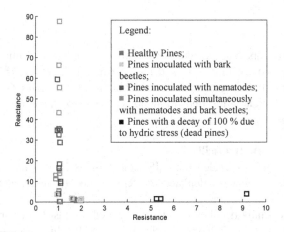

Fig. 9. Values of the impedance parameter Z_1/Z_{50} for each of the 24 pine trees. Note that there are represented two values for each pine.

The analysis of the obtained results shown that the healthy pines and the pines inoculated with bark beetles have similar Z_1/Z_{50} values. In fact, the bark beetles doesn't damage the inner structure of the trees, therefore it was expected that the impedance profiles were similar between healthy pines and pines inoculated with bark beetles.

On the other hand, Z_1/Z_{50} values for the pines inoculated with nematodes and also, for those inoculated simultaneously with nematodes and bark beetles, locates in the same region, different from the previous one, of the graph of Fig. 9. Those values present a relatively high dispersion in terms of reactance. It was later confirmed that higher reactance Z_1/Z_{50} values correspond to higher number of nematodes in the tree (see Fig. 10 from Sect. 4.2.3).

The pines that died due to hydric stress (decay of 100 %) were also studied and the Z_1/Z_{50} parameter present high resistance values in relation to all the other pines.

4.2.3   Relation Between the Number of Nematodes and Impedance Parameters

The counting of nematodes in the several cut sections revealed that the concentration of nematodes was higher in the cut sections (b) and (c) for the pines less watered (pines 1, 2 and 3) – see Table 2. It is known that the nematodes move toward watered regions along the trunk. For this reason, the concentration of nematodes in the lower parts of the trunks was much higher for the pines with less watering than for those with regular watering (pines 4, 5 and 6).

These results for the nematodes counting support the already referred results obtained for the Z_1/Z_{50} impedance parameter. In fact, it is observed a clear relation

Table 2. Number of nematodes in the trunks of pine trees per cut sections.

Tree	Cut section	Number of nematodes in 0.05 mL
1	a	1
	b	0
	c	133
2	a	43
	b	1
	c	0
3	a	0
	b	0
	c	112
4	a	4
	b	20
	c	0
5	a	0
	b	17
	c	0
6	a	0
	b	0
	c	14

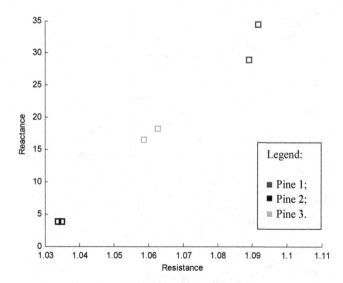

Fig. 10. Values of the impedance parameter Z_1/Z_{50} for pines inoculated with nematodes and with low watering (pines 1, 2 and 3 from the Table 2). Note that there are represented two values for each pine.

between the number of nematodes and the reactance dispersion for the Z_1/Z_{50} parameter, as Fig. 10 shows. The higher the number of nematodes is, the higher is the reactance value of Z_1/Z_{50}. It is considered that the dispersion in terms of resistance is not significant when compared with values from pines in other physiological condition – see Fig. 9 from Sect. 4.2.2.

4.2.4 EIS Monitoring During Pine Decay

There were monitored two healthy pines, one with low watering (2 min/day) and another with regular watering (5 min/day). After one week from the beginning of the monitoring, both pines were inoculated with nematodes. It was shown again a

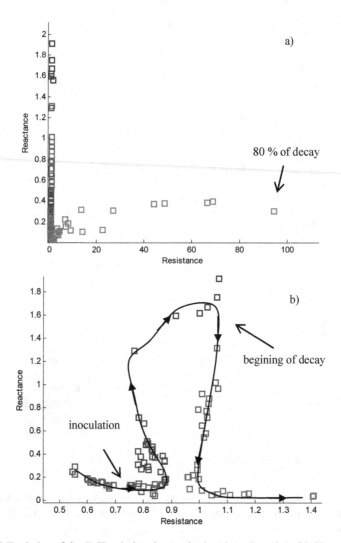

Fig. 11. (a) Evolution of the Z_1/Z_{50} during the monitoring time (8 weeks). (b) Closer view from Z_1/Z_{50}, showing a hysteresis-like behavior.

dispersion of the reactance values of the Z_1/Z_{50} parameter, as Fig. 11 shows. As time passed the reactance values became higher. The higher values of reactance were achieved for the pine with less watering. According to the previous presented results, it was expected that the number of nematodes increase in the below part of the trunk for the pines with less watering; and consequently, to observe a higher rising of the reactance of the Z_1/Z_{50} parameter. After the 6th week, pines start to decay strongly and it was observed a relevant decrease of the reactance and a significant increase of the resistance for the same parameter – see Fig. 11. The higher values of resistance were achieved for the pine with less watering, and also in a shorter period of time. At the end of the monitoring, the decay of the pines, evaluated by an expertise, was about 80% for the pine with regular watering and 100% for the pine with less watering.

From the Fig. 11(b), that represents a closer view of the Z_1/Z_{50} values for the monitoring, it is possible to observe that the path followed during the period of nematodes population increasing is different from the path followed during the period of decay, i.e., it is observed an hysteresis-like behavior.

5   Conclusions

The EIS system was developed in order to ensure a robust, efficient and fast bioimpedance analysis. The adaptability to different biological applications, the portability and the usage of easily accessible and affordable components, were preferred aspects taken into account. In this manner, the system allows the user to choose the settings of the analysis that best fit to a specific application. Furthermore, there were built two versions of the equipment: one OEM version for lab tests and a miniaturized version for field applications.

The system is able to perform AC scans within a frequency range from 1 kHz to 1 MHz. The frequency limits and the number of intervals of the scan can be selected at the user interface (developed with Matlab® tools). The type of signal used to excite de sample, voltage or current, can be preselected by an external switch. This allows the usage of the source with the best behavior in a concrete application.

The implemented PSD algorithm allows a very good phase shift assessment without the need to use a reference signal of amplitude equal to 1 and null-phase at the origin. In fact, the signal set as reference has undetermined phase and amplitude. All the algorithm tests revealed results analogous to the theoretical.

To overcome problems inherent to stray capacitive effects from cables, a driven shield technique is applied. The maximum phase shift reduction is estimated at 20.4 % for the current excitation mode and at 35.8 % for the voltage mode.

The biological application study aimed at discriminating between different pine tree physiological states.

The obtained results suggest that the implemented method may constitute a first innovative approach to the early diagnosis of plant diseases. In fact, the achieved impedance parameters allow discriminating three different physiological states: healthy trees, trees with PWD and trees in hydric stress.

The trees with PWD present Z_1/Z_{50} ratio with high values of reactance, suggesting that the current flows preferably trough the cytosol. In fact, the action of the nematodes inside the tree may destroy cell membranes. This means that membranes capacitor effect becomes less significant in the impedance measurement.

It was also shown that the number of nematodes and Z_1/Z_{50} impedance parameter are related. The higher the number of nematodes is, the higher the reactance of the ratio is.

The action of bark beetles seems not to interfere, at least in measurable terms, in the level of hydric stress of pine trees.

Healthy trees, with high values of hydric stress (decays above 80 %), and also trees with PWD at advanced stages, revealed low reactance and high resistance for the same studied parameter. The high values of resistance are justified due to the water loss in the tree. Consequently, it means that for this specific case, the method cannot distinguish between trees with PWD or trees with high level of hydric stress but with no disease. However, it is known that advanced stages of PWD promote high levels of hydric stress. This means that both cases represent, in practical terms, the same situation, i.e., the tree presents high probability to die. In addition, in the stages where the method is able to distinguish between healthy trees and trees with PWD, the decay was determined to round the 40 %. Therefore, if a cure is available, this diagnosis could help to administrate a treatment and reverse the disease evolution.

Hence, the main conclusion of the developed study is that the studied method could be used to assess physiological states of living pine trees, and that the Z_1/Z_{50} impedance parameter could be applied as a risk factor.

Acknowledgements. We acknowledge support from Fundação para a Ciência e Tecnologia, FCT (scholarship SFRH/BD/61522/2009).

References

1. Callegaro, L.: The metrology of electrical impedance at high frequency: A review. Meas. Sci. Technol. **20**, 022002 (2009)
2. Fukuma, H., Tanaka, K., Yamaura, I.: Measurement of impedance of columnar botanical tissue using the multielectrode method. Electron. Commun. Jpn. **3**, 1–11 (2001)
3. Repo, T., Zhang, G., Ryyppö, A., Rikala, R.: The electrical impedance spectroscopy of scots pine (Pinus sylvestris L.) shoots in relation to cold acclimation. J. Exp. Bot. **51**(353), 2095–2107 (2000)
4. Väinöla, A., Repo, T.: Impedance spectroscopy in frost hardiness evaluation of rhododendron leaves. Ann. Bot. **86**, 799–805 (2000)
5. Bauchot, A.D., Harker, F.R., Arnold, W.M.: The use of electrical impedance spectroscopy to assess the physiological condition of kiwifruit. Postharvest Biol. Technol. **18**, 9–18 (2000)
6. Dean, D.A., Ramanathan, T., Machado, D., Sundararajan, R.: Electrical impedance spectroscopy study of biological tissues. J Elesctrostat. **66**(3–4), 165–177 (2008)
7. Willis, J., Hobday, A.: Application of bioelectrical impedance analysis as a method for estimating composition and metabolic condition of southern bluefin tuna (Thumus maccoyii) during conventional tagging. Fish. Res. **93**, 64–71 (2008)
8. Kyle, U., et al.: Bioelectrical impedance analysis – Part I: Review of principles and methods. Clin. Nutr. **23**, 1226–1243 (2004)

9. Giouvanoudi, A.C., Spyrou, N.M.: Epigastric electrical impedance for the quantitative determination of the gastric acidity. Physiol. Meas. **29**, 1305–1317 (2008)

10. Vozáry, E., Mészáros, P.: Effect of mechanical stress on apple impedance parameters. In: IFMBE Proceedings, vol. 17 (2007)

11. Pliquett, U.: Bioimpedance: A review for food processing. Food Eng. Rev. **2**, 74–94 (2010)

12. Fang, Q., Liu, X., Cosic, I.: Bioimpedance study on four apple varieties. IFMBE Proc. **17**, 114–117 (2007)

13. Hayashi, T., Todoriki, S., Otobe, K., Sugiyama, J.: Impedance measuring technique for identifying irradiated potatoes. Biosci. Biotechnol. Biochem. **56**(12), 1929–1932 (1992)

14. Dejmek, P., Miyawaki, O.: Relantionship between the electrical and rheological properties of potato tuber tissue after various forms of processing. Biosci. Biotechnol. Biochem. **66**(6), 1218–1223 (2002)

15. Harker, F.R., Maindonald, J.H.: Ripening of nectarine fruit – changes in the cell wall, vacuole, and membranes detected using electrical impedance measurements. Plant Physiol. **106**, 165–171 (1994)

16. Ivorra, A.: Bioimpedance monitoring for physicians: An overview. Centre Nacional de Microelectrònica, Biomedical Applications Group (2003)

17. Grimnes, S., Martinsen, O.: Bioimpedance and Bioelectricity Basics, 2nd edn. Academic Press of Elsevier, London (2008)

18. Rafiei-Naeini, M., Wright, P., McCann, H.: Low-noise measurements for electrical impedance tomography. IFMBE Proc. **17**, 324–327 (2007)

19. Ross, A.S., Saulnier, G.J., Newell, J.C., Isaacson, D.: Current source design for electrical impedance tomography. Physiol. Meas. **24**, 509–516 (2003)

20. Yoo, P.J., Lee, D.H., Oh, T.I., Woo, E.J.: Wideband bio-impedance spectroscopy using voltage source and tetra-polar electrode configuration. J. Phys. **224**, 012160 (2010)

21. Saulnier, G.J., Ross, A.S., Liu, N.: A High-Precision Voltage Source for EIT. Physiol. Meas. **27**, S221–S236 (2006)

22. Seoane, F., Bragós, R., Lindecrantz, K.: Current source for multifrequency broadband electrical bioimpedance spectroscopy systems. In: Proceedings of the 28th IEEE on A Novel Approach, 1-4244-0033 (2006)

23. Yamamoto, T., Oomura, Y., Nishino, H., Aou, S., Nakano, Y.: Driven shield for multi-barrel electrode. Brain Res. Bull. **14**, 103–104 (1985)

24. He, C., Zhang, L., Liu, B., Xu, Z., Zhang, Z.: A digital phase-sensitive detector for electrical impedance tomography. In: IEEE Proceedings (2008)

Gold Standard Generation Using Electrooculogram Signal for Drowsiness Detection in Simulator Conditions

N. Rodríguez-Ibáñez[1]([⊠]), P. Meca-Calderón[2],
M.A. García-González[3], J. Ramos-Castro[3],
and M. Fernández-Chimeno[3]

[1] Ficosa International S.A, Can Magarola, Ctra. C-17 Km 13. Mollet Del Valles,
08100 Barcelona, Spain
noelia.rodriguez@ficosa.com
[2] Idneo Technologies S.L, Pol. Ind. Can Mitjans, Viladecavalls,
08232 Barcelona, Spain
pablo.meca@ficosa.com
[3] Group of Biomedical and Electronic Instrumentation of the Department
of Electronic Engineering of the Technical University of Catalonia (UPC),
08034 Barcelona, Spain
{miquel.angel.garcia,jramos,mireia.fernandez}@upc.edu

Abstract. The aim of this work is to generate a Gold Standard signal to assess the alertness state of drivers based on the Electrooculogram (EOG) dynamics derived from a polysomnography device. Different EOG patterns have been analyzed in order to determine the relation between ocular activity and sleep onset while doing complex tasks. More than 15 h of laboratory tests were analyzed in order to detect drowsiness while doing different cognitive activities. The proposed method has a sensitivity of 92.41 % and a Predictive Positive Value (VPP) of 93.41 % in detecting drowsiness in laboratory conditions. The results show that the proposed index may be promising to assess the alertness state of real drivers.

Keywords: Drowsiness detection · Gold standard · Driver monitoring · Electrooculogram · Electroencephalogram

1 Introduction

Drowsiness is one of the main causes of vehicle accidents. A recent study showed that 20 % of crashes and 12 % of near-crashes were caused by drowsy drivers [1]. The morbidity and mortality associated with drowsy-driving crashes are high, perhaps because of the higher speeds involved combined with delayed reaction time [2].

Driver behavior monitoring, and the reliable detection of drowsiness and fatigue is one of the leading objectives in the development of new Advanced Driver Assistance Systems (ADAS). Of the use of biomedical signal analysis to detect drowsiness in real vehicles appears the need of an objective gold standard to compare with the selected signals, in this case thoracic effort. The most objective signal to assess the sleep onset

© Springer-Verlag Berlin Heidelberg 2014
M. Fernández-Chimeno et al. (Eds.): BIOSTEC 2013, CCIS 452, pp. 74–88, 2014.
DOI: 10.1007/978-3-662-44485-6_6

phase is Electroencephalography (EEG). The problem associated to this signal is that, in real environments (i.e. vehicles) the actual devices used in hospital environment to acquire the data presents artefacts due to vibration and movements of the vehicle that masks the real EEG signal.

The aim of this work is to validate the EOG signal as a new Gold Standard and the EOG acquisition device as a good quality device to ensure the optimal quality of the data. The EOG signal is highly robust to artefacts signal when compared to EEG and valuable to compare with our drowsiness detection index based on thoracic effort variability (TEDD) in real environments [3].

2 Prior Work

2.1 EEG and EOG Signals as Gold Standard

During active wakefulness (i.e., when the person is awake and pursuing normal activities), the EEG is characterized by high frequencies (i.e., 16 to 25 Hz) and low voltage (i.e., 10 to 30 μV). EOG readings during wakefulness exhibit Rapid Eye movements (REM).

During relaxed wakefulness (i.e., when a person is awake but has his or her eyes closed and is relaxed), the EEG is characterized by a pattern of alpha waves with a frequency of 8 to 12 Hz and an amplitude of 20 to 40 μv. EOG readings show slow, rolling movements [4], increase of blinking frequency and lots of saccadic response at the transition to NREM sleep onset [5].

2.2 EEG and EOG Signals Acquisition in Real Driving Environments

The most important handicap in the field of drowsiness detection in real driving environments is the fact that the filtering of the low amplitude biomedical signals in order to eliminate vibration and movement artifacts is a very complex work that, in most cases, also affects the original signal of interest.

Hundreds of real vehicle tests have been made in the last three years with the objective of finding a biomedical signal robust to artifacts and also related to sympathetic-vagal system to provide drowsiness information in real vehicle tests.

The EEG signal has always been the most objective signal to define drowsiness in laboratory conditions but in real vehicle tests the EEG signal presents several problems as artifacts and the fact that the EEG codifications of the Rechtschaffen & Kale's method [6] is only recommended with closed eyes. According to the EEG-EOG studies there is a relation between EEG waves and EOG patterns that allows generating an objective Gold Standard signal for drowsiness detection from EOG signal.

For the first set of real vehicle tests the EEG and EOG signal was acquired with a Bitmed eXim Pro polysomnography device. The EOG signal quality was good before and after filtering the vehicle vibrations and movement artifacts but the EEG signal was lost in the filtering process due to the fact that the frequency of the vibrations was the same frequency that the waves of interest (theta and alpha waves).

Following this results, currently we have focused on finding new devices that avoids the problem of the artifacts in EEG signal. Two different tests have been made in real vehicles with two different polysomnography devices:

- Nicoletta wireless device
- Bionic EEG holter that provides active electrode technology

Although both systems show improvements in the EEG signal quality it hasn't enough quality to extract the drowsiness information. The filtering solution had the same problems that with other polysomnography devices.

Taking into account these results and the fact that the EEG and the EOG signals are physiologically related we recommend the use of EOG data as Gold Standard in real vehicle tests. This work proposes different indexes based on slow eye movement's detection, blinking frequency and saccade movement's inhibition.

3 Materials and Methods

3.1 Measurement Protocol

The participants in the test were 17 male and 6 female with ages between 20 and 29 years and no clinical conditions. These tests were designed and performed in laboratory conditions.

To perform these tests the setup was equipped with a biomedical monitor (Bitmed eXim Pro, BitMed) and a webcam. The biomedical signals selected as significant for this test were the external observer (video), Electrooculography (EOG) and thoracic effort. The thoracic effort signal was measured in all cases using an inductive band located at the middle trunk above the diaphragm. The EOG signal was measured with four Electromyography (EMG) single electrodes: two were located in the outer cantus of each eye in the case of the horizontal EOG setup, and two more electrodes located in the upper part and in the lower part of the right eye (Fig. 1). The EOG and the respiratory signal were sampled at 100 Hz.

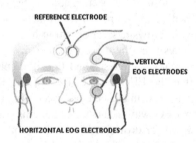

Fig. 1. EOG instrumentation.

Video signal was recorded to generate the external observer variable.

3.2   Test Design

The test was designed to classify the different eye movements and set a level of eye activity or eye inactivity (related to drowsiness). The test setup consists of a vehicle seat and a 19" inches monitor in front so the subject of the tests can see the patterns classification video seated on the vehicle seat. The test has two parts:

1. Patterns classification part

   Once the subjects are seated and connected to the acquisition systems the first part starts and they were asked to watch a 5 min video with the objective of following a point on the screen to force the movement of the eye for the following patterns of interest:
   - Saccadic movement
   - Compensation movements
   - Blinking
   - Fixed gaze
   - Seeking movements
   - Slow Eye Movements (SEM)

   The monitor has to be no more than 15 cm far from the face of the subject.
2. Drowsiness state classification part

   The subject rest relaxed in the seat for over 20 min with eyes open.

3.3   Patterns Classification

The patterns selected as indicative of drowsiness where the following (Fig. 2):

Saccadic Movements. Saccadic movements are defined as rapid symmetric eye movements with the objective of constantly changing the retinal focus from one point to the next point in the visual path.

There is a linear relation between the size of the saccade and the velocity of the ocular movement. The mean duration of saccadic movements ranges between 30 and 120 ms.

In an awake state these movements are mostly voluntary and they are used to redirect the gaze to the point of interest of the scene. In fatigue and drowsy states the saccadic speed decreases [7, 8] and the latent period between saccades increases.

Compensation Movement. Compensation movements are reflex movements that imply the coordination of both eyes. These movement works as an object fixation mechanism while moving head or body. The most important is the Vestibule-Ocular Reflex (VOR) with a response time of 16 ms (Fig. 3).

Blinking. Blinking is the rapid closing and opening of the eyelid that provides moisture to the eye by irrigation using tears and a lubricant that the eyes secrete. The mean frequency of blinks in a normal subject is 12 to 19 blinks per minute. This frequency can be influenced by internal or external factors. Fatigue and drowsiness decreases the blinking rate and increases the percentage of eye closure time (Fig. 4).

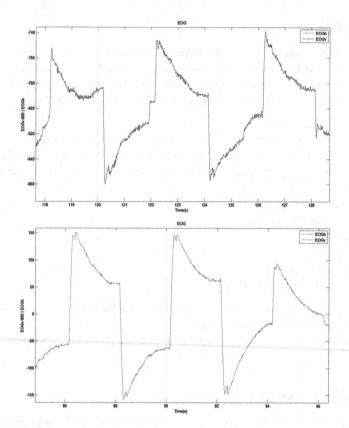

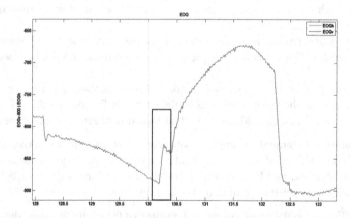

Fig. 2. EOGv saccades (up) and EOGh saccades (down). Filter: band-pass 0.2–30 Hz.

Fig. 3. Compensation movement in EOGv signal. Band-pass filter 0.2–30 Hz.

Fixed Gaze. The fixed gaze or ocular movement fixation can be a characteristic pattern of interest in one point or low cognitive activity depending on the duration of the pattern. In a normal context the fixed gaze duration ranges between 200 ms and 350 ms

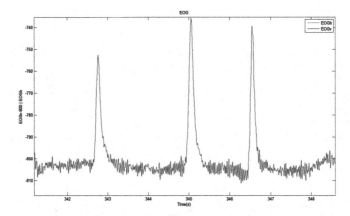

Fig. 4. Blinking pattern on EOGv signal. Band-pass filtering 0.2–30 Hz.

with open eyes. In phases of fatigue or drowsiness the fixed gaze time can reach 3 s becoming an ocular loss of activity [9, 10]. (Fig. 5)

Seeking Movements. Seeking movements are coordinated movements between two eyes with the purpose of following slow visual stimuli. Their function is to stabilize the dynamic visual image in the retina with velocities between 1 and 30°/s (Fig. 6).

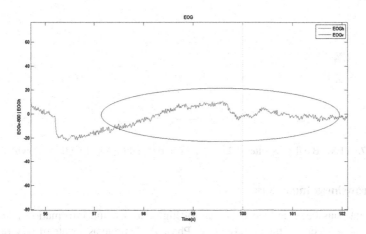

Fig. 5. EOGh fixed gaze pattern. Ban-pass filter 0.2–30 Hz.

Slow Eye Movements (SEM). Slow eye movements are eye movements with duration between 1 and 3 s mostly detected in the horizontal component of the EOG. This movement is characteristic of drowsiness states. It's characteristic of sleep onset with eyes closed but this pattern can also be seen with open eyes in drowsy drivers fighting not to fall sleep (Fig. 7).

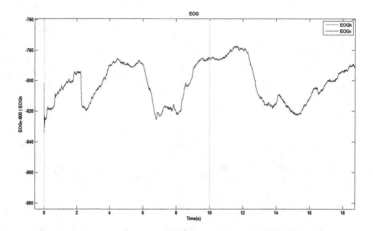

Fig. 6. Seeking movement in EOGv signal. Band-pass filter 0.2–30 Hz.

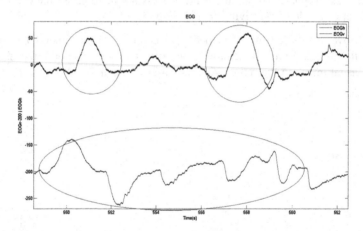

Fig. 7. SEM. Band pass filter 0.2–30 Hz of horizontal (up) and vertical (down) EOG.

3.4   Drowsiness Indicators

Awake state has been defined as a state of high activity and information interchange between the subject and the environment (Phase 0), Fatigue as a state of lack of energy and motivation (Phase 1) and Drowsiness as a state related to the sleep onset. Only some of the EOG patterns explained have direct relation with the sleep onset:

Blinking – An increase of the blinking frequency in addition to an increase of the percentage of eye closure are indicative of sleep onset.

Saccade – The number of saccades and the detection of fixations combined can be an index to estimate the ocular activity assuming saccades as activity and fixation as no activity. There is a direct relation between the reaction time of the subject and the velocity of the saccade movement (Fig. 8).

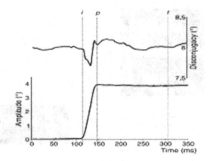

Fig. 8. Determination of the beginning and the final of the saccade movement. Binocular motor coordination during saccades and fixations while reading: A magnitude and time analysis [11].

Slow Eye Movements (SEM) – During the transition from awake to sleep stages, it is very common the appearance of slow eye movements (SEM), like pendulum low frequency (0.1–1 Hz) movements in the horizontal line of the eye.

Ocular Activity – This information combines the information given by all the previous patterns combined in a single index according to the state of the subject. In sleep onset periods the ocular activity diminishes according to the appearance of SEM and the decrease of blinks and saccades.

4   EOG Signal Processing

4.1   Preprocessing

A non linear filter preprocessing of the signal has been done. The filter used for the signal detrending was a non linear filter derived from the Hodrick-Prescott (1) filter with the objective of removing repeated oscillations in the signal. Cutoff frequency of [0.1, 30] at −6 dB.

$$H_{2,\lambda}(\omega) = \frac{1}{1 + 2^2 \lambda (\cos \omega - 1)^2} \qquad (1)$$

In secondly a band pass filter has been done. The high pass filter at 0.1 Hz filtered the baseline eliminating the electrode polarization effects and the movement artifacts. The low pass filter at 30 Hz removed the Electromyogram (EMG) artifacts of the signal.

4.2   Processing

As seen in the literature, the most representative EOG patterns used to estimate the sleep onset are saccade, blinks and slow eye movements. This investigation was focused on the analysis of blinking and saccade patterns as explained below.

Blinking Detectors. The analysis was divided in two blocks (Fig. 9): erosion and detection.

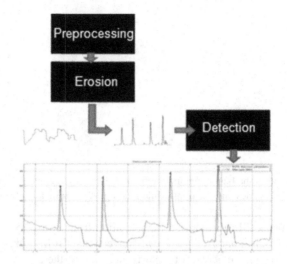

Fig. 9. Block diagram of the blinking detection algorithm.

First the signal passes the erosion block, where the abrupt swings are eliminated (Fig. 10). Then the filtered signal passes to the blinking detection module.

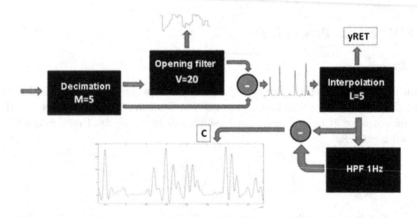

Fig. 10. Erosion block.

The objective of the erosion module is to stand out the blinking patterns from the rest of artifacts and saccade oscillations with the interpolation of the obtained "yRET" signal and its posteriors calculation of the very low frequency oscillations obtaining "FPA 1 Hz" signal. Finally the "FPA 1 Hz" signal is subtracted from the "yRET" signal to obtain C signal.

In the detection block C signal is processed with the objective of stand out the low frequency oscillations to avoid remaining artifacts. Finally the subtraction of yRET signal from C is done and the detection of peaks with a fixed threshold 'Um' gives the resultant signal with the blinks detected (Fig. 11).

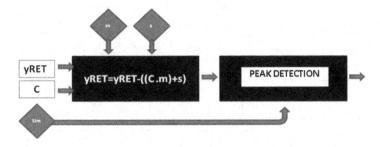

Fig. 11. Detection block.

Saccade Detectors. The saccade detection algorithm developed analyzes the horizontal EOG signal with an adaptation of the known Murty-Rangaraj method based on the detection of QRS segment in EKG signal [12].

As shown in the picture below (Fig. 12) the analysis is divided in three blocks: The preprocessing block explained in E.1, The Murty-Rangaraj adaptation block and the saccade detection block.

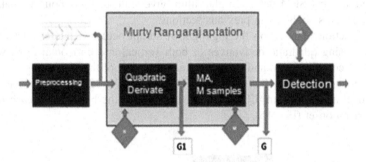

Fig. 12. Block diagram of the saccade detection algorithm.

Murthy-Rangaraj method consists in a pre-filtering of the signal followed by and estimation of the first weighted quadratic derive (2). The resulting signal was later filtered with a moving average filter (3) in order to smooth the obtained result.

$$g1(n) = \sum_{i=1}^{N} |x(n - i + 1) - x(n - i)|^2 (N - i + 1) \tag{2}$$

N: Window width smoothing

$$g(n) = \frac{1}{M} \sum_{j=0}^{M-1} g1(n - j) \tag{3}$$

Next step was the maximum and minimum identification of the signal in order to detect the position of the saccade using a fixed threshold (Fig. 13).

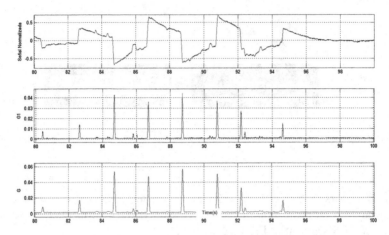

Fig. 13. Example of the saccade detection in horizontal EOG.

Ocular Activity Detectors. The ocular activity detector algorithm, comprising the same modules that SEM detection algorithm, gives information about the balance of the EOG patterns analyzed in previous sections.

The detection is based on the highlight of abrupt ocular events characteristics of awakeness using quadratic derivatives of both vertical and horizontal EOG with the objective of combine them and extract the activity level of the signal (Fig. 14). Three different frequency bands, [0.1–0.5] Hz, [3–8] Hz and [9–12] Hz, were analyzed for each vertical and horizontal EOG in order to compare the behavior of each one and relate to sleep onset (Fig. 15).

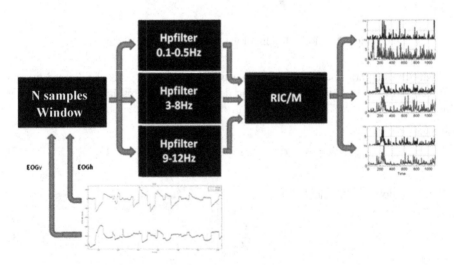

Fig. 14. Block diagram of the ocular activity detection algorithm.

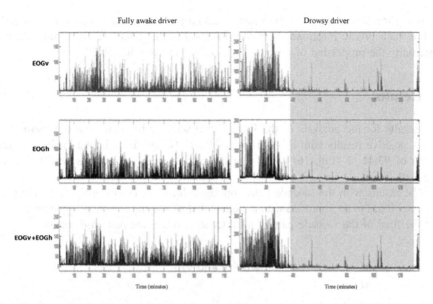

Fig. 15. Example of ocular activity detection combining vertical and horizontal EOG signal for an awake driver (left) and a drowsy driver (right). Marked in grey the drowsy zone.

5 Statistical Analysis

For each minute of recording, the phases obtained by the EOG different drowsiness detection algorithms were compared with the GS signal, in this case a combination of three external observers evaluating minute by minute the state of the subject using a video recording of the tests. To estimate the sensitivity and specificity of the different EOG methods a match signal was calculated having the number of false positives, false negatives, true positives and true negatives.

According to Table 1 [13], sensitivity and specificity for each phase is defined as:

Table 1. Sensitivity and Specificity definition.

		Gold Standard	
		PHASE 0	PHASE 2
EOG index	PH0	TN	FN
	PH2	FP	TP

Specificity = $\dfrac{TN}{TN + FP}$	Sensitivity = $\dfrac{TP}{TP + FN}$

Stone EA, et al. 2005.
Annu. Rev. Genomics Hum. Genet. 6:143–64

86   N. Rodríguez-Ibáñez et al.

being the Sensitivity the proportion of actual positives which are correctly identified as such giving information about how good is the detection algorithm, and the Specificity the proportion of negatives which are correctly identified.

6 Results

The results for the analysis of the EOG signal with de blinking detection algorithm shows positive results with a sensitivity of 92.41 % and a Predictive Positive Value (VPP) of 93.41 % (Fig. 16) comparing the results of the algorithms with the Gold Standard.

The results with the saccade detection algorithms shows also good results but, in this case, it has to be improved with a module that allows the detection of the beginning and the final of the saccade pattern in order to improve the pattern detection, yet the

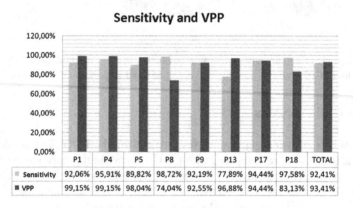

Fig. 16. Blinking detection algorithm results.

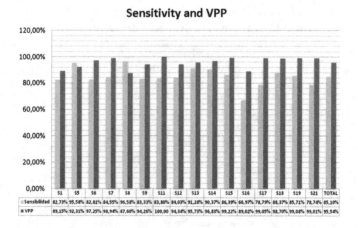

Fig. 17. Saccade detection algorithm results.

results are very promising for drowsiness detection porpoises with sensitivity values of 85.1 % and VPP values of 95.4 % (Fig. 17). The results of the Ocular Activity shows very promising results in drowsy subjects, further work will be done in order to avoid false positives in the detection due to movement artifacts [14].

7 Conclusions

The results confirmed the viability of the sleep onset detection using related to drowsiness patterns in the EOG signal as blinking frequency and saccade movements' appearance. Some misdetection of the algorithms may be due to the inter-subject variability mostly regarding the shape of the saccade pattern.

Future work will be focused in the improvement of the saccade detection algorithm by including the detection of initiation and end of the saccade pattern in order to make more specific the detection and accurate the calculation of the variable velocity of the saccade.

The future objective is to use the EOG signal as Gold Standard in vehicle tests replacing the EEG signal that shows low quality signal in real environments.

Acknowledgements. This work has been partially funded by the Spanish MINISTERIO DE CIENCIA E INNOVACIÓN. Project IPT-2011-0833-900000. Healthy Life style and Drowsiness Prevention-HEALING DROP.

References

1. NHTSA VSR "Chapter 3: objective 2, what are the environmental conditions associated with driver choice of engagement in secondary tasks or driving while drowsy? (2006)
2. Faber, J.: Detection of different levels of vigilance by EEG pseudospectra. Neural Netw. World 14(3–4), 285–290 (2004)
3. Rodrígue-Ibáñez, N., García-González, M.A., Fernández-Chimeno, M., Ramos-Castro, J.: Drowsiness detection by thoracic effort signal analysis in real driving environments. In: International Conference of the IEEE Engineering in Medicine and Biology Society, pp. 6055–6058 (2011)
4. Roehrs, T., Roth, T.: Sleep, sleepiness, and alcohol use. National Institute on Alcohol Abuse and Alcoholism of the National Institutes of Health (2011)
5. Dinges, D.F., Mallis, M.M., Maislin, G., Powell, J.W.: Evaluation of techniques for ocular measurement as an index of fatigue and the basis for alertness management (2005)
6. Rechtschaffen, A., Kales, A.: A manual of standardized terminology, technique and scoring system for sleep stages of human subjects los Angeles: Brain information services/Brain research institute, UCLA, pp. 1–12 (1968)
7. Galley, N.: Saccadic eye movement velocity as an indicator of (de)activation: a review and some speculations. J. Psychophysiol. 3, 229–244 (1989)
8. Sirevaag, E.J., Stern, J.A.: Ocular measures of fatigue and cognitive factors. In: Backs, R.W., Boucsein, W. (eds.) Engineering Psychophysiology, pp. 269–288. Lawrence Erlbaum Associates, Mahwah (2000)

9. Salthouse, T.A., Ellis, C.L.: Determinants of eye-fixation duration. Am J Psychol. **93**(2), 207–234 (1980)
10. Viviani, P.: Eye movements in visual search: cognitive, perceptual and motor control aspects. In: Kowler, E. (ed.) Eye Movements and Their Role in Visual and Cognitive Processes, pp. 353–394. Elsevier Science, Amsterdam (1990)
11. Vernet, M., et al.: Guiding binocular saccades during reading: a TMS Study of the PPC. Front Hum Neurosci., 5:14. Published online 2011, 7 February, 2011. doi:10.3389/fnhum. 2011.00014
12. Rangaraj, R.: Biomedical Signal Analysis, A Case Study approach. IEEE Press Series in Biomedical Engineering
13. Stone, E.A., Cooper, G.M., Sidow, A.: Trade-offs in detecting evolutionarily constrained sequence by comparative genomics. Annu. Rev. Genomics Hum. Genet. **6**, 143–164 (2005)
14. Calderón, P.M.: Estudio de la viabilidad de la señal de EOG como indicador de somnolencia. PFE. Master Oficial en Enginyeria Biomèdica. Department of Electronic Engineering of the Technical University of Catalonia (UPC), June 2011

Dielectric Barrier Discharge
Plasma for Endodontic Treatment

Avinash S. Bansode[1], Aumir Beg[2], Swanandi Pote[3],
Bushra Khan[2], Rama Bhadekar[3], Alok Patel[2],
S.V. Bhoraskar[1], and V.L. Mathe[1(✉)]

[1] Department of Physics, University of Pune, Pune 411007, India
{asb,svb,vlmathe}@physics.unipune.ac.in
[2] Department of Microbial Biotechnology,
Bharati Vidyapeeth University, Pune 411046, India
[3] Department of Pedodontics and Preventive Dentistry,
Bharati Vidyapeeth University, Pune 411046, India

Abstract. Recently Dielectric Barrier Discharge plasma is being explored for its application as an alternative to the conventional sterilization and disinfection techniques in medical sciences. Dielectric barrier discharge plasma torch has been developed to study the interaction of plasma and effect of its treatment on the growth rate of *S. aures*, *Escherichia coli*, *Bacillus subtilis* and *E. faecalis*. Plasma treated *E. faecalis* was compared with Chlorhexidine treated cultures and biofilms of *E. faecalis*. The results are analysed for significance ($P < 0.001$) using ANOVA and TUCKEY's test. In-situ optical emission spectroscopy has been employed to identify the plasma species interacting with the bioculture and biofilm. Based on its efficient disinfecting properties dielectric barrier discharge plasma has been proved to be a promising alternative to traditional techniques used for sterilization during the endodontic treatment.

Keywords: *DBD* plasma · Endodontic treatment · Plasma treatment on *E. faecalis*

1 Introduction

Recently, cold plasma is being explored for its application as an alternative to the conventional sterilization and disinfection techniques in medical field. Amongst the different types of plasmas, studies on Dielectric Barrier Discharge (DBD) plasma has gained importance in the medical field due to its property of being functional at room temperature and atmospheric pressure; unlike low pressure cold plasmas. The anti-microbial property of dielectric barrier discharge plasma has been demonstrated against several bacteria for example *Escherichia coli*, *Candida albicans*, *Streptococcus mutans*, *Bacillus subtilis* etc. [1].

In the field of dentistry, microorganisms and their by-products are considered to be the major cause of pulp and periradicular pathosis. The objective of endodontic therapy is to [i] remove diseased tissue, [ii] eliminate bacteria present in the canals as well as dentinal tubules and [iii] prevent post endodontic recontamination. Hence, a major

© Springer-Verlag Berlin Heidelberg 2014
M. Fernández-Chimeno et al. (Eds.): BIOSTEC 2013, CCIS 452, pp. 89–97, 2014.
DOI: 10.1007/978-3-662-44485-6_7

objective in root canal treatment is to disinfect the entire root canal system and to eliminate all the possible sources of infection. This can be accomplished by using mechanical instrumentation and chemical agents, in conjunction with medication of the root canal system between treatment sessions. In spite of these treatments, the survival of microorganisms in the apical portion of root filled teeth is still the major cause of endodontic failure. Studies have revealed that this increased resistance is due to the formation of biofilms by the microorganisms. Microorganisms like *Enterococcus faecalis, Streptococcus mutans, and Candida albicans* can adhere to the root canal walls and form communities organized in biofilm which makes them more resistant to phagocytosis, antibodies, and antimicrobials. This is due to the presence of exopoly-saccharides in comparison with non-biofilm producing organisms [2]. Also, due to the presence of dentin tubules; the disinfection of dentin posses special challenges during caries therapy. Conventionally disinfection is achieved by [i] invasive removal, [ii] use of chemicals like chlorhexidine. But these treatments do not disinfect the dental tubules completely. Contact dermatitis is a common adverse reaction to chlorhexidine. Chlorhexidine is also liable to cause desquamative gingivitis, discoloration of tongue and teeth or dysgeusia (distorted taste). Plasma being in the gaseous form would have a better reach in the confined and tortuous root canal system and may prove to be a better adjunct to root canal instrumentation for canal sterilization. Hence, inactivation of microorganisms using plasma has attracted much attention recently.

Numerous in vitro studies have been conducted on the effects of atmospheric plasma on pathogens like *Streptococcus mutans, Candida albicans, Chromobacterium violaceum* etc. Amongst the various oral pathogens *Enterococcus faecalis* is one of the primary organisms in patients with post treatment endodontic infection [3]. *E. faecalis* is known to form intracanal biofilms, periapical biofilms and biomaterial centered infection.

Inspite of being one of the important infection causing organism, relatively few reports are available that describe the efficacy of dielectric DBD plasma against *E. faecalis*. Hence, this study was aimed at evaluating the inhibitory effect of DBD plasma against *E. faecalis* culture and biofilms. Initially experiments were carried out with *S. aures, Escherichia coli, Bacillus subtilis* etc. which was found to be effective for destruction of these bacteria. Later on these bacteria were replaced by *E. faecalis* for further investigation. The paper reports the use of He (atmospheric pressure) as a plasma generating gas which enables the formation of active reactive species useful for bactericidal properties.

2 Materials and Methods

DBD plasma torch is found to be useful for varieties of applications. Its use for dental applications is an emerging field of research. In the present article we report on the development of a small plasma torch with helium as plasma forming gas. Figure 1 shows the schematic of the atmospheric pressure DBD plasma torch which consisted of a tungsten cathode in the form of thin wire axially fitted into a cylindrical glass tube with a fine nozzle. The anode was a stainless steel plate which formed the base below the microtitre plate. The anode is grounded. A 1.5 mm thick glass slab work as the dielectric material to obtain the DBD plasma. Helium was made to flow at a flow rate of 1 lpm through the torch system so as to reach the bacterial culture placed inside the microtitre plate. Pulsed dc voltage of 24 kV was used to operate the torch. The plasma plume 1–2 mm in diameter extended outside the nozzle up to a distance of 2–3 cm and reached the *E. faecalis* culture. Helium having a low ionization potential serves as a suitable ionizing medium and helps in extracting the plasma plume to larger distances. Compared to other gas plasmas it is easy to obtain plume of few centimeter by using helium gas as plasma forming gas [4]. There are several other advantages of using He as a plasma forming gas like, He gas flow propels radicals and metastables towards the surface to be treated. The flow diameter helps to determine the diameter of the treated portion of the sample and moreover it has a high thermal conductivity and thus acts like a coolant. Also the flow mixes with the atmospheric gases like H_2O, N_2, and O_2 at the downstream from the nozzle [5]; and help in producing reactive atomic oxygen and hydroxide (OH⁻) species which are well known to be bactericidal.

In order to examine the species responsible and to understand the mechanism of reaction of the bacteria and plasma species; the optical emission spectroscopic studies were carried out. The photon collection was facilitated with an optical fiber which was

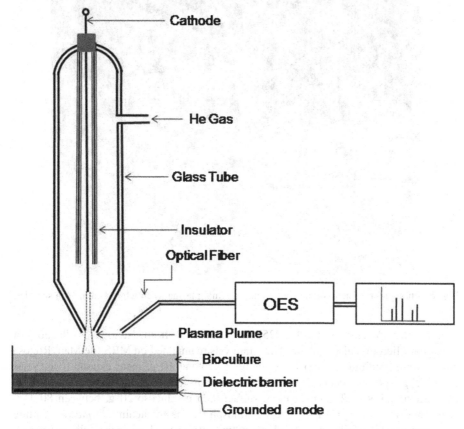

Fig. 1. Shows the schematic of the atmospheric pressure non thermal plasma torch.

directed towards the junction between the plasma plume and the upper meniscus of the bio-culture and bio-films. Ocean Optics spectrophotometer (model HR-4000) was employed for the measurements. A collection of species which were produced in the process has been useful to understand the reaction mechanism within the dye due to the plasma.

The glow in the plasma is due to the electron excitation, de-excitation and ionization of the gas atoms, so that it serves as a visual indicator of the presence of energetic electrons and photons. These energetic electrons generate radicals by dissociating gas molecules such as H_2O and O_2, or by generating metastable He ions that probably dissociate H_2O molecules. The air mixed with water molecules help in generating the reactive OH and O radicals which further assist in the killing of the *E. faecalis* bacteria (Fig. 2) [5].

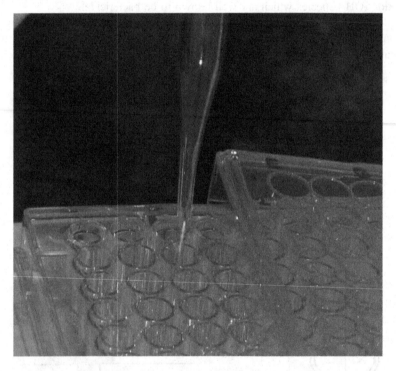

Fig. 2. Photograph of actual atmospheric non-thermal plasma torch while treating the samples.

Enterococcus faecalis (NCIM 5025) was procured from National Collection of Industrial Microorganisms, Pune. The strain was maintained on MRS (de Man, Rogosa and Sharpe) medium (composition per liter: peptone 10 g, meat extract 10 g, yeast extract 5 g, sodium acetate 5 g, dipotassium phosphate 2 g, ammonium citrate 2 g, magnesium sulfate 0.2 g, manganese sulfate 0.05 g, glucose 20 g, between 80 1 g) slants and stored at 4 °C. The effect of atmospheric plasma against *E. faecalis* culture was monitored as follows. 100 µl of culture was placed in each well of 96 well microtitre plate. The culture was then treated with atmospheric plasma for 2 min. The wells were grouped into 4 groups viz. group A - helium treated, group B, plasma treated and group C - chlorhexidine treated D - control, with a size of 30 wells per group. E. faecalis culture was treated with chlorhexidine for 2 min by adding 100 μl of 0.2 % chlorhexidine solution to the wells and subsequent removal of the solution after 2 min. The viability of the cells was checked by Triphenyl tetrazolium chloride (TTC) assay.

Preparation of the bio-films of E. faecalis was inoculated in MRS broth and incubated at 37 °C at 120 rpm for 24 h. 100 μl of culture (OD 1.0 according to McFarland's scale, colony forming unit (CFU) 4×10^8/ml) was added in each well of 96-well flat base microtitre plate. The plate was incubated at 37 °C for 1 h in order to allow the organisms to adhere to the surface of the wells. After 1 h, remaining culture from the wells was replaced by 100 μl of fresh MRS broth and the plate was further incubated at 37 °C for 24 h. After incubation, the excess medium from the wells was removed and the biofilms were then treated with DBD plasma for 2 min. The experiment was divided into 4 groups as mentioned above. S. aures, Escherichia coli and Bacillus subtilis were inoculated in Nutrient broth and incubated at 37 °C at 120 rpm for 24 h. The bioculture was treated with DBD plasma for 2 min. All the conditions are same as for E. Faecalis.

2.1 Viability Assay

The viability of the organisms was determined using the TTC viability assay described by C E Nwanyanwu with some modifications [6]. 100 μl of 1 % (w/v) solution of TTC prepared in sterile distilled water was added in the wells along with 100 μl of MRS broth following the treatments. The plates were incubated at 37 °C overnight. The color change was measured at 490 nm using Elisa plate reader (Bio Rad model no: IMark).

All the results are analysed for significance (P < 0.001) using ANOVA and TUCKEY's test. Optical Emission Spectrometer (OES) model HR-4000 (manufactured by Ocean Optics) was employed for the identification of active ionic species present in the plasma. This spectrometer had a linear CCD array detector (model TCD1304AP) and the range of detection extended from 200 nm to 1100 nm with a resolution of ∼0.03 nm (FWHM). In addition it has fiber optical input.

3   Results and Discussion

S. aures, B. subtilis, and E. coli cultures were exposed to atmospheric non-thermal helium plasma for 2 min. These organisms were also treated with chlorhexidine and helium. The results of the viability assay were then compared for all three processes as shown in Fig. 3. In all the four groups (Fig. 3), the mean bacterial count for three experiments has been shown. The control group is considered to be 100 % for bacterial survival. The group treated with helium showed same optical density as that of control; indicating that the killing effect of plasma is not due to presence of helium. Significant reduction in optical density was observed in the cultures exposed to chlorhexidine and plasma. In case of *S. aures* the optical density as compared to that of the control is reduced to be 25 % by chlorhexidine and 58 % after plasma treatment. In case of *B. subtilis* the optical density reduction was observed to be 2 % by cholorohexidine treatment while 37 % by plasma treatment. However in case of *E. coli* the reduction in optical density for chlorhexidine treated culture was found to be 9 % while 61 % for plasma treated culture. Effect of atmospheric plasma, helium and chlorhexidine on the culture plates was determined by TTC viability assay.

E. faecalis culture was exposed to DBD helium plasma for 2 min. Similarly, culture was also treated with chlorhexidine and helium for 2 min. The results of the viability assay were then compared for the three processes as shown in Fig. 4. It shows the mean bacterial count in terms of optical density in all the four groups. The wells in the control group and the wells treated with helium exhibited no change in the optical density (OD 1.0) indicating 100 % survival of the culture. However significant reduction in optical density was observed in the wells exposed to chlorhexidine (OD 0.7) and plasma (OD 0.6). Inhibitory effects of DBD plasma have also been reported on *E. faecalis* culture by Cao et al. [7]. The culture was exposed to helium-oxygen plasma for 5, 10 and 15 min. The post exposure viability was determined by measuring the zone of inhibition. Effects of atmospheric plasma, helium and chlorhexidine on *E. faecalis* biofilms adhered on wells of the microtitre plates were determined by TTC viability assay. The results were similar to those observed for the culture. Chlorhexidine and plasma treated biofilms showed decrease in optical density.

It was 0.5 for chlorhexidine, 0.47 for plasma as compared to 0.8 of control and helium treated biofilms. This clearly indicated reduction in viability of bacteria in biofilms (Fig. 6).

In order to confirm the presence of the reactive species we have carried out the spectral identification using optical emission spectroscopy. Figure 5 shows the OES

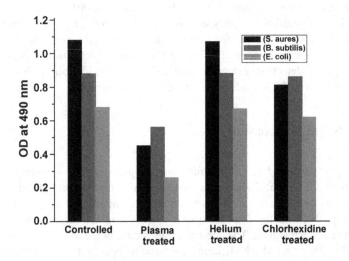

Fig. 3. Viability Assay of effect of plasma treatment and chlorhexidine treatment on different *organisms*.

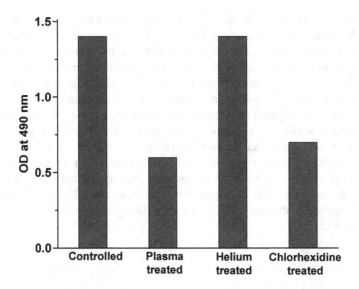

Fig. 4. Viability Assay of effect of plasma treatment and chlorhexidine treatment on *E. faecalis* culture.

spectrum recorded during the plasma exposure of the *E. faecalis* culture. Optical emission spectrum shows that the species present in the plasma include helium, atomic oxygen (O), Ozone (O_3), and OH radicals.

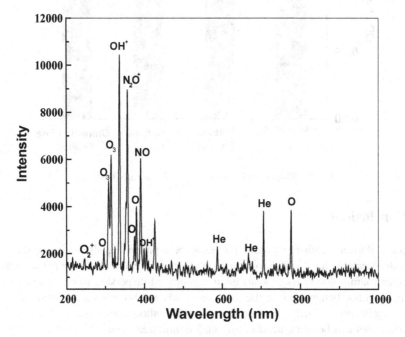

Fig. 5. Optical emission spectrum of helium plasma generated by dielectric barrier discharge plasma torch.

Figure 6 shows that helium gas alone did not show any inhibitory effect on bio-films. Inhibitory effects of DBD plasma have also been examined on *Chromobacterium violaceum* [8] and on *Streptococcus mutans* biofilms [9]. Du et al. [10] have also reported the reduction in viability of *E. faecalis* biofilms after treatment with DBD plasma. The authors also determined the viability using confocal laser scanning microscopy. Cao et al. [7] have reported antibacterial effects of atmospheric plasma on *E. faecalis* biofilms prepared on nitrocellulose membrane using SEM. Our results of 2 min exposure, leading to a significant reduction in the viability, suggest that this technique can be used as an adjunct for endodontic therapy in dentistry. Further research in this line could prove its potential as an alternative for traditional procedures of disinfection. However it has been obvious that the oxygen and OH radicals which are known to be bactericidal, most probably, react with the cell walls of the bacteria causing the rupture of the cell walls. As an effect the cytoplasm leaks out, causing the death of the bacteria.

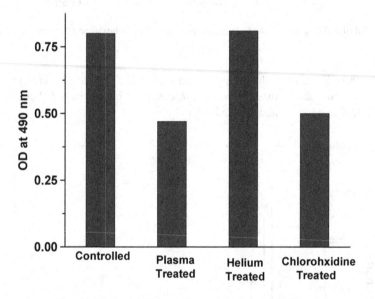

Fig. 6. Shows inhibitory effect on *E. faecalis* biofilms.

4   Conclusions

Dielectric Barrier Discharge plasma proves to be a promising alternative to the traditional disinfectants for disinfection during endodontic treatment. Using plasma will be beneficial mainly due to its gaseous nature, improved dispersion of disinfectant in the dentinal tubules, better reach in the tortuous canals and no dysgeusia (distorted taste). Hence, application of DBD plasma can be a quick method since post disinfection clean-up procedures can be minimized during endodontic treatments.

Acknowledgements. ASB is thankful to UGC for the award of Ph.D. fellowship. SVB is thankful to DST and CSIR, New Delhi for the financial support and award of ES fellowship.

References

1. Hong, Y.F., Kang, J.G., Lee, H.Y., Uhm, H.S., Moon, E., Park, Y.H.: Sterilization effect of atmospheric plasma on *Escherichia coli* and *Bacillus subtilis* endospores. Lett. Appl. Microbiol. **48**, 33–37 (2009)
2. Sabeena, M., Boopathi, T.: *Enterococcus faecalis* - an endodontic challenge. J. Ind. Acad. Dent. Spec. **1**(4), 46–50 (2010)
3. Mohammadi, Z., Abbott, P.V.: The properties and applications of chlorhexidine in endodontics. Int. Endod. J. **42**, 288–302 (2009)
4. Sladek, R.E.J., Stoffels, E., Walraven, R., Tielbeek, P.J.A., Koolhoven, R.A.: Plasma treatment of dental cavities: a feasibility study. IEEE Trans. Plasma Sci. **32**(4), 1540–1543 (2004)
5. Goree, J., Liu, B., Drake, D., Stoffels, E.: Killing of *S. mutans* bacteria using a plasma needle at atmospheric pressure. IEEE Trans. Plasma Sci. **34**(4), 1317–1323 (2006)
6. Nwanyanwu, C.E., Abu, G.O.: Influence of pH and inoculum size on phenol utilization by bacterial strains isolated from oil refinery effluent. Int. J. Nat. Appl. Sci. **7**(1), 8–15 (2011)
7. Cao, Y., et al.: Efficacy of atmospheric pressure plasma as an antibacterial agent against *Enterococcus faecalis* in vitro. Plasma Sci. Technol. **13**(1), 93–98 (2011)
8. Jonathan, C.J., Kwan, C., Abramzon, N., Vandervoort, K., Brelles-Marin, G.: Is gas-discharge plasma a new solution to the old problem of biofilm inactivation? Microbiology **155**, 724–732 (2009)
9. Sladek, R.E.J., Filoche, S.K., Sissons, C.H., Stoffels, E.: Treatment of *Streptococcus mutans* biofilms with a nonthermal atmospheric plasma. Lett. Appl. Microbiol. **45**(3), 318–323 (2007)
10. Du, T., Ma, J., Yang, P., Xiong, Z., Lu, X., Cao, Y.: Evaluation of antibacterial effects by atmospheric pressure non-equilibrium plasmas against *Enterococcus faecalis* biofilms in vitro. J. Endod. **38**(4), 545–549 (2011)

Characterization of an Acoustic Based Device for Local Arterial Stiffness Assessment

H.C. Pereira[1,2(✉)], J. Maldonado[3], T. Pereira[4], M. Contente[1],
V. Almeida[1], T. Pereira[1], J.B. Simões[1,2], J. Cardoso[1], and C. Correia[1]

[1] Physics Department, Instrumentation Centre (CI-GEI),
University of Coimbra, Coimbra, Portugal
{cpereira,vaniagalmeida,taniapereira,cardoso,correia}
@lei.fis.uc.pt, m_ines_c@hotmail.com
[2] ISA - Intelligent Sensing Anywhere, Coimbra, Portugal
jbasilio@isa.pt
[3] Instituto de Investigação e Formação Cardiovascular,
Aveleira, Penacova, Portugal
instituto@iifc.pt
[4] Escola Superior de Tecnologia da Saúde de Coimbra,
S.Martinho do Bispo, Coimbra, Portugal
telmo@estescoimbra.pt

Abstract. Arterial stiffness, recognized as an independent predictor of cardiovascular events, can be assessed non-invasively by regional and local methods. The present work proposes and describes a novel and low-cost device, based on a double-headed acoustic probe (AP), to assess local arterial stiffness, by means of pulse wave velocity (PWV) measurements. Local PWV is measured over the carotid artery and relies on the determination of the time delay between the signals acquired simultaneously by both acoustic sensors, placed at a fixed distance. The AP was characterized with dedicated test setups, in order to evaluate its performance concerning waveform analysis, repeatability, crosstalk effect and time resolution. Results show that AP signals are repeatable and crosstalk effect do not interfere with its time resolution, when the cross-correlation algorithm for time delay estimation is used. The AP's effectiveness in measuring higher PWV (14 m/s), with a relative error less than 5 %, when using two uncoupled APs, was also demonstrated. Finally, its clinical feasibility was investigated, in a set of 17 healthy subjects, in which local PWV and other hemodynamic parameters were measured. Carotid PWV yielded a mean value of 2.96 ± 1.08 m/s that is in agreement with the values obtained in other reference studies.

Keywords: Local pulse wave velocity · Double headed probe · Microphones · Test bench systems · in-Vivo measurements

1 Introduction

Arterial stiffness, which results from the progressive degeneration of the wall's elastic fibres, is a marker of cardiovascular risk that in the last few years has gained great relevance in the medical community due to its predictive value for cardiovascular

© Springer-Verlag Berlin Heidelberg 2014
M. Fernández-Chimeno et al. (Eds.): BIOSTEC 2013, CCIS 452, pp. 98–114, 2014.
DOI: 10.1007/978-3-662-44485-6_8

mortality and morbidity, target organ damage, coronary events and fatal strokes in patients with different levels of risk [1–4]. The most direct, simple and robust method of assessing arterial stiffness is pulse wave velocity (PWV), i.e. the velocity at which the pressure wave propagates along an artery. Carotid-femoral PWV is considered the 'gold standard' measurement of arterial stiffness, being supported by several clinical studies that highlight the relevant contribution of this parameter to the diagnosis, prognosis and follow-up of the general population/patient [5, 6].

The most prominent commercial devices require moderate technical expertise; however they have several drawbacks in PWV assessment. The practical solution used by these systems relies on the acquisition of pulse waves at the carotid and femoral arteries to determine the time delay measured between pressure upstroke at each site (t). The distance between the two acquisition locations (d) is usually assessed using a measuring tape and then PWV is automatically determined using the linear ratio between d and t [7, 8]. One source of error is related to time delay estimation and the lack of standardization on its determination. There are severable feasible methods to estimate it, however they can't be used interchangeably [9]. On the other hand, this solution not only negligence opposite wave propagation but also presents errors in estimating the distance between the recording locations (for example, the curvature of the arteries cannot be taken into account) [10–12]. Finally, it introduces a rough estimate of local properties of the artery, since it integrates different segments of arterial stiffness (carotid, aorta, iliac, femoral), being unable to differentiate between the muscular and elastic parts.

The possibility of assessing the local hemodynamics is in fact very useful, particularly in the carotid artery due to its predisposition to atherosclerotic plaques formation and its significance in the development of coronary and cerebrovascular diseases [13–16]. Several methods have already been investigated with the intention of evaluating indices of local arterial stiffness, such as distensibility, compliance, elastic modulus and local PWV. Most of them are based on the simultaneous assessment of diameter and pressure waveforms via ultrasound systems, however they present important limitations in what concerns the local pressure assessment that is often taken in a vessel distant (braquial) or at a time distant from where and when diameter waveform is acquired (carotid) [17–20]. To avoid inaccuracies related to local blood pressure measurement, two different approaches were further suggested. The first one proposed by Giannattasio et al. combines ecographic data with the pressure waveform captured using arterial tonometry in the contralateral artery, taking the ECG as reference point [21]. With this approach it is possible to perform a synchronized acquisition of diameter and pressure at similar regions. On the other hand, Hermeling et al. investigated an alternative method to assess local PWV that does not require the measurement of local blood pressure. This technique is based on the calculation of time delay between the several diameter waveforms acquired using piezoelectric elements of an ultrasound probe [22, 23].

In spite of the methodological advances that have been observed in local hemodynamic assessment, the devices available for the measurement of these variables require high technical expertise and also burdensome and specialized imaging technologies (i.e.: ultrasound and echo tracking techniques) limiting their generalized use in clinical practice [16]. These limitations show that the introduction of low-cost, non-invasive and easy accessible devices that identify local changes in arterial wall dynamics and also other hemodynamic parameters would be of great interest, mainly in CVDs surveillance and monitoring.

The present work intends to contribute to the achievement of the previous scenario, presenting and characterizing an efficient and low-priced tool based on a non-invasive device that is placed over the carotid artery and can be easily handled in diagnostic trials by an operator. Based on a previous work, where a double headed piezoelectric probe was designed and characterized in laboratory [24, 25], we developed a simpler and novel system that assesses local arterial stiffness by means of non-invasive PWV measurements and also other hemodynamic parameters, such as left ventricular ejection time (LVET), in a short length of an artery (less than 15 mm). The device is based on a double configuration of two acoustic sensors that are placed at a fixed distance, d, allowing simultaneous acquisition of two (sound) pulse waves. The measurement of time delay between the waves, Δt, allows local PWV to be determined, simply, as:

$$PWV(ms^{-1}) = \frac{d}{\Delta t} \qquad (1)$$

2 Materials

2.1 The Double Headed Probe

The developed probe, presented in Fig. 1, consists of two acoustic transducers (Pro-Signal, ABM-712-RC, microphone-solder pad) that are placed approximately 11 mm apart and fixed at the top of a plastic box (Multicomp, 77 mm × 49 mm × 26.6 mm). The transducers, based on 9.7 mm diameter electret condenser microphones with an operating frequency of 100 Hz–10 kHz and noise cancelling directivity, are placed at the minimum separating distance, while avoiding mechanical contact. These elements form an ergonomic configuration that allows a safe and effective way of collecting the pulse wave on the carotid artery for both the patient and the operator.

Fig. 1. Representation of the double headed acoustic probe. M_1 – Microphone 1; M_2 – Microphone 2; c1 - distance between transducers centres: ≈11 mm; d1: microphones diameter: 9.7 mm; h1: sensors external height: 2 mm.

The probe does not include any type of signal conditioning circuit, so the acoustic signals are acquired directly by a Personal Computer (PC) Sound Card. The AP uses parallel audio cable to convey the information obtained directly from the transducers, to the microphone input of the PC Sound Card. In circumstances in which the PC Sound Card does not have stereo input, the probe connects first to an external Sound Card (7.1 Sweex® USB Sound Card, 16-bit, 48 kHz Maximum Sampling Rate, 90 dB Signal to Noise Ratio) that then delivers the collected signals to the PC, via USB. The data acquisitions are displayed in real time, through a dedicated Matlab® Based Graphical User Interface (Cardiocheck GUI) and automatically stored in a non-commercial Microsoft Access® based database (Cardiocheck DB). The data is subsequently processed using different algorithms (detailed in Sect. 3.3) that aim the extraction of several hemodynamic parameters, namely the PWV and the LVET (Fig. 2).

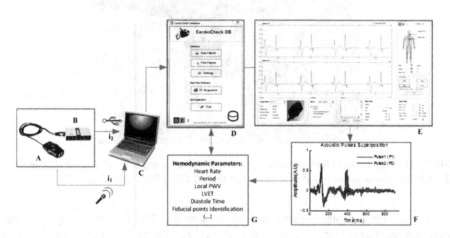

Fig. 2. Schematic representation of the overall system used in in-vivo measurements. A - Acoustic Probe; B - External USB Sound Card; C - PC; D - Cardiocheck Database; E - Cardiocheck Graphical User Interface; F - Data Pre-processing; G - Hemodynamic Parameters Extraction.

2.2 Test Setups

For characterizing the AP, as well as the various parameters extraction algorithms, it was developed two special purpose sets of test bench systems. The test setup I was designed to evaluate the ability of the probe in reproducing repeatedly different types of waveforms but also to evaluate the existence of crosstalk between both transducers. The setup uses a 700 μm stroke actuator, ACT, driven by a high voltage linear amplifier, HVA (Physik Instrumente GmbH P-287 and E-508, respectively) to generate a pressure wave that is fed to the acoustic probe by means of a "mushroom" shaped PVC interface (Fig. 3). This PVC interface (10 mm diameter), coupled to the ACT, is in mechanical contact with the transducer, without affecting the output voltage since the sensors does not respond to DC excitation. With this mechanical adapter it is possible to transmit the ACT's motion associated to the pressure wave, in such a way that the longitudinal forces are responsible for the transducer electric response. The waveforms are programmed and downloaded into an arbitrary waveform generator, AWG, Agilent 33220A that delivers the signal that is generated by the ACT and also the synchronism that triggers the data acquisition system, DAS (National Instruments®, USB6210). Although the AP is a prototype suitable for clinical tests, designed to be combined with a PC Sound Card, it was necessary to use a different DAS in test bench experiments, in order to acquire additional reference signals.

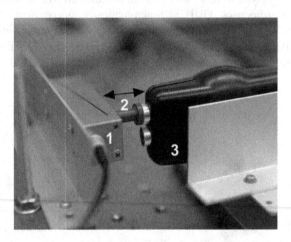

Fig. 3. Representation of the mechanical structure of the test setup I. 1 - ACT. 2 - PVC interface. 3 - AP. The arrow represents the movement of the ACT and PVC interface.

The test bench system II, schematically presented in Fig. 4, was developed aiming the assessment of the temporal resolution of the AP. This test bench emulates the main arterial pressure wave propagation features of the cardiovascular system, presenting an upgrade in relation to the experimental setup developed previously by Pereira et al. [26]. The main difference is based on the use of a natural rubber latex tube, considered to be a reliable material to simulate the compliance of a human artery and providing a higher distensibility than the silicon tube originally used. As in the test bench I, the pressure waveform is firstly generated using the AWG and then delivered to the ACT/HVA, that through a piston ("mushroom" shaped PVC, 15 mm diameter) - rubber membrane mechanism launches the wave into a 200 cm long latex tube (Primeline Industries, 7.9 mm inner diameter, 0.8 mm wall thickness), filled with water. The tube's sealing is made using a T-shaped scheme, guaranteeing geometric homogeneity. The wave is captured by the AP placed along the tube and by two pressure sensors PS1 and PS2 (Honeywell, 40PC015G1A), placed transversely and longitudinally to the tube. These sensors are used as the main reference for time delay/pulse wave velocity assessment but also monitor the DC pressure level of the tube, imposed by a piston P and a mass m at one of the tube's extremities. The range of DC pressure levels in the tube varies, approximately, from 30 mmHg to 400 mmHg, including (and exceeding) the pressure variations registered in a healthy and non-healthy human system. Although the variation in the DC pressure level interferes with the wave propagation velocity, the AP was tested at a constant DC pressure (≈66 mmHg), since it was not crucial for the present work having several wave propagation velocities.

To record simultaneously the different sensors response it was used the afore-mentioned DAS, triggered by the AWG.

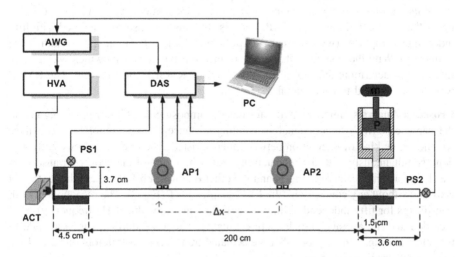

Fig. 4. Schematic representation of the test setup II.

3 Methods

3.1 Experimental Characterization

The experimental characterization of the AP consisted in the evaluation of its performance regarding three main aspects: repeatability in assessing pressure waveform, occurrence of crosstalk phenomenon and estimation of time resolution. Several pressure waveforms were programmed and used as inputs in these studies, including Gaussian-like and Cardiac-like pulses. The last ones, synthesized using a weighted combination of exponentially shaped sub-pulses [27], reproduce different states of arterial wall elasticity: type A and type B, correspond respectively to cases of pronounced and slight arterial stiffness (non-healthy subjects) and type C, commonly seen in healthy individuals, characterize elastic arteries [28].

In all the experiments, the data acquisition was performed through a dedicated acquisition module of National Instruments (NI© USB6210) and data logging was accomplished by NI LabView™ 2010 SignalExpress. All the signals were sampled at 5 kHz and stored for offline analysis using Matlab®. Data processing was undertaken in Matlab® 2009a and statistical analysis was performed using Microsoft Excel® 2010.

Waveform Analysis/Repeatability. The first part of this study aimed to examine the probe's response for different types of waveforms generated by the Agilent 33220A and exerted by the ACT. To obtain the best response of the transducer it was selected for each input signal, the best amplitude (3.5 V) and frequency (1 Hz). All the sensors signals were submitted to a 300 Hz low pass filter and to a band cut filter of 49 Hz–51 Hz, in order to avoid, respectively, the presence of the resonant frequency of the actuator ($\sim$380 Hz $\pm$ 20 %) and the 50 Hz power line interference. It was also performed an integration of the transducer signals using the Matlab® function *cumsum* to compare them with the original input signals.

In the second part, it was intended to measure the same waveform repeatedly and under the same conditions by the AP. For this study, each sensor was excited with fifty independent impulses (with the same amplitude and width (Gaussian, 1 s width, 1 Hz frequency). With those signals, it was determined the average signal which was used as reference to determine the root mean square error (RMSE), for each one. The RMSE was then computed to each signal.

Crosstalk Analysis. Since the two transducers composing the AP were incorporated in the same case with a very small separating distance, it was important to analyse whether some kind of interaction between them existed. The first part of this study was done simultaneously with the repeatability test, where one of the acoustic transducers was being actuated (microphone 1) and the other one was left free (microphone 2), that is to say without any contact with the PVC adapter/ACT (Fig. 3). The responses of both transducers for fifty independent impulses (Gaussian, 1 s width, 1 Hz frequency) were recorded and the average signals were estimated. This procedure was then applied to the other sensing element, such that the actuated transducer was microphone 2 and the free transducer was microphone 1.

The actuated transducers generated a typical differential signal with a good signal-to-noise ratio, while the free transducers generated a much lower amplitude signal, with a profile substantially opposite to that obtained for the actuated transducers (see results Sect. 4.1). Due to the characteristics of the signal obtained for the free transducers, it was necessary to perform an additional experiment to determine whether this transmission might interfere with one of the most important aspects of the probe: its time delay assessment. Thus, the second test consisted in the direct and simultaneous actuation of both transducers, with the purpose of time delay assessment. Both sensors were excited with three independent impulses (Gaussian, 1 s width, 1 Hz frequency) and for each acquisition it was determined the time delay between both transducers, using different algorithms yet to be described on Subsect. 3.3. In this particular experiment, the signals were sampled at 12.5 kHz, the same sampling frequency used in in vivo tests.

Time Resolution Evaluation. One of the main goals of the AP characterization was the evaluation of its ability of precisely assessing the time delay between two distinct points, separated from a very small distance.

In order to evaluate its time resolution, it was used two different APs (AP1 and AP2) that were placed on the tube of the test bench II, with the help of two external clamps (Fig. 4). One of the probes was kept fixed at the 50 cm position, while the other one was moving from 100 cm position to 54 cm position by 2 cm intervals. For each position, a Gaussian waveform (100 ms width, 10 Hz frequency) was delivered to the system, and then time delay and PWV were estimated between the first microphones of both probes and also between the pressure sensors (PS1 and PS2), attached at the extremities of the tube.

The relative errors between the reference PWV and the PWV obtained with the uncoupled transducers for each separating distance (Δx) were calculated, using the algorithms of Sect. 3.3. The test was repeated for more two times, for a constant DC pressure of $\approx$66 mmHg.

3.2 in-Vivo Measurements

Participants and Study Protocol. Seventeen young volunteers aged 22.12 $\pm$ 1.96 years were recruited and gave written informed consent prior to recording. Each participant was properly weighed and measured and after 5 m of rest of supine position, a blood pressure measurement was obtained from his right brachial artery, using an automatic clinically validated sphygmomanometer (MAM Colson BP 3AA1-2 ®; Colson, Paris). Next, a straight arterial segment of the right common carotid artery was identified by a skilled operator and a record of approximately 20 s–30 s was obtained with the probe longitudinally aligned to the artery. Data acquisition was performed with the dedicated real-time software Cardiocheck GUI and automatically stored in the Cardiocheck DB. Age, sex, weight, height, waist, systolic blood pressure (SBP) and diastolic blood pressure (DBP) were also stored in the same database.

All the signals were acquired at a sample rate of 12.5 kHz and were processed offline in Matlab® 2009a, aiming the extraction of carotid PWV and other hemodynamic parameters.

3.3 Signal Processing

In the first part of this work (experimental characterization), a set of dedicated algorithms have been developed aiming the estimation of time delay in two main situations: between the signals of the AP transducers and between the signals of pressure sensors (test setup II). After a common pre-processing, based on a low-pass filter with a cut-off frequency of 100 Hz to reduce high frequency noise, four different methods were used for time delay estimation: (a) maximum of cross-correlation function, (b) maximum and (c) minimum amplitude identification and (d) zero-crossing detection. The cross correlation method uses the xcorr function of Matlab's Signal Processing Toolbox to determine the peak of cross-correlogram that allows delay estimation by subtracting the peak time position from the pulse length. The other methods ensure an accurate detection of some fiducial points of the signal, such as the maximum, the minimum and the zero. As so, the methods of maximum and minimum amplitude identification use a 6th polynomial fit in the maximum and the minimum region of the signals, while the zero-crossing method applies a linear fit to the region where the signal crosses the zero. For all the methods, time delay is estimated between the maxima, minima and zero points detected in each set of signals.

In the last part of this work, the AP was used to assess PWV and other hemodynamic parameters in human carotid arteries. Since the acquisitions were constituted by several cardiac cycles, it was necessary to apply a dedicated segmentation routine, based on a minima detection approach to divide the data stream into single periods. Before applying the segmentation algorithm, the signals were filtered with the aforementioned 100 Hz low pass-filter and then heart rate was determined. For each cardiac cycle, the maximum of cross-correlation was used for carotid PWV estimation and an average value was obtained. Besides PWV, it was also possible to determine hemodynamic parameters, such as: LVET, defined as the period of time from the start of the pulse (aortic valve open) to the dicrotic notch (closure of the aortic valve) and diastole phase (DP), defined as the period of time from the dicrotic notch to the end of the pulse. These parameters were extracted based on the conviction that the onsets of the first and second carotid sounds (S1 and S2) coincide respectively with the onset and with the dicrotic notch of the carotid pulse waveform [29]. The onsets of carotid sounds S1 and S2 were identified as the maxima of the second time derivative of the acoustic signal. LVET and DP were calculated for each cardiac cycle. Data were expressed as mean ± SD.

4 Results and Discussion

4.1 Experimental Characterization

The first part of probe's experimental characterization consisted in the evaluation of the AP output to different waveforms and its repeatability. The response obtained by the AP for each one of the waveforms is presented in Fig. 5. The AP profiles are similar to those expected by a differentiator circuit; however it is not possible to precisely recover the original pressure waveform. When the acoustic signals are integrated there are noticeable similarities with the input signals, however the RMSE between both signals is quite high (approximately 13 % for each case). This performance was predictable, since the sensitivity of the acoustic sensors must be reduced for low frequencies that are below the microphone's 3 dB bandwidth (100 Hz–10 kHz). Since low frequencies are determinant for the precise reconstruction of arterial pressure waveform, the use of these acoustic sensors limits the possibility of the AP for waveform estimation purposes. Nevertheless, this fact does not disqualify the use of this probe this probe for its main purpose: local PWV estimation, once the method does not depend on the waveform accuracy.

The results regarding the repeatability test are shown in Table 1 and Fig. 6.

Although the microphone 2 (0.6198 ± 0.0298) exhibits a better performance than the microphone 1 (1.1781 ± 0.0345), the RMSE variance values obtained for both probes are identically low, evidencing the reliability of the system.

The second part of the AP's characterization intended to study the presence or absence of crosstalk effect in both transducers. The first results achieved in this study are illustrated in Fig. 7. The actuated sensors present a good signal-to-noise ratio and a typical profile when compared with the one obtained previously in the waveform analysis test (Fig. 5(a)).

The free transducers also present a slight profile but of much lower amplitude. Although the results suggest the existence of crosstalk effect, this phenomenon was seen as a mass inertial effect (transducer resistance to conserve its idle state), since the

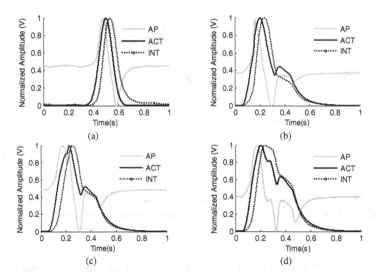

Fig. 5. Acoustic sensor responses to different excitation pressure waveforms. (a) Gaussian-like Pulse. (b) Type A Cardiac-like Pulse. (c) Type B Cardiac-like Pulse. (d) Type C Cardiac-like Pulse. AP - Acoustic Sensor Signal. ACT - Input Signal. INT - Integrated Sensor Signal.

Table 1. Statistics of the measurements obtained in the repeatability test.

Transducer	Mic 1	Mic 2
N° acquisitions	50	50
Mean (%)	1.1781	0.6198
Std. deviation (%)	0.0345	0.0298
Maximum (%)	1.2849	0.7379
Minimum (%)	1.1174	0.5890

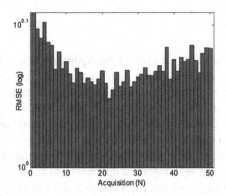

Fig. 6. Graphic representation of the RMSE distribution between the reference signal and the microphone 2 output, obtained in the repeatability test.

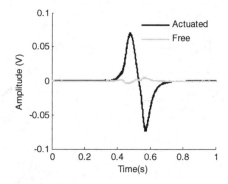

Fig. 7. Crosstalk phenomenon study: average response of both AP's transducers to fifty independent pulses. The actuated transducer is microphone 2 and the free transducer is microphone 1.

profile of the free transducer had an inversed shape relatively to the actuated one. This assumption could not be proven in the present work but it will be aim of futures studies. Nevertheless, and since the main purpose of this probe is the assessment of local PWV, it was employed a different approach, in order to determine if this "transmission" might interfere with the AP's time delay. For this purpose, both transducers were simultaneously actuated with three independent Gaussian waves and time delay was calculated using different algorithms. The results regarding this experiment are presented in Table 2 and Fig. 8.

Table 2. Time delay values obtained for each algorithm when both transducers are simultaneously actuated with three independent Gaussian impulses.

N	Time delay estimation method			
	xcorr	max	min	zc
1	8e-5	0.0134	0.0136	0.0037
2	8e-5	0.0109	0.0142	0.0038
3	8e-5	0.0103	0.0140	0.0035

The time delay obtained for each one of the algorithms is very different and actually surprising. It was not expected to obtain such a variable and elevated time delay values for maximum, minimum and zero crossing algorithms. In contrast, the cross-correlation algorithm presented a great performance, where time delay always matched the minimum detectable time, limited by the system, i.e.:, the sampling time (1/12500 Hz). In order to understand the achieved results, the AP's response was also analysed (Fig. 8). It is visible that the profiles obtained for each one of the transducers are identical; however, they present important differences in terms of amplitude and peaks correspondence. It was expected that the maxima and the minima of both signals were in agreement, but actually that didn't happen. These slight profiles difference can be justified with the experiment level of difficulty. It is extremely important that the simultaneous actuation of both transducers is made rigorously under the same

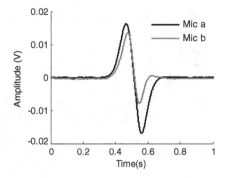

Fig. 8. Crosstalk phenomenon study: typical response of both AP's transducers to a Gauss ian pulse, simultaneously delivered to them.

conditions; otherwise the waveforms of each transducer can be affected. This also suggests that time delay algorithms that depend only on a fiducial point are more susceptible to error, especially if the waveforms don't have exactly the same profile. Finally and in what concerns to crosstalk effect, it can be concluded that the existence of a possible transmission between sensors does not affect the time delay, when the cross-correlation algorithm is used. In order to prove the effectiveness of the other algorithms, it will be necessary to proceed to additional experiments.

The last test concerning AP's experimental characterization intended to evaluate its time resolution. In this test, it was determined the PWV for two uncoupled AP's in successively smaller separation distances and the PWV reference obtained using the pressure sensors PS1 and PS2. The PWV results obtained for each algorithm and the relative errors between the reference PWV and the PWV obtained with the uncoupled transducers, for each separating distance and method are presented, respectively, in Figs. 9 and 10. The statistics of the measurements are synthesized in Table 3.

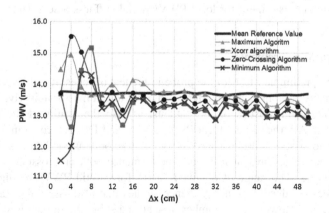

Fig. 9. Time resolution study: PWV values of uncoupled acoustic sensors and pressure sensors, yielded by the four algorithms. Each point is an average of three trials.

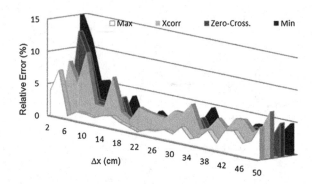

Fig. 10. Time resolution study: relative errors for each distance and method.

Table 3. Statistics of the measurements obtained in the time resolution test.

Algorithm	PWV(m/s)		Relative error mean (%)
	Pressure sensors (reference)	Acoustic sensors	
Maximum	13.856 ± 0.037	13.742 ± 0.372	2.083
Xcorr	13.716 ± 0.037	13.308 ± 0.524	4.246
ZC	13.702 ± 0.034	13.592 ± 0.560	2.863
Min	13.540 ± 0.039	13.176 ± 0.547	3.550

The algorithms with the best and worst general performance are the maximum and the cross-correlation with an average error less than 3 % and 5 %, respectively. However, and for the minimum distance achieved (2 cm) the magnitude of the errors is less than 1 %, when considering cross-correlation and zero-crossing algorithms.

The results obtained for AP time resolution for each algorithm, exhibit a very good performance suggesting that the AP have enough accuracy to be considered an interesting stand-alone instrument for local PWV/arterial stiffness assessment.

4.2   in-Vivo Measurements

Following the preliminary tests of the probe in the test benches, it was performed a set of measurements in human carotid arteries, in order to test the AP in in vivo conditions (Fig. 11). The characteristics of the patients, as also the results of the parameters assessed by the AP, (heart rate, local PWV, LVET and DP) are given in Table 4.

In order to assess pulse wave velocity, it was only used the cross-correlation algorithm, since it has presented the best performance both on crosstalk and on time resolution studies. The range of obtained values for carotid PWV are slightly lower than the values obtained by other reference studies that also assess the carotid PWV (≈4 m/s) [22, 30]. However, the number of analysed subjects not only is small as also include very young people (22.12 ± 1.96 years), which can justify a lower PWV mean (≈3 m/s), due to the high elasticity of young and healthier arteries. Although the

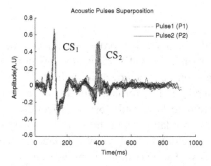

Fig. 11. Preliminary results of the AP acquiring data in a healthy young subject. CS_1 - First Carotid Sound. CS_2 - Second Carotid Sound.

Table 4. Main characteristics of the volunteers and AP parameters assessment.

Variable	Mean ± SD
Age, years	22.12 ± 1.96
N(Male/Female)	17(6/11)
Height, cm	166.82 ± 11.73
Weight, Kg	66.18 ± 18.87
BMI, Kg/m^2	23.50 ± 4.94
Waist, cm	75.35 ± 14.85
Brachial SBP, mmHg	114.35 ± 12.62
Brachial DBP, mmHg	70.94 ± 7.11
Heart Rate, bpm	67 ± 12.09
Local PWV(m/s)	2.96 ± 1.08
LVET (ms)	288.59 ± 21.42
DP (ms)	611.07 ± 148.84

obtained PWV variance is high (N = 17, ≈1 m/s), it is concordant with the PWV variance obtained in a recent study for a significant number of healthier subjects (N = 1774, ≈1.64 m/s) [30].

Nevertheless and in order to address more accurate results, it will be necessary to assess to a higher number of subjects, not only with a broader range of ages but also with pathologies, such as hypertension or atherosclerosis, where is expected to observe an increase of local PWV. The use of a reference method is also indispensable to validate the developed algorithms for AP hemodynamic parameters extraction. Currently, the probe presented a good performance in acquiring signals with a good bandwidth and signal-to-noise ratio in human carotid arteries, allowing the application of various algorithms that extract clinically relevant information.

Regarding the LVET values, we believe that is actually possible to determine this parameter as the time delay between the main onsets of each carotid sounds, since the estimated values are generally close to the expected for healthy subjects [31]. This parameter will be the subject of a further study, to evaluate the robustness of the algorithm.

5 Conclusions

A novel and low-cost doubled headed probe specifically designed to assess local stiffness, by means of non–invasive PWV measurements in a short segment of carotid artery, has been developed and characterized in dedicated test bench systems.

The probe demonstrated good performance on the dedicated test setups and results showed that its signals are repeatable and crosstalk effect do not interfere with its time resolution when the cross-correlation algorithm for time delay estimation is used.

It is also possible to conclude favorably towards the effectiveness of the AP in the measurement of local PWV. The maximum amplitude and the cross correlation algorithms exhibited the capability of measuring higher PWV ($\approx$14 m/s) with an error less than 10 %, for the several separating distances (50 cm–2 cm).

The natural follow-up to this work will be the continuation of the assessment of local PWV and other hemodynamic parameters in a significant numbers of patients (healthy and with various pathologies), under medical control. One of these studies would evaluate the agreement between the PWV values obtained with the AP and the ones obtained with the current 'gold-standard' system (Complior®). In general, lower values are expected for PWV measured over the carotid artery (local stiffness) than PWV measured over the carotid-femoral path (regional stiffness), nevertheless a correlation with aging and (cardiovascular) disease must be investigated. Another study will be designed to compare the experimental and in vivo performance of the AP probe with an improved version of the double headed piezoelectric probe [24], also developed in this research group, with the same purpose of local hemodynamics assessment.

Although studies to validate the clinical use of AP are still required, this device seems to be a valid alternative system, to local PWV stand-alone devices.

Acknowledgements. The authors acknowledge Fundação para a Ciência e Tecnologia for funding SFRH/BDE/15669/2007 and PTDC/SAU-BEB/100650/2008, project developed under the initiative of QREN, funding by UE/FEDER, through COMPETE - Programa Operacional Factores de Competitividade. The authors also thank to the company ISA-Intelligent Sensing Anywhere and the medical collaboration of Instituto de Investigação e Formação Cardiovascular, Clínica da Aveleira and Escola Superior de Tecnologia da Saúde de Coimbra.

References

1. Laurent, S., Boutouyrie, P., Asmar, R., Gautier, I., Laloux, B., Guize, L., et al.: Aortic stiffness is an independent predictor of all-cause and cardiovascular mortality in hypertensive patients. Hypertension **37**, 1236–1241 (2001)
2. Mattace-Raso, F., van der Cammen, T., Hofman, A., van Popele, N., Bos, M., Schalekamp, M., et al.: Arterial stiffness and risk of coronary heart disease and stroke the Rotterdam study. Circulation **113**, 657–663 (2006)
3. Mancia, G., De Backer, G., Dominiczak, A., Cifkova, R., Fagard, R., Germano, G., et al.: 2007 guidelines for the management of arterial hypertension: the task force for the management of arterial hypertension of the European society of hypertension (ESH) and of the European society of cardiology (ESC). J. Hypertens. **25**, 1105–1187 (2007)

4. Laurent, S., Katsahian, S., Fassot, C., Tropeano, A.I., Gautier, I., Laloux, B., et al.: Aortic stiffness is an independent predictor of fatal stroke in essential hypertension. Stroke **34**, 1203–1206 (2003)
5. Maldonado, J., Pereira, T., Polonia, J., Silva, J.A., Morais, J., Marques, M., et al.: Arterial stiffness predicts cardiovascular outcome in a low-to-moderate cardiovascular risk population: the EDIVA (Estudo de DIstensibilidade VAscular) project. J. Hypertens. **29**, 669–675 (2011)
6. Ben-Shlomo, Y., Spears, M., Boustred, C., May, M., Anderson, S., Boutouyrie, P., et al.: Prognostic value of carotid-femoral pulse wave velocity for cardiovascular events: an IPD meta-analysis of prospective observational data from 14 studies including 16,358 subjects. Artery Res. **5**, 138–139 (2011)
7. Asmar, R., Benetos, A., Topouchian, J., Laurent, P., Pannier, B., Brisac, A.M., et al.: Assessment of arterial distensibility by automatic pulse wave velocity measurement. Validation Clin. Appl. Stud. Hypertens. **26**, 485–490 (1995)
8. Boutouyrie, P., Briet, M., Collin, C., Vermeersch, S., Pannier, B.: Assessment of pulse wave velocity. Artery Res. **3**, 3–8 (2009)
9. Millasseau, S.C., Stewart, A.D., Patel, S.J., Redwood, S.R., Chowienczyk, P.J.: Evaluation of carotid-femoral pulse wave velocity: influence of timing algorithm and heart rate. Hypertension **45**, 222–226 (2005)
10. Rajzer, M.W., Wojciechowska, W., Klocek, M., Palka, I., Brzozowska-Kiszka, M., Kawecka-Jaszcz, K.: Comparison of aortic pulse wave velocity measured by three techniques: Complior. SphygmoCor Arteriograph. J. Hypertens. **26**, 2001–2007 (2008)
11. Weber, T., Rammer, M., Eber, B., O'Rourke, M.F.: Determination of travel distance for noninvasive measurement of pulse wave velocity: case closed? Hypertension **54**, e137 (2009)
12. Segers, P., Kips, J., Trachet, B., Swillens, A., Vermeersch, S., Mahieu, D., et al.: Limitations and pitfalls of non-invasive measurement of arterial pressure wave reflections and pulse wave velocity. Artery Res. **3**, 79–88 (2009)
13. Agabiti-Rosei, E., Mancia, G., O'Rourke, M.F., Roman, M.J., Safar, M.E., Smulyan, H., et al.: Central blood pressure measurements and antihypertensive therapy a consensus document. Hypertension **50**, 154–160 (2007)
14. Tu, S.T., Wang, I.W., Lin, H.F., Liao, Y.C., Lin, R.T., Liu, C.S., et al.: Carotid intima-media thickness and stiffness are independent risk factors for atherosclerotic diseases. J. Investig. Med. **58**, 786–790 (2010)
15. Gaszner, B., Lenkey, Z., Illyes, M., Sarszegi, Z., Horvath, I.G., Magyari, B., et al.: Comparison of aortic and carotid arterial stiffness parameters in patients with verified coronary artery disease. Clin. Cardiol. **35**, 26–31 (2012)
16. Laurent, S., Cockcroft, J., Van Bortel, L., Boutouyrie, P., Giannattasio, C., Hayoz, D., et al.: Expert consensus document on arterial stiffness: methodological issues and clinical applications. Eur. Heart J. **27**, 2588–2605 (2006)
17. Brands, P.J., Willigers, J.M., Ledoux, L.A.F., Reneman, R.S., Hoeks, A.P.G.: A noninvasive method to estimate pulse wave velocity in arteries locally by means of ultrasound. Ultrasound Med. Biol. **24**, 1325–1335 (1998)
18. Meinders, J.M., Hoeks, A.P.: Simultaneous assessment of diameter and pressure waveforms in the carotid artery. Ultrasound Med. Biol. **30**, 147–154 (2004)
19. Segers, P., Swillens, A., De Schryver, T., Vermeersch, S., Rietzschel, E., De Buyzere, M.: Assessing arterial distensibility using ultrasound wall-tracking diameter distension. In: 2009 IEEE International Ultrasonics Symposium (IUS), Rome, Italy, pp. 2460–2463 (2009)

20. Giannarelli, C., Bianchini, E., Bruno, R.M., Magagna, A., Landini, L., Faita, F., et al.: Local carotid stiffness and intima-media thickness assessment by a novel ultrasound-based system in essential hypertension. Atherosclerosis **223**, 372–377 (2012)
21. Giannattasio, C., Salvi, P., Valbusa, F.O., Kearney-Schwartz, A., Capra, A., Amigoni, M., et al.: Simultaneous measurement of beat-to-beat carotid diameter and pressure changes to assess arterial mechanical properties. Hypertension **52**, 896–902 (2008)
22. Hermeling, E., Reesink, K.D., Reneman, R.S., Hoeks, A.P.: Measurement of local pulse wave velocity: effects of signal processing on precision. Ultrasound Med. Biol. **33**, 774–781 (2007)
23. Hermeling, E., Reesink, K.D., Kornmann, L.M., Reneman, R.S., Hoeks, A.P.: The dicrotic notch as alternative time-reference point to measure local pulse wave velocity in the carotid artery by means of ultrasonography. J. Hypertens. **27**, 2028–2035 (2009)
24. Pereira, H.C., Pereira, T., Almeida, V., Borges, E., Figueiras, E., Simoes, J.B., et al.: Characterization of a double probe for local pulse wave velocity assessment. Physiol. Meas. **31**, 1449–1465 (2010)
25. Pereira, H.C., Simões, J.B., Malaquias, J.L., Pereira, T., Almeida, V., Borges, E., et al.: Double headed probe for local pulse wave velocity estimation - a new device for hemodynamic parameters assessment. In: BIODEVICES, pp. 444–447. SciTePress, Rome (2011)
26. Pereira, H.C., Cardoso, J.M., Almeida, V.G., Pereira, T., Borges, E., Figueiras, E., et al.: Programmable test bench for hemodynamic studies. In: Dössel, O., Schlegel, W. (eds.) World Congress on Medical Physics and Biomedical Engineering, Munich, Germany, September 7–12, 2009. IFMBE Proceedings, vol. 25/4, pp. 1460–1463. Springer, Heidelberg (2010)
27. Almeida, V., Pereira, T., Borges, E., Figueiras, E., Cardoso, J., Correia, C., et al.: Synthesized cardiac waveform in the evaluation of augmentation index algorithms - case study for a new wavelet based algorithm. In: BIOSIGNALS, pp. 385–388. INSTICC Press, Valencia (2010)
28. Murgo, J.P., Westerhof, N., Giolma, J.P., Altobelli, S.A.: Aortic input impedance in normal man: relationship to pressure wave forms. Circulation **62**, 105–116 (1980)
29. Hasegawa, M., Rodbard, D., Kinoshita, Y.: Timing of the carotid arterial sounds in normal adult men: measurement of left ventricular ejection, pre-ejection period and pulse transmission time. Cardiology **78**, 138–149 (1991)
30. Borlotti, A., Khir, A.W., Rietzschel, E.R., De Buyzere, M.L., Vermeersch, S., Segers, P.: Noninvasive determination of local pulse wave velocity and wave intensity: changes with age and gender in the carotid and femoral arteries of healthy human. J. Appl. Physiol. **113**, 727–735 (2012)
31. Willems, J.L., Roelandt, J.O.S., De Geest, H., Kesteloot, H., Joossens, J.V.: The left ventricular ejection time in elderly subjects. Circulation **42**, 37–42 (1970)

Bioinformatics Models, Methods
and Algorithms

Texture Classification of Proteins Using Support Vector Machines and Bio-inspired Metaheuristics

Carlos Fernandez-Lozano[1(✉)], Jose A. Seoane[2], Pablo Mesejo[3],
Youssef S.G. Nashed[3], Stefano Cagnoni[3], and Julian Dorado[1]

[1] Information and Communication Technologies Department,
Faculty of Computer Science, University of a Coruña,
Campus de Elviña s/n, 15071 A Coruña, Spain
{carlos.fernandez,julian}@udc.es
[2] MRC Centre for Causal Analyses in Translational Epidemiology, School
of Social and Community Medicine, University of Bristol,
Oakfield House, Oakfield Grove, Bristol BS82BN, UK
j.seoane@bristol.ac.uk
[3] Dipartimento Di Ingegneria Dell'Informazione, Universitá Degli Studi Di
Parma, Viale G. Usberti 181/a, 43100 Parma, Italy
{pmesejo,nashed,cagnoni}@ce.unipr.it

Abstract. In this paper, a novel classification method of two-dimensional polyacrylamide gel electrophoresis images is presented. Such a method uses textural features obtained by means of a feature selection process for whose implementation we compare Genetic Algorithms and Particle Swarm Optimization. Then, the selected features, among which the most decisive and representative ones appear to be those related to the second order co-occurrence matrix, are used as inputs for a Support Vector Machine. The accuracy of the proposed method is around 94 %, a statistically better performance than the classification based on the entire feature set. This classification step can be very useful for discarding over-segmented areas after a protein segmentation or identification process.

Keywords: Texture analysis · Feature selection · Electrophoresis · Support · Vector Machines · Genetic Algorithm · Proteomic imaging

1 Introduction

Proteomics is the study of protein properties in a cell, tissue or serum aimed at obtaining a global integrated view of disease, physiological and biochemical processes of cells, and regulatory networks. One of the most powerful techniques, widely used to analyze complex protein mixtures extracted from cells, tissues, or other biological samples, is two-dimensional polyacrylamide gel electrophoresis (2D-PAGE). In this method, proteins are classified by molecular weight (MWt) and iso-electric point (pI) using a controlled laboratory process and digital imaging equipment.

© Springer-Verlag Berlin Heidelberg 2014
M. Fernández-Chimeno et al. (Eds.): BIOSTEC 2013, CCIS 452, pp. 117–130, 2014.
DOI: 10.1007/978-3-662-44485-6_9

Since the beginning of proteomic research, 2D-PAGE has been the main protein separation technique, even before proteomics became a reality itself. The main advantages of this approach are its robustness and its unique ability to analyze complete proteins at high resolution, keeping them intact and being able to isolate them entirely [1]. However, this method has also several drawbacks like its very low effectiveness in the analysis of hydrophobic proteins, as well as its high sensitivity to dynamic range (i.e. the quantitative ratio between the rarest protein expressed in a sample and the most abundant one) and quantitative distribution issues [2]. The outcome of the process is an image like the ones showed in [3, 4].

Dealing with this kind of images is a difficult task because there is not a commonly accepted ground truth [3, 4]. Another aspect that makes the work difficult from a computer vision point of view is that both protein images and background noise seem to follow a Gaussian distribution [5]. The inter- and intra-operator variability of the outcome of manual analysis of these images is also a big drawback [6].

The aim of this paper is to demonstrate that there is enough texture information in 2D-electrophoresis images to discriminate proteins from noise or background. In this work the most representative group of textural features are selected using two meta-heuristics, namely Genetic Algorithms [7] and Particle Swarm Optimization [8].

2 Theoretical Background and Related Work

The method proposed in this work intends to assist 2D-PAGE image analysis by studying the textural information they carry. To do so, a novel combination of meta-heuristics and Support Vector Machines [9] is presented. In this section, the main techniques used in our approach are briefly introduced and explained.

2.1 Texture

One of the most important characteristics used for identifying objects or regions of interest in an image is texture, which is related with the spatial (statistical) distribution of the grey levels within an image [10]. Texture is a surface's property and can be regarded as an almost regular spatial organization of complex patterns, always present even if they could exist as a non-dominant feature. Different approaches to texture analysis have been applied which are extensively reviewed in [11, 12].

There are four major issues one can relate with texture: synthesis, classification, segmentation and shape from texture [12, 13].

- Texture synthesis: this is a subjective process which creates textures in synthetic images. It is important where the goal is to obtain object surfaces as realistic as possible [14].
- Texture classification: the goal is to classify regions of interest in the image according to the kind of texture they embed [15]. This particular issue is widely studied in medical imaging, quality control and remote sensing among others. Classification algorithms can rely on a quantitative measure of success. It is necessary to have a priori knowledge of the possible texture types in order to perform such a task.

- Texture segmentation: the principal task here is to find the texture boundaries in an image with a large number of textural types [16]. This is a blind process in the sense that there is no a priori information available about them or about how many different textures or types of textures are there.
- Shape from texture: the task is to reconstruct a three-dimensional object from a two-dimensional image based on textural information. Firstly proposed by Gibson [17].

This paper is focused on texture classification, where texture can be computed from the variations in the intensities within the image. As said before, there is no textural information in one pixel, so texture is a contextual property related with gray levels in a neighborhood. First-order statistics depend only on individual pixel values and can be computed from the histogram of pixel intensities in the image, but second-order statistics depend on pairs of grey values, concerning their relative position and spatial resolution. Commonly used second-order statistics can be derived from the so-called Grey Level Co-Occurrence Matrix (GLCM), first proposed by Haralick [9]. Common properties playing an important role with texture definition, as identified by Laws [18], are: density, coarseness, uniformity, roughness, regularity, linearity, directionality, direction, frequency, and phase. All these properties are often related [19].

2.2 Metaheuristics

Genetic Algorithms (GAs) are search techniques inspired by Darwinian Evolution and developed by Holland in the 1970 s [7]. In a GA, an initial population of individuals, i.e. possible solutions defined within the domain of a fitness function to be optimized, is evolved by means of genetic operators: selection, crossover and mutation. The selection operator ensures the survival of the fittest individuals, crossover represents the mating between individuals, and mutation introduces random modifications into the population. GAs possess effective capabilities to explore the search space in parallel, exploiting the information about the quality of the individuals evaluated so far [20]. Using the crossover operator, GAs combine the features of parents to produce new and better solutions, which preserve the parents' best characteristics. Using the mutation operator, new information is introduced in the population in order to explore new areas of the search space. The strategy known as elitism, which is a variant of the general process of constructing a new population, allows the best organisms from the current generation to survive to the next, remaining unaltered. At the end of the process, the population of solutions is expected to converge to the global optimum of the fitness function.

Particle Swarm Optimization [8] (PSO) is a bio-inspired optimization algorithm based on the simulation of the social behavior of bird flocks. In the last fifteen years PSO has been applied to a very large variety of problems [21] and numerous variants of the algorithm have been presented [22].

During the execution of PSO a set of particles moves within the function domain searching for the optimum of the function (best fitness value). The motion of each particle is driven by the best positions visited so far by the particle itself and by the entire swarm (gbest PSO) or by some pre-defined neighborhood of the particle (lbest PSO).

Consequently, each particle relies both on "individual" and on "swarm" intelligence, and its motion can be described by two simple discrete-time equations which regulate the particle's position and velocity.

2.3 Support Vector Machines

Vapnik introduced Support Vector Machines (SVMs) in the late 1970 s on the foundation of statistical learning theory [9]. The basic implementation deals with two-class problems in which data are separated by a hyperplane defined by a number of support vectors. This hyperplane separates the positive from the negative examples, maximizing the distance between the boundary and the nearest data point in each class; the nearest data points are used to define the margins, known as support vectors [23]. These classifiers have also proven to be exceptionally efficient in classification problems of high dimensionality [24, 25], because of their ability to generalize in high-dimensional spaces, such as the ones spanned by texture patterns. SVMs use different non-linear kernel functions, like polynomial, sigmoid and radial basis functions, to map the training samples from the input spaces into a higher-dimensional feature space through a mapping function [23].

2.4 Related Work

With respect to related work, the authors were not able to find any other work in the literature dealing with evolutionary computation in combination with texture analysis in 2D-electrophoresis images, while we did find one article describing a discriminant partial least squares regression (PLSR) method for spot filtering in 2D-electrophoresis [26]. The authors use a set of parameters to build a model based on texture, shape and intensity measurements using image segments from gel segmentation. As regards texture information, they focus on the descriptors related to the noisy surface texture of unwanted artifacts and conclude that their textural features allow them to distinguish noisy features from protein spots. In their work, five out of the eleven second-order textural features are used, along with five new textural features accounting for intensity relationships among sets of three pixels. They distinguish proteins in the image by using shape information, since cracks and artifacts in gel surface deviate from a circular shape. Besides that, a degree of Gaussian fit is calculated as an indicator of whether the image segment corresponds to a protein or to an artifact. Textural features are used for noise and crack detection and as a complement for spot segmentation. Finally, the 17 initial variables are reduced to five PLSR components to account for 85 % of the total variation with respect to the response factor, and 82 % of the total variation in the data matrix.

3 Materials

In order to generate the dataset, ten 2D-PAGE images, representative enough of different types of tissues and different experimental conditions, were used. These images are similar to the ones used by G.-Z. Yang (Imperial College of Science, Technology and Medicine, London), and are available for download at the webpage http://personalpages.manchester.ac.uk/staff/andrew.dowsey/rain/. It is important to notice that Hunt et al. [27] determined that 7–8 is the minimum acceptable number of samples for a proteomic study.

For each image, 50 regions of interest (ROIs) representing proteins and 50 representing non-proteins (noise, black non-protein regions, and background) were selected to build a training set with 1000 samples in a double-blind process in which two clinicians selected the fifty ROIs they considered and after that, within which they selected proteins which were representative of the different possible scenarios (isolated, overlapped, big, small, darker, etc.). For each ROI, as will be seen later, 296 texture features are computed.

The ROIs were selected taking into consideration that, for each manually selected protein, there is an area of influence surrounding it. It means that, once the clinician has selected a protein, the ROI is slightly larger than the visible dark surface of such a protein. This assumption is made because texture characterizes not only the darkest regions but also the lightest ones.

As said before, proteins seem to fit a Gaussian peak, and ideally the center of the protein is in the darkest zone of that peak. This approach prevents the loss of information caused by neglecting the lowest values of the inverted protein (grey levels closest to white) that also fit the Gaussian peak. This information could be useful to classify a ROI as including a protein or to discard it.

4 Proposed Method

This paper goes further than related work in texture analysis of 2D-electrophoresis images, studying the ability of textural features to discriminate not only cracks from proteins but background and non-protein dark spots as well.

The first step in texture analysis is texture feature extraction from the ROIs. With a specialized software called Mazda [28], 296 texture features are computed for each element in the training set. Various approaches have demonstrated the effectiveness of this software in extracting textural features from different types of medical images [29–33].

These features [34], reported in Table 1, are based on:

- Image histogram.
- Co-occurrence matrix: information about the grey level value distribution of pairs of pixels with a preset angle and distance between each other.
- Run-length matrix: information about sequences of pixels with the same grey level values in a given direction.
- Image gradients: spatial variation of grey level values.
- Autoregressive models: description of texture based on the statistical correlation between neighboring pixels.
- Wavelet analysis: information about the image frequency content at different scales.

Thus, within each ROI, texture information was analyzed by extracting first and second-order statistics, spatial frequencies, co-occurrence matrices and two other statistical methods as autoregressive model and wavelet based analysis, preserving the original gray-level and spatial resolution in all runs. Histogram-related measures conform the first-order statistics proposed by Haralick [10] but second-order statistics are those derived from the Grey Level Co-occurrence Matrices (GLCM). Additionally, a group of features derived from the textural ones is also calculated, such as the area of the ROI, but cannot be used for texture characterization.

Table 1. Textural features extracted and used in this work.

Group	Features
Histogram	Mean, variance, skewness, kurtosis, percentiles 1 %, 10 %, 50 %, 90 % and 99 %
Absolute Gradient	Mean, variance, skewness, kurtosis and percentage of pixels with nonzero gradient
Run-length Matrix	Run-length non-uniformity, grey-level non-uniformity, long-run emphasis, short-run emphasis and fraction of image in runs
Co-ocurrence Matrix	Angular second moment, contrast, correlation, sum of squares, inverse difference moment, sum average, sum variance, sum entropy, entropy, difference variance and difference entropy
Autoregressive Model	Theta: model parameter vector, four parameters; Sigma: standard deviation of the driving noise
Wavelet	Energy of wavelet coefficients in sub-bands at successive scales; max four scales, each with four parameters

All these feature sets were included in the dataset. The normalization method applied was the one set by default in Mazda: image intensities were normalized in the range from 1 to $Ng = 2^k$, where k is the number of bits per pixel used to encode the image under analysis.

Two solutions are available for decreasing the dimensionality: extraction of new features derived from the existing ones and selection of relevant features to build robust models. In order to extract a feature set from the problem data, principal component analysis (PCA) has been commonly used. In this work, we use GAs to find the smallest feature subset able to yield a fitness value above a threshold. Besides optimizing the complexity of the classifier, feature selection may also improve the classifiers' quality. In fact, classification accuracy could even improve if noisy or dependent features are removed.

The use of GAs for feature selection were first proposed by Siedlecki and Skalansky [35]. Many studies have been done on GAs for feature selection since then [36], concluding that they are suitable for finding optimal solutions to large problems with more than 40 features to select from. GAs for feature selection could be used in combination with a classifier such as SVM, k-nearest neighbor (KNN) or artificial neural networks (ANN), optimizing its performance. In terms of classification accuracy with imaging problems, SVMs have shown to yield good performance with textural features [37–39], but also KNN [40]; hybrid approaches which use a combination of both classifiers [41] have obtained good results. Other techniques use GAs to optimize both feature selection and classifier parameters [42, 43].

In our method, based on both GAs and SVMs, there is not a fixed number of variables. As GAs continuously reduce the number of variables that characterize the samples, a pruned search is implemented. Each individual in the genetic population is described by p genes (using binary encoding). The fitness function (1) considers not only the classification results but also the number of variables used for such a classification, so it is defined as the sum of two terms, one related to the classification results and another to the number of variables selected. In (1) the number of genes with a true binary value (feature selected) is represented by *numberActiveFeatures*. Regarding classification results, taking into account the F-measure apparently gives better results than only using the accuracy obtained with image features [44, 45]. F-measure (2) is a function made up of the recall (true positives rate or sensitivity: proportion of actual positives which are correctly identified) and precision (or positive predictive value: proportion of positive test results that are true positives) measurements.

$$Fitness = (1 - F) + \frac{numberActiveFeatures}{numberTotalFeatures} \qquad (1)$$

$$F = 2 . \frac{precision.recall}{precision + recall} \qquad (2)$$

Therefore individuals with fewer active features (genes) are favored. Once the reduced feature dataset is generated, a statistical parametric test is made to evaluate the adequacy of the feature selection process.

5   Experimental Results

The test set is composed of ten representative images for the different types of proteomic available images, and for each one of them, 50 protein and 50 non-protein ROIs have been extracted to generate a dataset with 1000 elements, that was divided randomly in 800 elements, of which 600 elements are used for training and 200 elements are used for validation (inside the GA feature selection process) and finally, 200 elements for test. Once the GA finishes, the best individual found (the one with lowest fitness value) is tested, using a 10-fold cross validation (10-fold CV), to calculate the error of the proposed model using the full and reduced datasets. Then, a test set is used in order to evaluate the adequacy of the reduction process.

After a preliminary experimental study of the values suggested in the literature, the parameters used in the feature selection process were empirically set the population size to 250 individuals, with no elitism, a 95 % crossover probability, a 2 % mutation rate, with scattered crossover, tournament selection and uniform mutation.

To evaluate the performance of this method, there are several number of well-known accuracy measures for a two-class classifier in the literature such as: classification rate (accuracy), precision, sensitivity, specificity, F-score, Area Under an ROC Curve (AUC), Youden's index, Cohen's kappa, likelihoods, discriminant power, etc. The ROC curve is a graphical plot of the sensitivity against 1-specificity as the detector threshold, or a parameter which modifies the balance between sensitivity and specificity. An experimental comparison of performance measures for classification could be found in [46]. In [47], the authors proposed that AUC is a better measure in general than accuracy when comparing classifiers and in general. The most common measures used for their simplicity and successful application are the classification rate and Cohen's kappa measures. Tables 3a and 3b in the appendix shows the results for classification rate (accuracy), AUC, F-measure, Youden's and unweighted Cohen's Kappa for each kernel. For this problem, all the measures consider the same ranking, and the best kernel function is the linear one. For each kernel, Table 4 in the Appendix section shows the selected features in their textural membership group.

Among others, Mazda computes the area for each ROI. This feature merely indicates the number of pixels used to compute the textural features and, since it has nothing to do with the description of textures, it cannot be used for texture characterization. With linear, polynomial (order 3), and RBF (C = 100 and sigma = 10) kernels, non- textural features are selected for classification. The results obtained seem to indicate that the textural group with more representatives in 2D-PAGE images is the Co-occurrence matrix Group (second-order statistics).

As the proposed work intends to evaluate the textural information present in a 2D-PAGE image, the RBF(2) kernel function is selected as the most appropriate for solving this problem, since only textural features where selected for classification with this kernel, and it yields the best accuracy.

In order to compare the GA-based feature selection results, a binary Particle Swarm Optimization implementation for feature selection is used [48]. PSO is an optimization algorithm inspired by the organized behavior of large groups of simple animals and, like GAs and other Evolutionary Computation techniques; it is a derivative-free global optimum solver. Firstly proposed by Kennedy and Eberhart [8] and used optimization of non-linear functions [49].

The experiments were performed with the same final combination of common parameters (population size, stall conditions, etc.), and the same elements for training and validation separated with the same seed in order to reproduce experimental conditions with the RBF(2) kernel function. Results are shown in Table 2.

The results show that GA-based have better results, improving the AUC-ROC score of the PSO and is able to reduce to 6 features whilst the PSO is only able to reduce to 58 features. Both techniques reach the stall condition in a very close number of generations.

Table 2. Results for the GA and PSO approaches with RBF(2) kernel.

	Accuracy	AUC	F-Measure	Number of variables	Generations
GA	0.88	**0.88**	0.89	6	45
PSO	0.83	0.86	0.85	58	44

We evaluated the reduced textural feature set on the 200 patterns of the validation dataset using the RBF (2) kernel, by calculating the F-measure and the areas under the receiver operating characteristic curves and a 10-fold CV, using the Libsvm classifier implementation [50] in Weka [51] and comparing the results with the same classifier using the full dataset. Thus, we have obtained samples composed by 10 AUC measures. AUC area can be seen as the capacity to be sensitive and specific at the same time, in the sense that the larger is the AUC, the more accurate is the model. We use the RBF kernel with different gamma values (the parameter controls the width of the kernel) to check if there is a significant improvement when the reduced dataset is used.

In order to use a parametric test, it is necessary to check the independence, normality and heteroscedasdicity [52]. In statistics, two events are independent when the occurrence of one does not modify the probability of the other one. An observation is normal when its behavior follows a normal or Gaussian distribution with a certain value of mean and variance. The heteroscedasticity indicates the existence of a violation of the hypothesis of equality of variances [53].

With respect to the independence condition, we separate the data using 10-fold CV. We perform a normality analysis using the Shapiro-Wilk test [54] with a level of confidence alpha = 0.05, for the Null Hypothesis that the data come from a normally distributed population, and such null hypothesis was rejected. The observed data fulfill the normality condition, a Bartlett test [55] is performed in order to evaluate the heteroscedasticity with a level of confidence alpha = 0.05.

A corrected paired Student's t-test could be performed in Weka [51], with a level of confidence alpha = 0.05, for the Null Hypothesis that there are no differences between the average values obtained by both methods. Results in average, with standard deviation in brackets for AUC-ROC are 0.94 (0.07) for the reduced dataset, and 0.55 (0.34) for the full dataset and the corrected paired Student's t-test determines that there is a significant improvement in using the reduced dataset. The reduced dataset has better accuracy result than the full dataset. Even more, the corrected paired Student's t-test evaluates this improvement as significant with an alpha = 0.05.

6 Summary and Conclusions

To the best of our knowledge, this is the first work in which protein classification in two-dimensional gel electrophoresis images is tackled using Evolutionary Computation, Support Vector Machines and Textural Analysis. In fact, this paper demonstrates the existence of enough textural information to discriminate proteins from noise and background, as well as to show the potential of SVMs in proteomic classification problems.

A new dataset with six features, starting from the 296 original ones, is created without loss of accuracy, and the most representative textural group has shown to be the one related to the Co-occurrence matrix Group (second-order statistics). A proper statistical test has determined that there is a significant improvement in using this reduced feature set with respect to the full feature set.

Acknowledgements. This work is by "Development of new image analysis techniques in 2D Gel for biomedical research" (Ref. 10SIN105004PR), CN2102/217, CN2011/034 and CN2012/130 by Xunta de Galicia. Jose A. Seoane acknowledges Medical Research Council Project Grant G1000427. Pablo Mesejo and Youssef S.G. Nashed are funded by the European Commission (MIBISOC Marie Curie Initial Training Network, FP7 PEOPLE-ITN-2008, GA n. 238819).

Appendix

Table 3a. Results with different SVM Gaussian kernels.

Kernel Type	Measure	Value
RBF(1)	TP	90
	FN	10
	FP	18
	TN	82
	Accuracy	0.86
	AUC	0.86
	F-Measure	0.86
	Youden's	0.72
	Kappa	0.72
	Nvar	8
RBF(2)	TP	94
	FN	6
	FP	17
	TN	83
	Accuracy	0.88
	AUC	0.88
	F-Measure	0.89
	Youden's	0.77
	Kappa	0.77
	Nvar	6
RBF (100;10)	TP	94
	FN	6
	FP	18
	TN	82
	Accuracy	0.88
	AUC	0.88
	F-Measure	0.88
	Youden's	0.76
	Kappa	0.76
	Nvar	8

Table 3b. Results with different SVM polynomial kernels.

Kernel Type	Measure	Value
Linear	TP	95
	FN	5
	FP	11
	TN	89
	Accuracy	0.92
	AUC	0.92
	F-Measure	0.92
	Youden's	0.85
	Kappa	0.85
	Nvar	6
Poli (3)	TP	87
	FN	13
	FP	19
	TN	81
	Accuracy	0.84
	AUC	0.84
	F-Measure	0.84
	Youden's	0.68
	Kappa	0.68
	Nvar	16

Table 4. Study of texture parameters between best SVM kernels in accuracy.

	Histogram	Absolute gradient	Run-length matrix	Co-occurence matrix	Wavelet	Non-textural features
RBF(1)				S(2,0)InvDfMom S(0,3)SumAverg S(0,3)DifVarnc S(4,-4)Contrast S(0,5)SumEntrp S(0,5)DifEntrp S(5,5)SumEntrp S(5,-5)Entropy		
RBF(2)	Perc.01%			S(2,-2)DifEntrp S(5,0)Correlat S(5,0)InvDfMom S(0,5)DifVarnc S(5,5)SumEntrp		
Linear	Skewness			S(2,2)Correlat S(4,0)InvDfMom S(5,0)Contrast		Area_S(0,4) Area_S(5,-5)
Poli(3)	Kurtosis	GrKurtosis	45dgr_GLev NonU	S(1,-1)Contrast S(1,-1)DifEntrp S(0,2)DifEntrp S(0,4)SumAverg S(4,-4)Correlat S(4,-4)SumVarnc S(5,0)InvDfMom S(0,5)SumOfSqs S(0,5)InvDfMom S(0,5)SumEntrp	WavEnLH_s-2 WavEnLH_s-4	AreaGr
RBF(100;10)			Horzl_GLev NonU	S(2,0)InvDfMom S(5,0)InvDfMom S(0,5)InvDfMom S(5,-5)DifEntrp	WavEnLH_s-4	Area_S(0,1) Area_S(5,0)

References

1. Rabilloud, T., Chevallet, M., Luche, S., Lelong, C.: Two-dimensional gel electrophoresis in proteomics: past, present and future. J. Proteomics **73**, 2064–2077 (2010)
2. Zhang, J., Tan, T.: Brief review of invariant texture analysis methods. Pattern Recogn. **35**, 735–747 (2002)
3. Marten Lab Proteomics Page. http://www.umbc.edu/proteome/image_analysis.html
4. Center for Cancer Research Nanobiology Program (CCRNP). http://www.ccrnp.ncifcrf.gov/users/lemkin
5. Tsakanikas, P., Manolakos, E.S.: Improving 2-DE gel image denoising using contourlets. Proteomics **9**, 3877–3888 (2009)
6. Millioni, R., Sbrignadello, S., Tura, A., Iori, E., Murphy, E., Tessari, P.: The inter- and intra-operator variability in manual spot segmentation and its effect on spot quantitation in two-dimensional electrophoresis analysis. Electrophoresis **31**, 1739–1742 (2010)
7. Holland, J.H.: Adaptation in Natural and Artificial Systems: An Introductory Analysis with Applications to Biology, Control, and Artificial Intelligence. University of Michigan Press, Ann Arbor (1975)
8. Kennedy, J., Eberhart, R.: Particle swarm optimization. In: IEEE International Conference on Neural Networks, Proceedings, vol. 1944, pp. 1942–1948 (1995)
9. Vapnik, V.N.: Estimation of dependences based on empirical data [in Russian]. English translation Springer Verlang, 1982, Nauka (1979)
10. Haralick, R.M., Shanmugam, K., Dinstein, I.: Textural features for image classification. IEEE Trans. Syst. Man Cybern. smc **3**, 610–621 (1973)
11. Materka, A., Strzelecki, M.: Texture analysis methods-A review. Technical University of Lodz, Institute of Electronics. COST B11 report (1998)
12. Tuceryan, M., Jain, A.: Texture analysis. Handbook of pattern recognition and computer vision, vol. 2. World Scientific Publishing Company, Incorporated (1999)
13. Levina, E.: Statistical Issues in Texture Analysis. University of California, Berkeley (2002)
14. Peitgen, H.O., Saupe, D., Barnsley, M.F.: The Science of Fractal Images. Springer-Verlag, New York (1988)
15. Pietikainen, K.: Texture Analysis in Machine Vision. World Scientific Publishing Company (Incorporated), River Edge (2000)
16. Mirmedhdi, M., Xie, X., Suri, J.S.: Handbook of Texture Analysis. Imperial College Press, London (2008)
17. Gibson, J.J.: The Perception of the Visual World. Houghton Mifflin, Boston (1950)
18. Laws, K.I.: Textured Image Segmentation. University of Southern California, Los Angles (1980)
19. Tomita, F., Tsuji, S.: Computer Analysis of Visual Textures. Kluwer Academic Publishers, Boston (1990)
20. Goldberg, D.: Genetic Algorithms in Search, Optimization, and Machine Learning. Addison-Wesley Professional, Upper Saddle River (1989)
21. Poli, R.: Analysis of the publications on the applications of particle swarm optimisation. J. Artif. Evol. App. **2008**, 1–10 (2008)
22. Banks, A., Vincent, J., Anyakoha, C.: A review of particle swarm optimization. Part I: Backgr. Dev. **6**, 467–484 (2007)
23. Burges, C.J.C.: A tutorial on support vector machines for pattern recognition. Data Min. Knowl. Disc. **2**, 121–167 (1998)

24. Moulin, L.S., Da Silva, A.P.A., El-Sharkawi, M.A., Marks Ii, R.J.: Support vector machines for transient stability analysis of large-scale power systems. IEEE Trans. Power Syst. **19**, 818–825 (2004)
25. Chapelle, O., Haffner, P., Vapnik, V.N.: Support vector machines for histogram-based image classification. IEEE Trans. Neural Netw. **10**, 1055–1064 (1999)
26. Rye, M.B., Alsberg, B.K.: A multivariate spot filtering model for two-dimensional gel electrophoresis. Electrophoresis **29**, 1369–1381 (2008)
27. Hunt, S.M.N., Thomas, M.R., Sebastian, L.T., Pedersen, S.K., Harcourt, R.L., Sloane, A.J., Wilkins, M.R.: Optimal replication and the importance of experimental design for gel-based quantitative proteomics. J. Proteome Res. **4**, 809–819 (2005)
28. Szczypinski, P.M., Strzelecki, M., Materka, A.: MaZda - A software for texture analysis, pp. 245–249
29. Szymanski, J.J., Jamison, J.T., DeGracia, D.J.: Texture analysis of poly-adenylated mRNA staining following global brain ischemia and reperfusion. Comput. Methods Programs Biomed. **105**, 81–94 (2012)
30. Harrison, L., Dastidar, P., Eskola, H., Järvenpää, R., Pertovaara, H., Luukkaala, T., Kellokumpu-Lehtinen, P.L., Soimakallio, S.: Texture analysis on MRI images of non-Hodgkin lymphoma. Comput. Biol. Med. **38**, 519–524 (2008)
31. Mayerhoefer, M.E., Breitenseher, M.J., Kramer, J., Aigner, N., Hofmann, S., Materka, A.: Texture analysis for tissue discrimination on T1-weighted MR images of the knee joint in a multicenter study: Transferability of texture features and comparison of feature selection methods and classifiers. J. Magn. Reson. Imaging **22**, 674–680 (2005)
32. Bonilha, L., Kobayashi, E., Castellano, G., Coelho, G., Tinois, E., Cendes, F., Li, L.M.: Texture analysis of hippocampal sclerosis. Epilepsia **44**, 1546–1550 (2003)
33. Létal, J., Jirák, D., Suderlová, L., Hájek, M.: MRI 'texture' analysis of MR images of apples during ripening and storage. LWT - Food Sci. Technol. **36**, 719–727 (2003)
34. Szczypiski, P.M., Strzelecki, M., Materka, A., Klepaczko, A.: MaZda-A software package for image texture analysis. Comput. Methods Programs Biomed. **94**, 66–76 (2009)
35. Siedlecki, W., Sklansky, J.: A note on genetic algorithms for large-scale feature selection. Pattern Recogn. Lett. **10**, 335–347 (1989)
36. Kudo, M., Sklansky, J.: A comparative evaluation of medium- and large-scale feature selectors for pattern classifiers. Kybernetika **34**, 429–434 (1998)
37. Li, S., Kwok, J.T., Zhu, H., Wang, Y.: Texture classification using the support vector machines. Pattern Recogn. **36**, 2883–2893 (2003)
38. Kim, K.I., Jung, K., Park, S.H., Kim, H.J.: Support vector machines for texture classification. IEEE Trans. Pattern Anal. Mach. Intell. **24**, 1542–1550 (2002)
39. Buciu, I., Kotropoulos, C., Pitas, I.: Demonstrating the stability of support vector machines for classification. Sig. Process. **86**, 2364–2380 (2006)
40. Jain, A.: Feature selection: evaluation, application, and small sample performance. IEEE Trans. Pattern Anal. Mach. Intell. **19**, 153–158 (1997)
41. Zhang, H., Berg, A.C., Maire, M., Malik, J.: SVM-KNN: Discriminative nearest neighbor classification for visual category recognition. pp. 2126–2136. (Year)
42. Huang, C.L., Wang, C.J.: A GA-based feature selection and parameters optimization for support vector machines. Expert Syst. Appl. **31**, 231–240 (2006)
43. Manimala, K., Selvi, K., Ahila, R.: Hybrid soft computing techniques for feature selection and parameter optimization in power quality data mining. Appl. Soft Comput. J. **11**, 5485–5497 (2011)
44. Müller, Meinard, Demuth, Bastian, Rosenhahn, Bodo: An evolutionary approach for learning motion class patterns. In: Rigoll, Gerhard (ed.) DAGM 2008. LNCS, vol. 5096, pp. 365–374. Springer, Heidelberg (2008)

45. Tamboli, A.S., Shah, M.A.: A Generic Structure of Object Classification Using Genetic Programming. In: 2011 International Conference on Communication Systems and Network Technologies (CSNT), pp. 723–728 (2011)
46. Ferri, C., Hernádez-Orallo, J., Modroiu, R.: An experimental comparison of performance measures for classification. Pattern Recogn. Lett. **30**, 27–38 (2009)
47. Huang, J., Ling, C.X.: Using AUC and accuracy in evaluating learning algorithms. IEEE Trans. Knowl. Data Eng. **17**, 299–310 (2005)
48. Chen, S.: Another Particle Swarm Optimization Toolbox. Ontario (2003)
49. Perez, R.E., Behdinan, K.: Particle swarm approach for structural design optimization. Comput. Struct. **85**, 1579–1588 (2007)
50. Chang, C.C., Lin, C.J.: LIBSVM: a library for support vector machines. ACM Trans. Intell. Syst. Technol. **2**, 1–27 (2011)
51. Hall, M., Frank, E., Holmes, G., Pfahringer, B., Reutemann, P., Witten, I.H.: The WEKA data mining software: an update. SIGKDD Explor. Newsl. **11**, 10–18 (2009)
52. Sheskin, D.J.: Handbook of Parametric and Nonparametric Statistical Procedures. Taylor and Francis, Boca Raton (2011)
53. García, S., Fernández, A., Luengo, J., Herrera, F.: A study of statistical techniques and performance measures for genetics-based machine learning: accuracy and interpretability. Soft. Comput. **13**, 959–977 (2009)
54. Shapiro, S.S., Wilk, M.B.: An analysis of variance test for normality (complete samples). Biometrika **52**, 591–611 (1965)
55. Bartlett, M.S.: Properties of sufficiency and statistical tests. Proc. R. Soc. Lond. Ser. A Math. Phys. Sci. **160**, 268–282 (1937)

Formal Analysis of Gene Networks Using Network Motifs

Sohei Ito[1]([✉]), Takuma Ichinose[2], Masaya Shimakawa[2], Naoko Izumi[3], Shigeki Hagihara[2], and Naoki Yonezaki[2]

[1] National Fisheries University, Shimonoseki, Japan
ito@fish-u.ac.jp
[2] Tokyo Institute of Technology, Tokyo, Japan
[3] Jumonji University, Niiza, Japan

Abstract. We developed a theoretical framework to analyse gene regulatory networks. Our framework is based on the *formal methods* which are well-known techniques to analyse software/hardware systems. Behaviours of gene networks are abstracted into transition systems which has discrete time structure. We characterise possible behaviours of given networks by linear temporal logic (LTL) formulae. By checking the satisfiability of LTL formulae, we analyse whether all/some behaviours of given networks satisfy given biological properties. Due to the complexity of LTL satisfiability checking, analyses of large networks are generally intractable in this method. To mitigate this computational difficulty, we proposed approximate analysis method using network motifs to circumvent the computational difficulty of LTL satisfiability checking. Experiments show that our approximate method is surprisingly efficient.

Keywords: Gene regulatory network · Temporal logic · Formal methods · Network motif

1 Introduction

Formal methods are useful in analysing and developing software/hardware systems in which we mathematically model the system, give the property and mathematically analyse whether the system satisfies the property. Although formal methods are mainly studied and applied in computer science, they are also useful in bioinformatics and systems biology.

We developed a theoretical framework to analyse gene regulatory networks based on formal methods [1]. Our framework is closely related with verification of *reactive system* specifications [2–4] in which the specification of the system is described in linear temporal logic (LTL) and check properties of it. The specification of a reactive system is intended to cover all possible behaviours of it. Based on this method, we characterise possible behaviours of gene networks in LTL and check whether they satisfy given biological properties.

In our method, behaviours are captured as transition systems using propositions for gene states (ON or OFF) and thresholds on gene activation/inhibition.

© Springer-Verlag Berlin Heidelberg 2014
M. Fernández-Chimeno et al. (Eds.): BIOSTEC 2013, CCIS 452, pp. 131–146, 2014.
DOI: 10.1007/978-3-662-44485-6_10

We characterise possible behaviours of networks by specifying changes in concentration levels of gene products and changes in gene states using LTL. The constraints are intended to cover all possible behaviours of networks. Expected biological properties such as reachability, stability and oscillation are also described in LTL. We check satisfiability of these formulae to verify whether some or all behaviours satisfy the corresponding biological property.

Our method is purely qualitative. We do not need any quantitative information such as kinetic parameters. Thus our method is useful when we do not have precise kinetic parameters but are interested in checking some qualitative property, e.g. 'is a certain gene oscillates'? If such property is computationally possible, biologists are motivated to check whether the property is really observed.

The bottleneck of our method is the computational difficulty of LTL satisfiability checking which is PSPACE-complete [5]. Known algorithms have exponential time complexity with respect to the length of an input formula. The length of a formula specifying possible behaviours of a network is proportional to the size of the network in our method, and thus analyses of large networks are generally intractable.

To mitigate this computational difficulty, we developed approximate analysis method to enable analysis of large networks in our framework. We approximate the set of possible behaviours by simple specifications. It is not trivial to find approximate specifications for any network. Thus we consider some common network patterns which can be used for many gene networks and give approximate specifications for them. Network motifs are such common patterns in gene networks [6]. The motifs we studied are negative auto-regulation, coherent type 1 feed-forward loops, incoherent type 1 feed-forward loops, single-input modules and multi-output feed-forward loops. Experimental results show that our approximate analysis method is signifincantly more efficient than the non-approximate method.

This paper is organised as follows. Section 2 introduces the logical structure which describes abstract behaviours of gene regulatory networks. In Sect. 3, we show how networks are qualitatively analysed by satisfiability checking of LTL. Most part of Sects. 2 and 3 is based on our previous work [1], but we modify some behaviour descriptions and introduce two manners in behaviour descriptions. In Sect. 4, we present the approximate analysis method and show some experimental results. The final section offers some conclusions and discusses future directions.

2   Abstracting Behaviours as Transition Systems

In gene regulation, a regulator is often inefficient below a threshold concentration, and its effect rapidly increases above this threshold [7]. The sigmoid nature of gene regulation is shown in Fig. 2, where gene u activates v and inhibits w (Fig. 1). Each axis represents the concentration of products for each gene.

Important landmark concentration values for u are, (1) the level u_v at which u begins to affect v, and (2) the level u_w at which u begins to affect w. In this

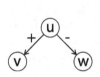

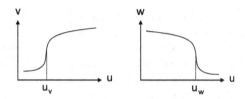

Fig. 1. Gene u activates v and inhibits w.

Fig. 2. Regulation effect.

case, whether genes are active or not can be specified by the expression levels of their regulator genes. If the concentration of u exceeds u_v then v is active (ON), and if the concentration of u exceeds u_w then w is not active (OFF). We exploit this switching view of genes to capture behaviours of gene networks in transition systems.

We now illustrate how we capture behaviours of gene regulatory networks as transition systems using a simple example network (Fig. 3) in which gene x activates gene y and gene y activates gene z, and its behaviour depicted in Fig. 4 where x_y is the threshold of x for y and y_z that of y for z.

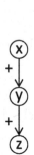

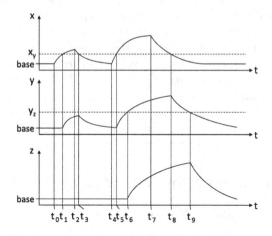

Fig. 3. Simple example.

Fig. 4. Change of concentrations with time.

To obtain a symbolic representation of behaviours of this network, we introduce logical propositions that represent whether genes are active or not (ON or OFF) and whether concentrations of products of genes exceed threshold values. In this network, we introduce the propositions on_x, on_y, on_z, x_y and y_z. Propositions on_x, on_y, on_z mean whether or not gene x, y or z is active, x_y whether gene x is expressed *beyond* the threshold x_y[1], and y_z whether gene y is expressed *beyond* the threshold y_z.

---

[1] Note that the symbol x_y is used for both the threshold and proposition but we can clearly distinguish them from the context.

Using these propositions, we discretise the above behaviour to the sequence of states (called *transition system*) shown in Fig. 5, where $0, \ldots, 10$ are states, arrows represent state transitions that abstract the temporal evolution of the system, and the propositions below each state are true in that state.

State 0 represents the interval $[0, t_0)$, state 1 represents the interval $[t_0, t_1), \ldots$ and state 10 represents $[t_9, \infty)$.

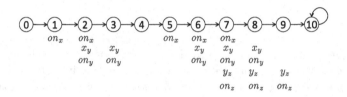

Fig. 5. State transition system corresponding to Fig. 4.

A single state transition can represent any length of time, since the actual duration of the transition (in real time) is immaterial. Therefore, the difference between $t_2 - t_0$ and $t_7 - t_4$, the duration of the input signal to x, in Fig. 4 is not captured directly. Figure 5 captures whether the concentration of y exceeds y_z; that is, we can infer that the latter duration is sufficiently long for x to activate y by comparing the propositions in state 1 to 3 and those in state 5 to 9. Moreover, the real values of thresholds are irrelevant. Propositions such as x_y merely represent the fact that the concentration of x is above the level at which x affects y.

In our abstraction, behaviours are identified with each other if they have the same transition system. Such logical abstraction preserves essential qualitative features of the dynamics [7,8].

3   Formal Analysis of Gene Regulatory Networks in LTL

3.1   Linear Temporal Logic (LTL)

Here we introduce LTL which is a logic to reason about linear time structures. If A is a finite set, A^{ω} denotes the set of all infinite sequences on A. The i-th element of $\sigma \in A^{\omega}$ is denoted by $\sigma[i]$. Let AP be a set of propositions. A *time structure* is a sequence $\sigma \in \mathfrak{P}(AP)^{\omega}$ where $\mathfrak{P}(AP)$ is the powerset of AP.

The formulae of LTL is defined as follows. Atomic propositions in AP are formulae. If ϕ and ψ are formulae, then $\neg\phi, \phi \wedge \psi, \phi \vee \psi$, and $\phi U \psi$ are also formulae. We introduce the following abbreviations: $\bot \equiv p \wedge \neg p$ for some $p \in AP$, $\top \equiv \neg\bot$, $\phi \rightarrow \psi \equiv \neg\phi \vee \psi$, $\phi \leftrightarrow \psi \equiv (\phi \rightarrow \psi) \wedge (\psi \rightarrow \phi)$, $F\phi \equiv \top U\phi$, $G\phi \equiv \neg F\neg\phi$, and $\phi W \psi \equiv (\phi U \psi) \vee G\phi$.

The semantics of LTL are given below. Let σ be a time structure and ϕ be a formula. We write $\sigma \models \phi$ for 'ϕ is true in σ'. The satisfaction relation $\models$ is defined inductively as follows:

$$\sigma \models p \quad \text{iff } p \in \sigma[0] \text{ for } p \in AP$$
$$\sigma \models \neg\phi \quad \text{iff } \sigma \not\models \phi$$
$$\sigma \models \phi \wedge \psi \text{ iff } \sigma \models \phi \text{ and } \sigma \models \psi$$
$$\sigma \models \phi \vee \psi \text{ iff } \sigma \models \phi \text{ or } \sigma \models \psi$$
$$\sigma \models \phi U \psi \text{ iff } (\exists i \geq 0)(\sigma^i \models \psi \text{ and } \forall j (0 \leq j < i)\sigma^j \models \phi)$$

where $\sigma^i = \sigma[i]\sigma[i+1]\ldots$, the i-th suffix of σ.

Intuitively, $\neg\phi$ means 'ϕ is not true', $\phi \wedge \psi$ 'both ϕ and ψ are true', $\phi \vee \psi$ 'ϕ or ψ is true', $\phi U \psi$ 'ϕ continues to hold until ψ holds', $\perp$ a false proposition, $\top$ a true proposition, $F\phi$ 'ϕ holds at some future time', $G\phi$ 'ϕ holds globally', and $\phi W \psi$ is the 'weak until' operator in that ψ is not obliged to hold, in which case ϕ must always hold.

An LTL formula ϕ is *satisfiable* if there exists a time structure σ such that $\sigma \models \phi$. A time structure σ such that $\sigma \models \phi$ is called a *model* of ϕ.

3.2 Analysis of Gene Regulatory Networks by Satisfiability Checking of LTL

As we can see in Sect. 2, a behaviour of a gene regulatory network can be seen as a time structure on atomic propositions for the network. Let AP be the set of propositions for a network. Formally, a behaviour of a network is an element of $\mathfrak{P}(AP)^\omega$. However, not all of the sequences in $\mathfrak{P}(AP)^\omega$ are possible behaviours. For example, in the network in Fig. 3, y cannot be ON before x becomes ON when y completely depends on x since y is activated only by x. We characterise the possible behaviours of a network in LTL[2] considering the meaning of gene activation/inhibition and changes of expression levels of genes.

Assume that we obtain a formula ϕ which characterises possible behaviours of a network. We also specify a biological property of interest in LTL and call it ψ. Then we can check whether some possible behaviour satisfies a given biological property by checking whether $\phi \wedge \psi$ is *satisfiable* which means there exists a behaviour which is possible in the network (satisfying ϕ) and satisfies a biological property (satisfying ψ). Also, we can check whether all possible behaviours satisfy a given biological property by checking whether $\phi \wedge \neg\psi$ is *not satisfiable* which means if a sequence σ is possible in the network (satisfying ϕ), then it is impossible that σ violates a biological property ψ.

3.3 Specification of Behaviours in LTL

We see how we specify ϕ for a given network in this section. As in Sect. 2, we assume that we have the following propositions:

- on_u for each gene u in a given network.
- u_v for each regulation from gene u to gene v in a given network.

---

[2] This contrasts with the framework in which behaviours are described in ordinary differential equations.

Additionally, we may introduce other propositions representing landmark concentration values that are not thresholds for other nodes (say, representing 'low level', 'maximum' and so on).

The basic idea of specifying possible behaviours of a network is the following qualitative principle:

- Genes are ON when their activators express over some threshold.
- Genes are OFF when their inhibitors express over some threshold.
- If genes are ON, the concentrations of their products increase.
- If genes are OFF, the concentrations of their products decrease.

Thus we specify the above rules in LTL using the propositions introduced above. The switching conditions for gene u can be specified by regulators $x, y, \ldots$ using their threshold values $x_u, y_u, \ldots$. The concentration increase or decrease for some gene u relates to the propositions $u_v, u_w, \ldots$, that is, the threshold values that u has. For this, the total order of threshold values must be fixed.

We show how we specify the above rules in LTL. The specification is written so that the behaviours that satisfy it are as large as possible.

Conditions for Gene Activation and Inhibition. First we consider the simple case in which a gene is regulated by a single gene. For example, let gene v be regulated only by u. If the effect of u on v is positive, then v is turned on when the concentration of u exceeds the threshold u_v. We have two choices for description in LTL. One is $G(u_v \rightarrow on_v)$ and the other is $G(u_v \leftrightarrow on_v)$. The former allows on_v to be true when u_v is not, but the latter does not. The former specification takes hidden activators or external regulation for v into account. The choice of the specifications depends on the system or the situation. If the effect of u on v is negative, this case is described by $G(u_v \rightarrow \neg on_v)$. Similarly we may write $G(u_v \leftrightarrow \neg on_v)$ depending on the system or situation.

Now we consider a gene that is regulated by multiple genes. In general, the multivariate regulation functions of organisms are unknown [6]. Thus we only describe the trivial facts. For example, we assume that genes u and v activate x and w inhibits x. In this example, the following two facts hold trivially.

- If u and v exceed u_x and v_x respectively, and w does not exceed w_x, then x is ON. This is described as $G((u_x \wedge v_x \wedge \neg w_x) \rightarrow on_x)$.
- If u and v do not exceed u_x and v_x respectively, and w exceeds w_x, then x is OFF. This is described as $G((\neg u_x \wedge \neg v_x \wedge w_x) \rightarrow \neg on_x)$.

If we know more information, such as the positive effect of u and v on x is conjunctive; that is, both u and v need to exceed their thresholds, or the negative regulation effect of w is dominant and overpowers other positive effects, then we can reflect these conditions in the specification.

In gene regulation, some genes regulate not genes but the regulation effect itself (e.g. when some gene's product intercepts another gene's product). Let us consider a case where x inhibits y and z inhibits the regulation effect of x on y. In this case y is turned OFF when x affects y but z does not affect the regulation.

To describe this, we introduce a threshold z_x above that z inhibits the effect of x. We can describe this as $G((x_y \land \neg z_x) \to \neg on_y)$. In this case, z_x may not be a fixed value but a function that takes the concentration of x and returns the threshold of z. The proposition z_x simply says that z influences the regulation effect of x and the real value of the concentration of z does not matter.

To capture alternative splicing we can use multiple (virtual) genes to represent one gene with multiple states. If a gene has two states (namely one produces A and the other B), we use propositions on_A and on_B and individually have the switching conditions and concentration changes for them.

Total Order of Threshold Values. We now specify the fixed total order of threshold values. Assume that u regulates $x_1, x_2, \ldots, x_m$ and the threshold values for them are in this order. This order relation can be described in LTL as follows:

$$\bigwedge_{1 \leq i < m} G(u_{x_{i+1}} \to u_{x_i}).$$

Concentration Changes When Genes are ON. If gene u is ON the concentration of its product increases with time. We have two kinds of specification on this principle: a strong one and a weak one.

In what follows, we assume that gene u has threshold values $u_1, u_2, \ldots, u_m$ in this order.

First we introduce the strong specification:

$$G((on_u \to F(\neg on_u \lor u_1)) \land \tag{1}$$
$$((on_u \land u_1) \to (u_1 U(\neg on_u \lor u_2))) \land \tag{2}$$
$$((on_u \land u_2) \to (u_2 U(\neg on_u \lor u_3))) \land \tag{3}$$

$$\vdots$$

$$((on_u \land u_{m-1}) \to (u_{m-1} U(\neg on_u \lor u_m))) \land \tag{4}$$
$$((on_u \land u_m) \to (u_m W \neg on_u))). \tag{5}$$

To see what the above formula says, suppose that u is ON and its concentration is between u_2 and u_3 at the beginning. Recall that u_i means the concentration of u exceeds u_i. Thus the left-hand sides of (1)–(3) in the above formula hold. From the total order of threshold values, u_3 implies u_1 and u_2, and u_2 implies u_1. Accordingly, (1)–(3) may be summed up as the statement that the concentration of u is not less than u_2 until v is turned OFF, or it exceeds u_3. Behaviours that satisfy this constraint with a starting concentration of u between u_2 and u_3 is such that in some future state the concentration of u exceeds u_3 but until that time it remains above u_2. The exception is that u is turned OFF before reaching u_3, so u may not exceed u_3. Behaviours in which u falls below u_2 while being ON are excluded. Moreover, u is not allowed to remain between u_2 and u_3 indefinitely although it is ON.

From this observation, we understand that the above formula says that the concentration of u does not decrease as long as u is ON and must increase (unless u is greater than u_m) if u is always ON.

Next we introduce the weak specification:

$$G((on_u \to F(\neg on_u \vee u_1)) \wedge$$
$$((on_u \wedge u_1) \to (u_1 W \neg on_u)) \wedge$$
$$\vdots$$
$$((on_u \wedge u_m) \to (u_m W \neg on_u))).$$

The difference compared with the strong specification is that behaviours in which u keeps its concentration although it is always ON are allowed; that is, the concentration does not have to increase strictly. This represents a situation where generation and degradation are equilibrated.

Concentration Changes When Genes are OFF. This is symmetric to the case when genes are ON. We again assume that gene u has threshold values $u_1, u_2, \ldots, u_m$ in this order. We also have both a strong specification and a weak one which are shown in Figs. 6 and 7 respectively.

$$G((\neg on_u \to F(on_u \vee \neg u_m)) \wedge$$
$$((\neg on_u \wedge \neg u_m) \to (\neg u_m U(on_u \vee \neg u_{m-1}))) \wedge$$
$$\vdots$$
$$((\neg on_u \wedge \neg u_2) \to (\neg u_2 U(on_u \vee \neg u_1))) \wedge$$
$$((\neg on_u \wedge \neg u_1) \to (\neg u_1 W on_u)))$$

Fig. 6. Strong specification.

$$G((\neg on_u \to F(on_u \vee \neg u_m)) \wedge$$
$$((\neg on_u \wedge \neg u_m) \to (\neg u_m W on_u)) \wedge$$
$$\vdots$$
$$((\neg on_u \wedge \neg u_1) \to (\neg u_1 W on_u)))$$

Fig. 7. Weak specification

In the strong specification, it is not possible that u keeps its concentration when it is always OFF but this is possible in the weak specification.

Remark 1. The choice between a strong and weak specification is made for both the ON and OFF behaviour of each gene. Thus, there are two options (i.e., strong or weak) for the ON behaviour and two for the OFF behaviour. For example, if there are two genes, there are $2^4 = 16$ possible combinations of specifications.

3.4   Biological Properties in LTL

Many biologically interesting properties can be described in temporal logic. For example, the property 'the system eventually reaches a state in which gene x is active but gene y is not active' is a type of *reachability* described as $F(on_x \wedge \neg on_y)$. The property 'the concentration of x is always above x_y' is a type of *stability* described as Gx_y. *Oscillation*, where 'some property ϕ is alternately true and false indefinitely', is described as $G((\phi \to F\neg\phi) \wedge (\neg\phi \to F\phi))$. Conditional properties can also be specified. For example, 'if gene x is always OFF then the property ϕ holds' is described as $(G\neg on_x) \to \phi$. These are just examples of properties expressible in LTL. We can use full LTL to specify properties of interest.

3.5 Satisfiability of LTL

To check the satisfiability of LTL, we construct a Büchi automaton [9] which is equivalent to a given LTL formula.

Theorem 1 ([10]). *Given an LTL formula ϕ, one can construct a Büchi automaton $\mathcal{A}_\phi = \langle Q, \Sigma, \delta, q_I, F \rangle$ such that $|Q|$ is in $2^{O(|\phi|)}$, $\Sigma = \mathfrak{P}(Prop(\phi))$ and $L(\mathcal{A}_\phi) = \{\sigma \in \mathfrak{P}(Prop(\phi))^\omega \mid \sigma \models \phi\}$.*

Corollary 1. *An LTL formula ϕ is satisfiable if and only if $L(\mathcal{A}_\phi) \neq \emptyset$.*

This corollary says that the problem of LTL satisfiability checking is reduced to empty-testing of a Büchi automaton. Empty-testing of Büchi automaton is known to be proportional in the size of automaton. As a result, we have exponential-time algorithm to check LTL satisfiability. As we can see from Sect. 3.3, the length of a formula specifying possible behaviours of a network is proportional to the size of the network, and accordingly, analyses of large networks are generally intractable in our method. To tackle this issue, we develop an approximate analysis which is discussed in the next section.

4 Approximate Analysis

In this section, we describe approximate analysis method and introduce approximate specifications for network motifs. The key of approximation is to reduce the number of propositions. By omitting some propositions, we approximately specify the possible behaviours of networks.

To guarantee the correctness of analysis, if approximate specifications are satisfiable or unsatisfiable, so be the original ones. These conditions can be assured by the fact that the set of behaviours of approximate specifications are smaller or larger than that of original ones. Thus there are two ways in approximation.

Now we formally state our framework of approximate analysis. Let A and B be finite sets such that $A \subseteq B$ and suppose $\sigma \in \mathfrak{P}(B)^\omega$. The sequence $\sigma|_A$ denotes $(\sigma[0] \cap A)(\sigma[1] \cap A) \cdots \in \mathfrak{P}(A)^\omega$. Let $L \subseteq \mathfrak{P}(B)^\omega$. The set $L|_A$ denotes $\{\sigma|_A \mid \sigma \in L\}$, the restriction of L to A.

Definition 1. *Let ϕ and ψ be LTL formulae such that $Prop(\psi) \subseteq Prop(\phi)$. We say ψ is a lower approximation of ϕ just if $L(\mathcal{A}_\psi) \subseteq L(\mathcal{A}_\phi)|_{Prop(\psi)}$. We say ψ is an upper approximation of ϕ just if $L(\mathcal{A}_\psi) \supseteq L(\mathcal{A}_\phi)|_{Prop(\psi)}$.*

Note that approximate formulae have less propositions than the original ones. We have the following theorem about approximations.

Theorem 2. *Let ϕ and ψ be LTL formulae. We have (i) if ψ is a lower approximation of ϕ then ψ is satisfiable implies ϕ is satisfiable, and (ii) if ψ is an upper approximation of ϕ then ψ is unsatisfiable implies ϕ is unsatisfiable.*

Proof. (i) Suppose that ϕ is not satisfiable. By Corollary 1, $L(\mathcal{A}_\phi) = \emptyset$. Then $L(\mathcal{A}_\phi)|_{Prop(\psi)} = \emptyset$. By Definition 1, $L(\mathcal{A}_\psi) = \emptyset$, that is to say, ψ is not satisfiable. (ii) Suppose that ϕ is satisfiable. By Corollary 1, $L(\mathcal{A}_\phi) \neq \emptyset$. Thus there exists $\sigma \in L(\mathcal{A}_\phi)$. By Definition 1, there exists $\rho \in L(\mathcal{A}_\psi)$ such that $\sigma|_{Prop(\phi)} = \rho$. Thus $L(\mathcal{A}_\psi) \neq \emptyset$, that is to say, ψ is satisfiable. □

Now we have the approximate analysis method. Let $\phi = \phi_1 \wedge \cdots \wedge \phi_n$ be a behavioural specification and ψ be a biological property. When we check the satisfiability of $\phi \wedge \psi$, we replace ϕ_i by ψ_i such that ψ_i is a lower approximation of ϕ_i and check the satisfiability of $\psi_1 \wedge \cdots \wedge \psi_n \wedge \psi$. If it is satisfiable, then $\phi \wedge \psi$ is also satisfiable by Theorem 2. When we check the unsatisfiability of $\phi \wedge \neg\psi$, we replace ϕ_i by ψ_i such that ψ_i is an upper approximation of ϕ_i and check the unsatisfiability of $\psi_1 \wedge \cdots \wedge \psi_n \wedge \neg\psi$. If it is unsatisfiable, then $\phi \wedge \neg\psi$ is also unsatisfiable by Theorem 2.

It is notable that the correctness of the approximate analysis is theoretically proved, which is not possible in ordinary differential equation approach.

To find approximate specifications for any network is not a trivial task. However, there is a small set of recurring regulation patterns, called network motifs [6], in gene regulatory networks. We, therefore, present approximate specifications for network motifs since we can apply them in analyses of many gene networks. In this paper we consider five motifs: negative auto-regulation, coherent type 1 feed-forward loops, incoherent type 1 feed-forward loops, single-input modules and multi-output feed-forward loops. The reason why we focus on these five motifs is that they have certain functions. So the approximate specifications are given by considering their functions.

Remark 2. It is worth noting that the weak specification is an upper approximation of the strong specification (recall these specifications from Sect. 3.3).

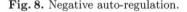

Fig. 8. Negative auto-regulation.   **Fig. 9.** Feed-forward loop.

In the following, lower approximations are given for strong specifications and upper approximations are given for weak specifications. From the above remark, lower approximations for strong specifications are also lower approximations for weak specifications, and upper approximations for weak specifications are also upper approximations for strong specifications.

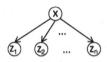

Fig. 10. Single-input module.

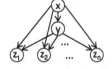

Fig. 11. Multi-output feed-forward loop.

Negative Auto-Regulation. Negative auto-regulation is depicted in Fig. 8. This motif has the function of response acceleration. In our abstraction, this function cannot be described since we cannot refer to an actual response time in LTL; that is, accelerated behaviours and non-accelerated behaviours cannot be distinguished. Therefore, we may ignore negative auto-regulation in our analysis. We now present the following lower approximation in which negative auto-regulation of x is ignored:

$$G(\qquad (in_x \leftrightarrow on_x) \qquad \land$$
$$(on_x \rightarrow F(x_{out} \lor \neg on_x)) \qquad \land$$
$$((on_x \land x_{out}) \rightarrow (x_{out} W \neg on_x)) \qquad \land$$
$$(\neg on_x \rightarrow F(\neg x_{out} \lor on_x)) \qquad \land$$
$$((\neg on_x \land \neg x_{out}) \rightarrow (\neg x_{out} W on_x)) \qquad).$$

The abstracted proposition is x_x. Here for simplicity we assume that there is one input and one output for x but this is easily generalised.

It is difficult to present a meaningful upper approximation for the weak specification since the behaviour principles prescribed in Sect. 3.3 will be violated by weakening the specification.

Coherent Type 1 Feed-Forward Loop. A feed-forward loop (FFL) is a pattern consisting of three nodes as depicted in Fig. 9. There are 8 patterns in FFL depending on regulation effects of three edges. The coherent type 1 FFL (C1-FFL) is the pattern in which all edges represent activation. There are two types of input function (AND/OR) for z that merge the influence of x and y. For the AND function, C1-FFL shows a delay after stimulation, but no delay when stimulation stops. For the OR function, the FFL has the opposite effect to the AND case; that is, it shows no delay after stimulation but shows a delay when stimulation stops. These functions are real-time properties and do not make a difference in our abstraction. For lower approximation, we ignore y and consider them as simple regulations. Thus the difference between AND and OR does not occur in the approximate formula. Although the original specifications for this motif depend on the orderings of the thresholds x_y and x_z, we can present a single lower approximation as follows (i.e. the difference of the ordering vanishes):

$$G(\quad (on_x \rightarrow F(on_z \lor \neg on_x)) \qquad \land$$
$$((on_x \land on_z) \rightarrow (on_z W \neg on_x)) \qquad \land$$

$$(\neg on_x \to F(\neg on_z \vee on_x)) \qquad \wedge$$
$$((\neg on_x \wedge \neg on_z) \to (\neg on_z W on_x)) \quad).$$

The abstracted propositions are x_y, x_z, y_z and on_y.

It is also difficult to present a consistent upper approximation since that would allow behaviours violating the behaviour principles. For example, we may weaken the constraint concerning activation of z and inactivation of z by allowing z not to be turned ON when x is ON or not to be turned OFF when x is OFF. However, this amounts to regarding z to be independent of x.

Incoherent Type 1 Feed-Forward Loop. In incoherent type 1 FFL (I1-FFL), x activates y and z but y inhibits z in Fig. 9. Assume that the threshold of x for z is higher than that of x for y. When x becomes ON, z will be turned ON. After some time y becomes ON. At that time y inhibits z, so z becomes OFF. As a result, this motif generates pulse-like dynamics on z. We specify this pulse-like dynamics as the following lower approximation.

$$G(\quad (on_x \wedge \neg on_y) \to F(on_z \vee \neg on_x)) \qquad \wedge$$
$$((on_x \wedge on_z) \to (on_z U on_y)) \qquad \wedge$$
$$((on_x \wedge on_y) \to ((on_y \wedge \neg on_z)W \neg on_x)) \wedge$$
$$(\neg on_x \to (\neg on_z \wedge \neg on_y)) \qquad).$$

The abstracted propositions are x_y, x_z and y_z.

It is difficult to give an upper approximation representing this pulse-like dynamics since this dynamics is a part of the possible behaviours of I1-FFL.

Single-Input Module. A single-input module is a pattern in which one regulator (called the master gene) regulates a group of target genes (Fig. 10). All regulations from the master gene are of the same type (positive or negative). We only consider the positive case but the negative case is similar.

The function of this motif is a last-in first-out (LIFO) temporal order on expressions of target genes. Assume that the thresholds for $z_1, z_2, \ldots, z_n$ occur in this ascending order. When the master regulator x is ON, the regulated genes $z_1, z_2, \ldots, z_n$ are turned ON in this order. When x is turned OFF, $z_n, z_{n-1}, \ldots z_1$ are turned OFF in this order.

We first present a lower approximation. For simplicity we set $n = 2$ but this is easily generalised:

$$G(\qquad (on_{z_2} \to on_{z_1}) \qquad \wedge$$
$$(on_x \to (on_x U on_{z_2})) \qquad \wedge$$
$$((on_x \wedge on_{z_1}) \to (on_{z_1} W \neg on_x)) \quad \wedge$$
$$((on_x \wedge on_{z_2}) \to (on_{z_2} W \neg on_x)) \quad \wedge$$
$$(\neg on_x \to (\neg on_x U \neg on_{z_1})) \qquad \wedge$$
$$((\neg on_x \wedge \neg on_{z_2}) \to (\neg on_{z_2} W on_x)) \wedge$$
$$((\neg on_x \wedge \neg on_{z_1}) \to (\neg on_{z_1} W on_x)) \quad).$$

The abstracted propositions are x_{z_1} and x_{z_2}. Behaviours that satisfy this formula are such that once x is turned ON it remains ON until all target genes become active, and once x is turned OFF it remains OFF until all target genes become inactive. Therefore, behaviours such that x is turned OFF before all target genes become active or x is turned ON before all target genes become inactive are eliminated from the possible behaviours obtained using the original specification.

We now present an upper approximation:

$$G(\qquad\qquad (on_{z_2} \to on_{z_1}) \qquad\qquad\qquad \wedge$$
$$((on_x \wedge on_{z_1}) \to (on_{z_1} W \neg on_x)) \quad \wedge$$
$$((on_x \wedge on_{z_2}) \to (on_{z_2} W \neg on_x)) \quad \wedge$$
$$((\neg on_x \wedge \neg on_{z_2}) \to (\neg on_{z_2} W on_x)) \wedge$$
$$((\neg on_x \wedge \neg on_{z_1}) \to (\neg on_{z_1} W on_x)) \quad).$$

Propositions x_{z_1} and x_{z_2} are ignored. This upper approximation says that the temporal order of activation and inactivation of target genes is preserved but some genes may not be activated when x is turned ON or inactivated when x is turned OFF. Such behaviours are also allowed in the weak specification but we can specify the same constraint with less propositions.

Multi-Output Feed-Forward Loop. A multi-output feed-forward loop is a generalisation of a feed-forward loop with n target genes (Fig. 11). The function of this motif is interesting when each input function for z_i is OR and the threshold orders for x and y are inverted, that is, $x_{z_1} < x_{z_2} < \cdots < x_{z_n}$ and $y_{z_1} > y_{z_2} > \cdots > y_{z_n}$. In this case this motif can generate a first-in first-out (FIFO) temporal order on expression of target genes. The activation order is $z_1 z_2 \ldots z_n$ and the inactivation order is the opposite. The position of threshold x_y does not matter. Focusing on this property we have the following lower approximation (we set $n = 2$ but this is easily generalised):

$$G(\qquad\qquad (on_x \to (on_{z_2} \to on_{z_1})) \qquad\qquad \wedge$$
$$(\neg on_x \to (\neg on_{z_2} \to \neg on_{z_1})) \qquad \wedge$$
$$((on_x \to (on_x U on_{z_1}))) \qquad\qquad \wedge$$
$$((on_x \wedge on_{z_1}) \to ((on_x \wedge on_{z_1}) U (on_x \wedge on_{z_2}))) \wedge$$
$$((on_x \wedge on_{z_2}) \to (on_{z_2} W \neg on_x)) \qquad \wedge$$
$$(\neg on_x \to (\neg on_x U \neg on_{z_1})) \qquad \wedge$$
$$((\neg on_x \wedge \neg on_{z_1}) \to$$
$$((\neg on_x \wedge \neg on_{z_1}) U (\neg on_x \wedge \neg on_{z_2}))) \qquad \wedge$$
$$((\neg on_x \wedge \neg on_{z_2}) \to (\neg on_{z_2} W on_x)) \qquad).$$

The abstracted propositions are x_y, x_{z_1}, x_{z_2}, on_y, y_{z_1}, and y_{z_2}. This formula says that when x is turned ON, z_1 and z_2 are activated in this order, and when x is turned OFF, z_1 and z_2 are inactivated in the opposite order.

Note that this motif can generate other temporal orders on expressions of target genes even if the threshold ordering is as assumed. Thus there are many possible behaviours, and we cannot give an upper approximation without violating the behaviour principles.

4.1   Experiments

We demonstrate the approximate analysis using networks depicted in Figs. 12, 13 and 14, and compare the approximate specifications for the original specifications. Lower approximation is used in this experiments.

The first example is the network depicted in Fig. 12. There are two motifs in Fig. 12, one a negative auto-regulation and the other a single-input module.

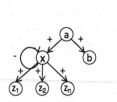

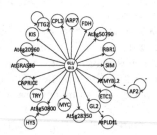

Fig. 12. A simple example network.   **Fig. 13.** A network in *Arabidopsis*.

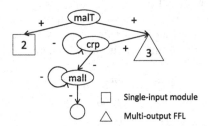

Fig. 14. The network from *E. coli* involving the *malT* gene.

The second example is a network in *Arabidopsis thaliana* depicted in Fig. 13. This network is obtained from ReIN[3]. In this network, we can find a single-input module which has 18 target genes in the motif. In this network, there are some genes which have regulators other than master gene. Thus we cannot use the approximate specification introduced above directly. But the modification is not difficult. We do not omit propositions x_y if y has a regulator other than the

---

[3] http://arabidopsis.med.ohio-state.edu/REIN/

master gene x, and specify the conditions for activation and inhibition of y the same as the original ones.

The third example is a network in *Escherichia coli* involving the *malT* gene. The numbers in the box and the triangle are the numbers of target genes in each motif. In this network we approximate two negative auto-regulations, one single-input module and one multi-output feed-forward loop. We show the result of satisfiability checking of these specifications with our satisfiability checker. We used our implementation (in OCaml) of the LTL satisfiability checker based on the algorithm of Aoshima [11]. The results are shown in Table 1[4].

Table 1. Results of satisfiability checking.

	Specification	Number of propositions	Size of formula	Time
Fig. 12	Original	12	233	0.020s
	Approximate	8	154	0.016s
Fig. 13	Original	45	821	17.357s
	Approximate	23	644	1.128s
Fig. 14	Original	28	575	94.454s
	Approximate	16	368	0.340s

In all cases, the cost of analysis is improved. Especially, in the last case, the improvement is drastic.

5 Conclusions

In this paper, we have presented a method for analysing the dynamics of gene regulatory networks using LTL satisfiability checking. To ease analysis of large networks, we developed the approximate analysis method and showed how it works well.

We presented approximate specifications for five network motifs. For further development, it is important to find approximate specifications for more network patterns. The other direction is to utilise *modular analysis method* [12] in combination with approximate method. In modular analysis method, we decompose a network into a few subnetworks, check them individually, and then integrate them. The modular analysis method is applicable to arbitrary network. So we will have no difficulty in utilising approximate method and modular method in combination, and the efficiency will be improved further.

---

[4] The following computational environment was used: openSUSE 11.0, Intel(R) Pentium(R) D CPU 3.00 GHz and 2 GB of RAM.

References

1. Ito, S., Izumi, N., Hagihara, S., Yonezaki, N.: Qualitative analysis of gene regulatory networks by satisfiability checking of linear temporal logic. In: 10th IEEE International Conference on Bioinformatics & Bioengineering, pp. 232–237. IEEE Computer Society (2010)
2. Pnueli, A., Rosner, R.: On the synthesis of a reactive module. In: 16th ACM SIGPLAN-SIGACT Symposium on Principles of Programming Languages, pp. 179–190. ACM Press (1989)
3. Abadi, M., Lamport, L., Wolper, P.: Realizable and unrealizable specifications of reactive systems. In: Ausiello, G., Dezani-Ciancaglini, M., Rocca, S.R.D. (eds.) Automata, Languages and Programming. LNCS, vol. 372, pp. 1–17. Springer, Heidelberg (1989)
4. Mori, R., Yonezaki, N.: Several realizability concepts in reactive objects. In: Information Modeling and Knowledge Bases IV, pp. 407–424 (1993)
5. Sistla, A.P., Clarke, E.M.: The complexity of propositional linear temporal logics. J. ACM **32**, 733–749 (1985)
6. Alon, U.: Network motifs: theory and experimental approaches. Nat. Rev. Genet. **8**, 450–461 (2007)
7. Thomas, R., Kauffman, M.: Multistationarity, the basis of cell differentiation and memory. II. logical analysis of regulatory networks in terms of feedback circuits. Chaos **11**, 180–195 (2001)
8. Snoussi, E., Thomas, R.: Logical identification of all steady states: the concept of feedback loop characteristic states. Bull. Math. Biol. **55**, 973–991 (1993)
9. Farwer, B.: ω-Automata. In: Grädel, E., Thomas, W., Wilke, T. (eds.) Automata Logics, and Infinite Games: A Guide to Current Research, pp. 3–20. Springer-Verlag New York, Inc., New York (2002)
10. Vardi, M.Y., Wolper, P.: Reasoning about infinite computations. Inf. Comput. **115**, 1–37 (1994)
11. Aoshima, T., Sakuma, K., Yonezaki, N.: An efficient verification procedure supporting evolution of reactive system specifications. In: 4th International Workshop on Principles of Software Evolution, pp. 182–185. ACM, New York (2001)
12. Ito, S., Ichinose, T., Shimakawa, M., Izumi, N., Hagihara, S., Yonezaki, N.: Modular analysis of gene networks by linear temporal logic. J. Integr. Bioinform. **10**, 216 (2013)

Word Match Counts Between Markovian Biological Sequences

Conrad Burden[1]([✉]), Paul Leopardi[1], and Sylvain Forêt[2]

[1] Mathematical Sciences Institute,
Australian National University, Canberra, ACT 0200, Australia
conrad.burden@anu.edu.au
[2] Research School of Biology, Australian National University,
Canberra, ACT 0200, Australia
http://wwwmaths.anu.edu.au/~burden

Abstract. The D_2 statistic, which counts the number of word matches between two given sequences, has long been proposed as a measure of similarity for biological sequences. Much of the mathematically rigorous work carried out to date on the properties of the D_2 statistic has been restricted to the case of 'Bernoulli' sequences composed of identically and independently distributed letters. Here the properties of the distribution of this statistic for the biologically more realistic case of Markovian sequences is studied. The approach is novel in that Markovian dependency is defined for sequences with periodic boundary conditions, and this enables exact analytic formulae for the mean and variance to be derived. The formulae are confirmed using numerical simulations, and asymptotic approximations to the full distribution are tested.

Keywords: Word matches · Biological sequence comparison

1 Introduction

The D_2 statistic is defined as the number of exact word matches of pre-specified length k between two sequences of letters from a finite alphabet $\mathcal{A}$. This statistic [1], and its many variants [2–5] have been proposed as a measures of similarity between biological sequences in cases where the more commonly used alignment methods may not be appropriate. The distributional properties of the D_2 statistic under the null hypothesis of sequences composed of independently and identically distributed (i.i.d.) letters have been studied extensively [1,6–10].

Analysis of the k-mer spectra of the genomes of several species provides strong evidence that genomic sequences are more appropriately modelled as having a Markovian dependence [11]. In the current work existing exact analytic results results for the mean, variance and an empirical distribution of D_2 for i.i.d. sequences is extended to the case of Markovian sequences.

A previous study of this problem, with some approximations, has been carried out by Kantorovitz et al. [12] in the process of developing a method for detecting regulatory modules in genomic sequences. The current study differs in that we

© Springer-Verlag Berlin Heidelberg 2014
M. Fernández-Chimeno et al. (Eds.): BIOSTEC 2013, CCIS 452, pp. 147–161, 2014.
DOI: 10.1007/978-3-662-44485-6_11

consider sequences with periodic boundary conditions (PBCs), for which we introduce a new definition of Markovian sequences. The restriction to periodic sequences simplifies calculations of the mean and variance, enabling an exact analytic formula for the variance for first order Markovian sequences which is rapidly computable to double precision accuracy for arbitrary sequence lengths. In biological applications of the analogous results for i.i.d. sequences [9, 10] we have found generally that the PBCs are not an impediment, as they can simply be imposed on the sequences prior to calculating D_2 without without seriously affecting its efficacy as a measure of sequence similarity.

2   Definitions

Definition 1. *Consider a sequence* $\mathbf{x} = x_1, x_2 \ldots$ *of letters from an alphabet* $\mathcal{A}$ *of size* d. *We say that* $\mathbf{x}$ *has periodic boundary conditions (PBCs) and is of length* m *if* $x_{i+m} = x_i$ *for all* $i = 1, 2, \ldots$.

A sequence $\mathbf{X} = X_1, X_2 \ldots$ *of random letters has an* θ-*th order Markovian dependence if*

$$\text{Prob} \left((X_{i+\theta} = b | (X_i, \ldots, X_{i+\theta-1} = (a_1, \ldots, a_\theta)) \right.$$
$$= M(a_1, \ldots, a_\theta; b), \tag{1}$$

for a specified $d^\theta \times d$ *matrix* M *satisfying*

$$0 \leq M(a_1, \ldots, a_\theta; b) \leq 1; \quad \sum_{b \in \mathcal{A}} M(a_1, \ldots, a_\theta; b) = 1, \tag{2}$$

for all $a_1, \ldots, a_\theta, b \in \mathcal{A}$.

As a shorthand notation, we will write a string of length θ in bold italics:

$$\boldsymbol{x} = (x_1, \ldots x_\theta), \tag{3}$$

and write any substring of $\mathbf{X}$ of length θ in a similar fashion, labelled by the index of the first element:

$$\boldsymbol{X}_i = (X_i, \ldots X_{i+\theta-1}), \tag{4}$$

Thus (1) is written more compactly as

$$\text{Prob} \left(X_{i+\theta} = b | \boldsymbol{X}_i = \boldsymbol{a} \right) = M(\boldsymbol{a}; b). \tag{5}$$

Following the notation of ref. [13], define a $d^\theta \times d^\theta$ square matrix $\mathbb{M}$ as

$$\mathbb{M}(\boldsymbol{a}, \boldsymbol{b}) = \begin{cases} M(\boldsymbol{a}; b_\theta) & \text{if } (a_2, \ldots, a_\theta) = (b_1, \ldots b_{\theta-1}), \\ 0 & \text{otherwise.} \end{cases} \tag{6}$$

Then the Markovian dependency can be written as a first order Markovian dependency as

$$\text{Prob} \left(\boldsymbol{X}_{i+1} = \boldsymbol{b} | \boldsymbol{X}_i = \boldsymbol{a} \right) = \mathbb{M}(\boldsymbol{a}, \boldsymbol{b}). \tag{7}$$

2.1   Markov Sequences with PBCs

Definition 2. *Given an order θ Markov transition matrix M, we define a Markov sequence with PBCs of length n to be a random sequence $\mathbf{X} = X_1, X_2 \ldots, X_n$ for which the probability of observing the sequence $\mathbf{x} = x_1, x_2 \ldots, x_n \in \mathcal{A}^n$ is*

$$\operatorname{Prob}(\mathbf{X} = \mathbf{x}) = \frac{\mathbb{M}(\boldsymbol{x_1}, \boldsymbol{x_2}) \mathbb{M}(\boldsymbol{x_2}, \boldsymbol{x_3}) \ldots \mathbb{M}(\boldsymbol{x_m}, \boldsymbol{x_1})}{\operatorname{tr}(\mathbb{M}^m)}, \tag{8}$$

where $\mathbb{M}$ is the equivalent first order transition matrix defined by (6) [14].

It is shown in ref. [10] that the following algorithm gives a practical way of generating such a sequence

Algorithm 1

Step 1: *Generate $\boldsymbol{X_1} = X_1, \ldots X_\theta$ from the uniform distribution $\operatorname{Prob}(\boldsymbol{X_1} = \boldsymbol{x}) = 1/d^\theta$ for all $\boldsymbol{x} \in \mathcal{A}^\theta$.*
Step 2: *Generate $X_{\theta+1}, \ldots, X_{\theta+n}$ using (5).*
Step 3: *If $\boldsymbol{X_{n+1}} = \boldsymbol{X_1}$, accept the sequence $\mathbf{X} = X_1, X_2 \ldots, X_n$, otherwise repeat from Step 1 until an accepted sequence is obtained.*

Note that, counter-intuitively, it is important that the initial θ-mer is chosen from a uniform distribution and not the stationary distribution of the Markov model in order to generate the correct distribution.

3   Strand Symmetry and the Transition Matrix M

To model genomic DNA sequences, the transition matrix M is generally estimated from observed word counts in a genome or part of a genome via the asymptotic maximum likelihood estimator for infinitely long sequences

$$\hat{M}(\boldsymbol{a}; b) = \frac{N_{\boldsymbol{a}b}}{N_{\boldsymbol{a}}} \tag{9}$$

where $N_{\boldsymbol{a}b}$ is the number of occurrences of the $(\theta+1)$-mer $(a_1 \ldots a_\theta b)$ and $N_{\boldsymbol{a}} = \sum_{c \in \mathcal{A}} N_{\boldsymbol{a}c}$ is the number of occurrences of the θ-mer $\boldsymbol{a}$ [13].

Most genomic sequences, when examined on a sufficiently large scale, are observed to have the property of strand symmetry [15]. That is, the number of occurrences of any given k-mer is, to a good approximation, equal to the number of occurrences of its reverse complement. In the interests of reducing the number of free parameters one would like to build this property into genomic Markov models.

To give a more mathematical framework to this statement, let us assume that the alphabet size d is even and each letter $a \in \mathcal{A}$ has a complement $\bar{a}$ such that $\bar{\bar{a}} = a$ and $\bar{a} \neq a$. In general the alphabet splits into $d/2$ 'purines' and $d/2$ 'pyrimidines'. For the usual nucleotide alphabet, A and G are purines, C and T are pyrimidines and $\bar{A} = T$, $\bar{G} = C$. We wish to determine what practical restrictions are placed on the estimate $\hat{M}$ of the transition matrix by the strand-symmetry restriction

$$\text{Prob}\,((X_1,\ldots,X_n) = (x_1,\ldots,x_n)) = \text{Prob}\,((X_1,\ldots,X_n) = (\bar{x}_n,\ldots,\bar{x}_1)). \tag{10}$$

To this purpose it is more convenient to work with the square matrix M defined by (6). Define a matrix $\mathbb{N}_{ab}$, $\boldsymbol{a},\boldsymbol{b} \in \mathcal{A}^\theta$, from the count matrix $N(\boldsymbol{a},b)$ in a manner analogous to (6). Then M is estimated by normalising the rows of $\mathbb{N}$ to add to 1:

$$\hat{\mathbb{M}}(\boldsymbol{a},\boldsymbol{b}) = \frac{\mathbb{N}_{ab}}{\sum_{c \in \mathcal{A}^\theta} \mathbb{N}_{ac}}. \tag{11}$$

Now rearrange the order of columns of $\mathbb{N}$ to form a new matrix Q so that if the rows are labelled by the complete set of θ-mers $\boldsymbol{w}_1, \boldsymbol{w}_2, \ldots \boldsymbol{w}_{d^\theta}$ the columns are labelled in the order $\bar{\boldsymbol{w}}_1, \bar{\boldsymbol{w}}_2, \ldots \bar{\boldsymbol{w}}_{d^\theta}$, where $\bar{\boldsymbol{w}}_i = \overline{(w_{i1}\ldots w_{i\theta})} = (\bar{w}_{i\theta}\ldots\bar{w}_{i1})$ is the reverse complement of the ith θ-mer $\boldsymbol{w}_i$. Strand symmetry implies that Q will be symmetric because the probability of making the transition $(a_1\ldots a_\theta) \to (b_1\ldots b_\theta)$ will be the same as the probability of making the transition $(\bar{b}_\theta\ldots\bar{b}_1) \to (\bar{a}_\theta\ldots\bar{a}_1)$.

The problem of determining the most general form of transition matrix is equivalent to answering the following question: how many independent non-zero elements does the matrix Q have, given the restrictions

1. $Q_{ab} = Q_{\bar{b}\bar{a}}$
2. Q_{ab} is zero unless $(a_2\ldots a_\theta) = (\bar{b}_\theta\ldots\bar{b}_2)$?

Consider first any diagonal element Q_{aa}. It will be zero unless $(a_2\ldots a_\theta) = (\bar{a}_\theta\ldots\bar{a}_2)$. If θ is even, this requires $a_{\theta/2+1} = \bar{a}_{\theta/2+1}$ which is impossible since no letter of the alphabet is its own complement. If θ is odd this condition can be satisfied by independently specifying the letters $a_1,\ldots,a_{(\theta+1)/2}$. Thus the number of non-zero diagonal elements of Q is

$$\begin{array}{ll} 0 & \text{if } \theta \text{ is even,} \\ d^{(\theta+1)/2} & \text{if } \theta \text{ is odd.} \end{array} \tag{12}$$

Now consider the off-diagonal elements of Q. The total number of non-zero off-diagonal elements is

$$\begin{array}{ll} d^{\theta+1} - 0 & \text{if } \theta \text{ is even,} \\ d^{\theta+1} - d^{(\theta+1)/2} & \text{if } \theta \text{ is odd.} \end{array} \tag{13}$$

Since the matrix is symmetric, exactly half of these are independent:

$$\begin{array}{ll} \frac{1}{2}d^{\theta+1} & \text{if } \theta \text{ is even,} \\ \frac{1}{2}d^{(\theta+1)/2}(d^{(\theta+1)/2} - 1) & \text{if } \theta \text{ is odd.} \end{array} \tag{14}$$

Finally, adding the number of independent diagonal elements gives the number of independent elements of M as

$$\begin{array}{ll} \frac{1}{2}d^{\theta+1} & \text{if } \theta \text{ is even,} \\ \frac{1}{2}d^{(\theta+1)/2}(d^{(\theta+1)/2} + 1) & \text{if } \theta \text{ is odd.} \end{array} \tag{15}$$

Table 1. The number of independent $\theta + 1$-mer word counts N_{ab} needed to estimate a θ-order strand-symmetric transition matrix for an alphabet of $d = 4$ letters.

θ	# of independent elements of Q	# of non-zero elements of $\mathbb{M}$
1	10	16
2	32	64
3	136	256
4	512	1024
5	2080	4096
6	8192	16384

The number of independent elements of Q for an alphabet of size $d = 4$ is listed in Table 1.

In order to estimate a strand-symmetric transition matrix from a given observed sequence, it is sufficient to extend the definition of the count matrix N_{ab} to include a count of the number of occurrences of the $(\theta + 1)$-mer $(a_1 \ldots a_\theta b)$ in the sequence plus the number of occurrences of the same $(\theta+1)$-mer in its reverse complement. Such a matrix will automatically have the same number of independent elements as the corresponding Q, and the matrix $\hat{M}(a, b)$ constructed according to (9) will then have the required symmetry.

4   The D_2 Statistic

We now consider statistical properties of the alignment-free sequence similarity measure known as the D_2 statistic. The distributional properties of this statistic presented here apply to any higher order Markov model with or without the constraint of strand symmetry.

Definition 3. *Given two sequences* $\mathbf{X}$ *and* $\mathbf{Y}$ *with PBCs of length* m *and* n *respectively, the* D_2 *statistic is defined as the number of* k-word matches, *including overlaps, between* $\mathbf{X}$ *and* $\mathbf{Y}$:

$$D_2(k, M) = \sum_{i=1}^{m} \sum_{j=1}^{n} I_{ij}, \tag{16}$$

where I_{ij} *is the word match indicator random variable for words length* k *positioned at site* i *in sequence* $\mathbf{X}$ *and site* j *in sequence* $\mathbf{Y}$:

$$I_{ij} = \begin{cases} 1 & \text{if } (X_i, \ldots, X_{i+k-1}) = (Y_j, \ldots, Y_{j+k-1}), \\ 0 & \text{otherwise.} \end{cases} \tag{17}$$

More specifically, we are interested in the case where the two sequences are Markovian. From (6) and (7) it is easy to see that any ensemble of pairs of random sequences $(\mathbf{X}, \mathbf{Y})$ generated by an θ-order transition matrix M is in one-to-one correspondence with an ensemble of pairs of random sequences $(\mathbb{X}, \mathbb{Y})$ of letters from a d^θ-letter alphabet generated by the equivalent sparse $d^\theta \times d^\theta$ matrix $\mathbb{M}$. Furthermore, any k-mer match between $\mathbf{X}$ and $\mathbf{Y}$ corresponds to a $(k - \theta + 1)$-mer match between $\mathbb{X}$ and $\mathbb{Y}$. It follows that the distributional properties of D_2 for Markovian sequences can be determined in terms of the properties of D_2 for an equivalent first order system. In particular, for the mean and variance:

$$E(D_2(k, M)) = E(D_2(k - \theta + 1, \mathbb{M})),$$
$$\mathrm{Var}\,(D_2(k, M)) = \mathrm{Var}\,(D_2(k - \theta + 1, \mathbb{M})), \qquad k \geq \theta. \qquad (18)$$

Therefore, to calculate $E(D_2(k, M))$ and $\mathrm{Var}\,(D_2(k, M))$ for any $k \geq \theta$ it is sufficient to derive formulae for a first order Markov model. These formulae are given below. An R implementation [16] of the mean and variance for $k \geq \theta$ and a formula for the mean when $k < \theta$ will be published elsewhere [10].

4.1 D_2 Mean for $\theta = 1$

The mean of D_2 for $\theta = 1$ is

$$E(D_2) = \frac{mn}{\mathrm{tr}\,(M^m)\mathrm{tr}\,(M^n)}\mathrm{tr}\,[(M^{m-k+1} \circ M^{n-k+1})(M \circ M)^{k-1}], \qquad (19)$$

where the Hadamard product $A \circ B$ of two matrices A and B is defined as the matrix whose (α, β)-th element is

$$(A \circ B)_{\alpha\beta} = A_{\alpha\beta}B_{\alpha\beta}. \qquad (20)$$

Proof. We have that

$$E(D_2) = \sum_{i=1}^{m}\sum_{j=1}^{n} E(I_{ij}) = \sum_{i=1}^{m}\sum_{j=1}^{n} \mathrm{Prob}\,(I_{ij} = 1), \qquad (21)$$

where

$$\mathrm{Prob}\,(I_{ij} = 1) = \sum_{w \in \mathcal{A}^k} \mathrm{Prob}\,(X_i \ldots X_{i+k-1} = w)\mathrm{Prob}\,(Y_j \ldots Y_{j+k-1} = w). \qquad (22)$$

To calculate $\mathrm{Prob}\,(X_i \ldots X_{i+k-1} = w)$ we sum (8) over all sequences $\mathbf{x}$ such that $(x_i \ldots x_{i+k-1}) = w$. Thus

$$\mathrm{Prob}\,(X_i \ldots X_{i+k-1} = w) =$$
$$\frac{M^{m-k+1}(w_k, w_1)M(w_1, w_2)\ldots M(w_{k-1}, w_k)}{\mathrm{tr}\,(M^m)}, \qquad (23)$$

where the factor $M^{m-k+1}(w_k, w_1)$ arises from summing over the letters $x_1, \ldots, x_{i-1}, x_{i+k}, \ldots, x_m$. Similarly we have

$$\mathrm{Prob}\,(Y_j \ldots Y_{j+k-1} = w) =$$
$$\frac{M^{n-k+1}(w_k, w_1)M(w_1, w_2)\ldots M(w_{k-1}, w_k)}{\mathrm{tr}\,(M^n)}. \qquad (24)$$

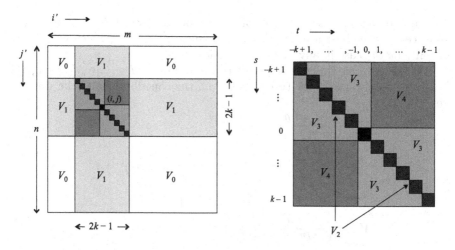

Fig. 1. Contributions to Var (D_2) via the sum in (27). The left-hand diagram shows the (i', j')-plane for a fixed value of (i, j), shown as the black square. The right-hand diagram is an expanded view of the 'accordion' region $-k + 1 \leq s, t \leq k - 1$, where $t = i' - i$ and $s = j' - j$ up to PBCs.

Substituting (23) and (24) into (22) gives

$$\text{Prob} \left(I_{ij} = 1 \right) = \frac{\text{tr} \left[(M^{m-k+1} \circ M^{n-k+1})(M \circ M)^{k-1} \right]}{\text{tr} \left(M^m \right) \text{tr} \left(M^n \right)}. \tag{25}$$

Equation (21) then gives the required result.   □

4.2 D_2 Variance for $\theta = 1$

The exact variance of D_2 for Markovian sequences with PBCs requires an extensive calculation. Here we give a summary of the result, which is valid for $m, n \geq 2k$. Full technical details of the derivation will be published elsewhere [10].

We have

$$\text{Var} \left(D_2 \right) = E(D_2{}^2) - E(D_2)^2. \tag{26}$$

The second term can be calculated from (19). The first term is a sum of contributions obtained from (16) by partitioning a sum over words beginning at positions i and i' in sequence $\mathbf{X}$ and beginning at j and j' in sequence $\mathbf{Y}$,

$$
\begin{aligned}
E(D_2{}^2) &= \sum_{i,i'=1}^{m} \sum_{j,j'=1}^{n} E(I_{ij} I_{i'j'}) \\
&= \sum_{i,i'=1}^{m} \sum_{j,j'=1}^{n} \text{Prob} \left(I_{ij} = 1, I_{i'j'} = 1 \right) \\
&= V_0 + V_1 + V_2 + V_3 + V_4.
\end{aligned}
\tag{27}
$$

The partitioning reflects the degree of overlap between words in each of the two sequences, and is illustrated in Fig. 1. We assume $m, n \geq 2k$, which will almost certainly be the case in any biological application.

We will write a Hadamard product of q factors, $M \circ \ldots \circ M$, using the shorthand notation $M^{\circ q}$. With this notation, the contributions to the variance are:

$$V_0 = \frac{mn}{\operatorname{tr}(M^m)\operatorname{tr}(M^n)} \times$$

$$\sum_{r=0}^{m-2k} \sum_{s=0}^{n-2k} \operatorname{tr}\left[(M^{r+1} \circ M^{s+1})(M \circ M)^{k-1} \times \right.$$

$$\left. (M^{m-2k-r+1} \circ M^{n-2k-s+1})(M \circ M)^{k-1}\right],$$

$$V_1 = \frac{mn}{\operatorname{tr}(M^m)\operatorname{tr}(M^n)} \times \tag{28}$$

$$\left\{\sum_{s=0}^{n-2k} \left[\operatorname{tr}\{[(M \circ M \circ M)^{k-1} \circ (M^{s+1})^T]\times\right.\right.$$

$$\left. (M^{m-k+1} \circ M^{n-2k-s+1})\}\right.$$

$$+ 2\sum_{r=1}^{k-1} \operatorname{tr}\{(M \circ M)^r \times$$

$$[(M \circ M \circ M)^{k-r-1} \circ (M^{s+1})^T]\times$$

$$\left.\left. (M \circ M)^r (M^{m-k-r+1} \circ M^{n-2k-s+1})\}\right]\right.$$

$$\left.+ \text{ the same with } m \text{ and } n \text{ interchanged.}\right\}, \tag{29}$$

$$V_2 = \frac{mn}{\operatorname{tr}(M^m)\operatorname{tr}(M^n)} \times$$

$$\left\{\operatorname{tr}\left[(M^{m-k+1} \circ M^{n-k+1})(M \circ M)^{k-1}\right]\right.$$

$$+ 2\sum_{t=1}^{k-1} \operatorname{tr}\left[(M^{m-k-t+1} \circ M^{n-k-t+1})\times\right.$$

$$\left.\left. (M \circ M)^{k+t-1}\right]\right\}, \tag{30}$$

$$V_3 = \frac{2mn}{\operatorname{tr}(M^m)\operatorname{tr}(M^n)} \times$$

$$\sum_{t=1}^{k-1}\sum_{s=0}^{t-1} \operatorname{tr}\left[(M \circ M)^s Q(M \circ M)^s \times\right.$$

$$(M^{m-k-t+1} \circ M^{n-k-s+1} +$$

$$\left. M^{n-k-t+1} \circ M^{m-k-s+1})\right], \tag{31}$$

where

$$
Q = \begin{cases}
(M^{\circ(2\nu+3)})^{\rho-1} \circ [(M^{\circ(2\nu+1)})^{t-s-\rho+1}]^T \\
\qquad\qquad\qquad\qquad \text{if } \rho > 0, \\
(M^{\circ(2\nu+1)})^{t-s-1} \circ (M^{\circ(2\nu-1)})^T \\
\qquad\qquad\qquad\qquad \text{if } \rho = 0,
\end{cases}
\tag{32}
$$

and

$$
\nu = \left\lfloor \frac{k-s}{t-s} \right\rfloor, \qquad \rho = (k-s) \mod (t-s).
\tag{33}
$$

Finally,

$$
V_4 = \frac{2mn}{\operatorname{tr}(M^m)\operatorname{tr}(M^n)} \sum_{r,t=1}^{k-1} \operatorname{tr} U,
\tag{34}
$$

where

$$
U =
\begin{cases}
\left\{ (M^{\circ(2\nu+1)})^{t-1} \circ (M^{m-k-t+1})^T \right\} M^{\circ 2\nu} \times \\
\quad \left\{ (M^{\circ(2\nu+1)})^{r-1} \circ (M^{n-k-r+1})^T \right\} M^{\circ 2\nu} \\
\qquad\qquad\qquad\qquad \text{if } \zeta = 0, \\
\left\{ (M^{\circ(2\nu+1)})^{r-\zeta+1} \circ M^{m-k-t+1} \right\} \times \\
\quad (M^{\circ(2\nu+2)})^{\zeta-1} \times \\
\qquad \left\{ (M^{\circ(2\nu+1)})^{t-\zeta+1} \circ M^{n-k-r+1} \right\} \times \\
\qquad (M^{\circ(2\nu+2)})^{\zeta-1} \\
\qquad\qquad\qquad\qquad \text{if } 0 < \zeta \le r,t, \\
\left\{ (M^{\circ(2\nu+3)})^{\zeta-r-1} \circ (M^{m-k-t+1})^T \right\} \times \\
\quad (M^{\circ(2\nu+2)})^r \left\{ (M^{\circ(2\nu+1)})^{t-\zeta+1} \circ M^{n-k-r+1} \right\} \\
\quad \times (M^{\circ(2\nu+2)})^r \\
\qquad\qquad\qquad\qquad \text{if } r < \zeta \le t, \\
\{\text{as above with } m \text{ and } n \text{ interchanged} \\
\quad \text{and } r \text{ and } t \text{ interchanged}\} \\
\qquad\qquad\qquad\qquad \text{if } t < \zeta \le r, \\
\left\{ (M^{\circ(2\nu+3)})^{\zeta-r-1} \circ (M^{m-k-t+1})^T \right\} \times \\
\quad (M^{\circ(2\nu+2)})^{t+r-\zeta+1} \times \\
\qquad \left\{ (M^{\circ(2\nu+3)})^{\zeta-t-1} \circ (M^{n-k-r+1})^T \right\} \times \\
\qquad (M^{\circ(2\nu+2)})^{t+r-\zeta+1} \\
\qquad\qquad\qquad\qquad \text{if } r,t < \zeta,
\end{cases}
$$

and

$$
\nu = \left\lfloor \frac{k}{r+t} \right\rfloor, \qquad \zeta = k \mod (r+t).
\tag{35}
$$

5   Numerical Simulations

For short sequences and small alphabets the distribution of the D_2 statistic can be computed by enumerating all possible sequences. We have confirmed the accuracy of the formulae for the mean and variance given in Sect. 4 to 11 significant figures by generating the complete distribution of D_2 using double precision arithmetic for sequences up to length $m = n = 9$ for $k = 3$, $d = 2$ and up to length $m = n = 7$ for $k = 2$, $D = 3$. The Markov matrices M are generated randomly by choosing each element from a uniform distribution on the interval $[0, 1]$ and then normalising each row sum to 1. Two examples of the exact D_2 distribution are shown in Fig. 2. Note that the introduction of

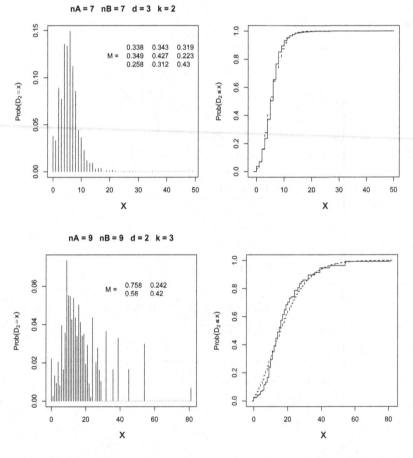

Fig. 2. The exact distribution of the D_2 statistic for short sequences of length n_A, n_B and words of length k from an alphabet of size d. The Markov matrix M has been generated randomly in each case. Also shown (dashed curve) is the cumulative distribution of the Pólya-Aeppli distribution with mean and variance set to the theoretical values using the formulae of Sect. 4.

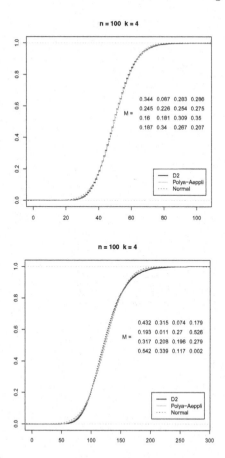

Fig. 3. Two examples of empirical cumulative distribution of the D_2 statistic estimated from 10,000 independently generated random sequences of length $m = n = 100$ for words of length $k = 4$ and an alphabet of size $d = 4$. The Markov matrix M has been generated randomly in each case. Also shown are the cumulative distribution of the normal and Pólya-Aeppli distributions with mean and variance set to the theoretical values using the formulae of Sect. 4.

random Markov matrices is to enable an efficient check of the above formulae for a range of M, and is not intended to have any biological meaning. Maximum likelihood estimates of Markov transition matrices from various genomes have been published, for instance, by Chor et al. [11], which can be used in biological applications. We note that the Chor estimates are close to satisfying the strand-symmetry condition restrictions of Sect. 3 (data not shown).

For longer sequences of realistic biological length, the distribution of D_2 can be estimated from a Monte Carlo ensemble of random sequences generated from the algorithm described in Sect. 2.1. Examples of cumulative distribution functions for $d = 4$, $k = 4$ estimated from ensembles of 10,000 pairs of independently

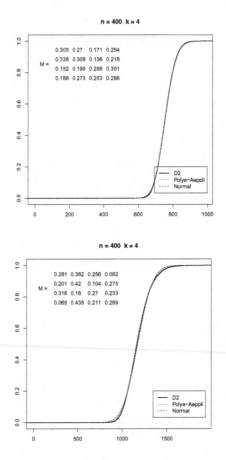

Fig. 4. The same as Fig. 3, except $m = n = 400$.

generated random sequences of length $m = n = 100$ and 400 are shown in Figs. 3 and 4 respectively. The Markov matrix is again generated randomly, and it is interesting to note that the mean of the distribution can vary considerably with M. We have made a number of simulations, and find that in roughly the expected proportion of times the mean and variance calculated from the formulae of Sect. 4 lie within the 95 % confidence intervals computed from the ensemble.

For the case of sequences composed of i.i.d. letters certain rigorous results are known for the asymptotic distribution of D_2 as the sequence lengths $m, n \to \infty$. For $m = n$, it has been shown that the limiting distribution is normal in the regime $k < 1/2 \log_b n + \text{const.}$ [17] and Pólya-Aeppli in the regime $k > 2 \log_b n + \text{const.}$ [1]. Here $b = 1/\sum_{a \in \mathcal{A}} p_a^2$ where p_a is the probability of occurrence of letter a. A Pólya-Aeppli random variable is the sum of a Poisson number of geometric random variables, and is therefore an example of a compound Poisson random variable. It often arises in the study of random word counts as a Poisson number of clumps of overlapping words, each clump containing a geometric number of

k-words [13]. Although the asymptotic results for D_2 are not proved for Markovian sequences, it is a reasonable experiment to compare our numerical simulations with these distributions as they may potentially provide an accurate estimate of p-values in biological applications.

One would not expect the asymptotic distributions to be an accurate fit for the short sequences considered in Fig. 2. Nevertheless we have included the Pólya-Aeppli distribution function and find it to be surprisingly close for the $d = 3$ case. Disagreement arises in the tail of the distribution because, for combinatoric reasons, certain values of D_2 within the range 0 to mn do not occur, whereas the Pólya-Aeppli has support over the whole range (and also out to ∞, albeit with very low probability).

If one were dealing with i.i.d. sequences with a uniform letter distribution, then the parameters $m = n = 100$ or 400, $k = 4$ used for the simulations in Figs. 3 and 4 would inhabit the region between the normal and Pólya-Aeppli asymptotic regimes described above. Both asymptotic distributions are superimposed on the empirical distribution functions in Figs. 3 and 4. We observe that the normal and Pólya-Aeppli do not differ greatly from one another, though the Pólya-Aeppli does appear to give a better fit, particularly in the important tail of the distribution relevant to estimating p-values.

6   Conclusions

This paper introduces the concept of periodic boundary conditions for Markovian sequences as an elegant mathematical construct which avoids the inconvenience of boundary effects in analytic calculations. We have demonstrated that the mean and variance of the D_2 word match statistic can be calculated analytically and readily computed to any desired accuracy through formulae involving only traces of products of matrices. Calculation of the mean and variance is fast as powers of Hadamard products need only be calculated once for a given Markovian model, and only need to be calculated up to the point of convergence. For biological applications such as measuring sequence similarity or identifying regions of regulatory motifs, sequences lengths tend to be of at least a few hundred letters. In these cases loss of information about boundary effects is unlikely to be a serious impediment. For instance, in previous studies of a database of cis-regulatory modelled as a set of i.i.d. sequences was successfully studied using the D_2 statistics simply by imposing PBCs on the sequences prior to calculating the D_2 [9,18].

The current work is a preliminary study designed to illustrate the computational effectiveness of imposing periodic boundary conditions when calculating the D_2 statistic. In ongoing work we are testing the agreement between the theoretical Markovian distributions studied herein and empirical distributions from genomic DNA. In general, we find that the empirical distribution tends to have heavier left and right tails, suggesting the existence of a subset of k-mers which are over- or under-represented within the genomes studied [10].

Further work also needs to be done on extending the results to more viable variants of the D_2 statistic. It has been argued that a potential shortcoming of the D_2 statistic is that the signal of sequence similarity one is trying to detect maybe hidden by its variability due to noise in each of the single sequences, and that to overcome this problem one should instead calculate a 'centred' version of D_2 in which word count vectors are replaced with those centred about their mean [1,3]. There also exist 'standardised' versions of D_2 [4,19] designed to account for biases arising from the fact that some words are naturally overrepresented, and 'weighted' versions [5] designed to account for higher substitution rates of chemically similar amino acids in protein sequences. Extension of the mathematical formalisms developed herein to these D_2 variants, as well as a more compete study of the accuracy of approximating p-values with asymptotic distributions, will be the subject of future work.

Acknowledgements. This work was supported in part by ARC Discovery Grant DP120101422 and NHMRC grant 525453.

References

1. Lippert, R.A., Huang, H., Waterman, M.S.: Distributional regimes for the number of k-word matches between two random sequences. Proc. Natl. Acad. Sci. USA **99**, 13980–13989 (2002)
2. Vinga, S., Almeida, J.: Alignment-free sequence comparison-a review. Bioinformatics **19**, 513–23 (2003)
3. Reinert, G., Chew, D., Sun, F., Waterman, M.S.: Alignment-free sequence comparison (I): statistics and power. J. Comput. Biol. **16**, 1615–1634 (2009)
4. Göke, J., Schulz, M., Lasserre, J., Vingron, M.: Estimation of pairwise sequence similarity of mammalian enhancers with word neighbourhood counts. Bioinformatics **28**, 656–663 (2012)
5. Jing, J., Wilson, S.R., Burden, C.J.: Weighted k-word matches: A sequence comparison tool for proteins. ANZIAM J. **52**(CTAC2010), 172–189 (2011)
6. Forêt, S., Kantorovitz, M.R., Burden, C.J.: Asymptotic behaviour and optimal word size for exact and approximate word matches between random sequences. BMC Bioinformatics **7**(Suppl 5), S21 (2006)
7. Kantorovitz, M.R., Booth, H.S., Burden, C.J., Wilson, S.R.: Asymptotic behavior of k-word matches between two uniformly distributed sequences. J. Appl. Probab. **44**, 788–805 (2006)
8. Forêt, S., Wilson, S.R., Burden, C.J.: Empirical distribution of k-word matches in biological sequences. Pattern Recogn. **42**, 539–548 (2009)
9. Forêt, S., Wilson, S.R., Burden, C.J.: Characterizing the $D2$ statistic: Word matches in biological sequences. Stat. Appl. Genet. Mol. Biol. **8**(1), Article No. 43 (2009)
10. Burden, C.J., Leopardi, P., Forêt, S.: The distribution of word matches between Markovian sequences with periodic boundary conditions. J. Comput. Biol. **21**(1), 41–63 (2014)
11. Chor, B., Horn, D., Goldman, N., Levy, Y., Massingham, T.: Genomic DNA k-mer spectra: models and modalities. Genome Biol. **10**(10), R108.10 (2009)
12. Kantorovitz, M.R., Robinson, G.E., Sinha, S.: A statistical method for alignment-free comparison of regulatory sequences. Bioinformatics **23**, i249–55 (2007)

13. Reinert, G., Schbath, S., Waterman, M.: Statistics on words with applications to biological sequences. In: Lotharie, M. (ed.) Applied Combinatorics on Words. Cambridge University Press, Cambridge (2005)
14. Percus, J., Percus, O.: The statistics of words on rings. Commun. Pure Appl. Math. **59**, 145–160 (2006)
15. Baisnée, P.F., Hampson, S., Baldi, P.: Why are complementary DNA strands symmetric? Bioinformatics **18**, 1021–1033 (2002)
16. R Core Team.: R: A Language and Environment for Statistical Computing. R Foundation for Statistical Computing, Vienna, Austria (2012)
17. Burden, C.J., Kantorovitz, M.R., Wilson, S.R.: Approximate word matches between two random sequences. Ann. Appl. Probab. **18**, 1–21 (2008)
18. Burden, C.J., Jing, J., Wilson, S.R.: Alignment-free sequence comparison for biologically realistic sequences of moderate length. Stat. Appl. Genet. Mol. Biol. **11**(1), 1–28 (2012). Article No. 3
19. Liu, X., Wan, L., Li, J., Reinert, G., Waterman, M.S., Sun, F.: New powerful statistics for alignment-free sequence comparison under a pattern transfer model. J. Theoret. Biol. **284**, 106–116 (2011)

Systematic Analysis of Homologous Tandem Repeat Family in the Human Genome

Woo-Chan Kim[✉] and Dong-Ho Cho

Department of Electrical Engineering, Korea Advanced Institute
of Science and Technology, Daejeon, South Korea
wckim@comis.kaist.ac.kr, dhcho@kaist.ac.kr
http://umls.kaist.ac.kr

Abstract. The vast majority of the human genome consists of repetitive elements that form many complex but highly-ordered patterns. In particular, tandem repeats, whose repeat units are placed adjacent to each other, form highly structured patterns in the human genome when homologous tandem repeats are close together. Herein, the structure of the homologous tandem repeat family (HTRF) is assessed using systematic analysis. In the proposed system for analyzing HTRF, the original tandem repeat units are derived using the characteristics of homology of HTRF, and represented in a diagram in order to show the structure of HTRF easily. The analysis results of the four HTRFs in the human genome are shown here and the proposed algorithm may be seen to be very efficient for analyzing HTRF via the comparison of three conventional algorithms.

Keywords: Repetitive element · Tandem repeat · Homologous tandem repeat array · Systematic analysis · Human genome

1 Introduction

There are many repeated DNA sequences in the genome of most organisms, which is called *repetitive element*. The two major classes of repetitive elements are interspersed repeats and tandem repeats. Interspersed repeats are usually present as single copies and distributed widely throughout the genome, whereas tandem repeats are DNA sequences of which repeat units are placed next to each other. Although the functions of many repetitive elements have not yet been known, their impact and importance on genomes is evident. Mobile repeat elements have been a critical factor in gene evolution [1,2]. Also, some tandem repeats cause a number of genetic diseases [3] and they have been used as genetic markers for human identity testing [4]. Therefore, analyzing repetitive elements is very important and we study tandem repeats especially in this paper.

Tandem repeats are classified into three types, which are satellite, minisatellite, and microsatellite. Satellites form arrays of 1,000 to 10 million repeat units particularly in the heterochromatin of chromosomes. Minisatellite form arrays

© Springer-Verlag Berlin Heidelberg 2014
M. Fernndez-Chimeno et al. (Eds.): BIOSTEC 2013, CCIS 452, pp. 162–175, 2014.
DOI: 10.1007/978-3-662-44485-6_12

of several hundred repeat units of 7 to 100 bp in length. They are present everywhere with an increasing concentration toward the telomeres. Microsatellites are composed of units of one to six nucleotides, repeated up to a length of 100 bp or more.

Although tandem repeats have been characterized by some features, which are the position in the genome, sequence, size, number of copies, and presence or absence of coding regions, there are much more complex tandem repeats in the human genome. In [5], the authors researched complex pattern structures of variable length tandem repeat (VLTR) and multi-periodic tandem repeat (MPTR). Also, our previous studies to find and visualize all repetitive elements in the genomes showed that the structure of the unknown as well as known repetitive elements is very complex but highly organized [6,7]. We, here, focus on the structure of multiple tandem repeats, which is called HTRF (Homologous Tandem Repeat Family).

HTRFs, which consist of multiple homologous tandem repeats dispersed throughout specific sequence regions, are abundant in the genomes of human and mouse [6,7]. Despite of lack of research of HTRF, we expect that HTRF plays an important role involving biological functions from its abundance and unique structure. Also, we can easily find a consensus tandem repeat unit of an HTRF array since two or more tandem repeats are homologous. By getting a consensus tandem repeat unit, we can find out how much the original HTRF are broken, which can be used as an evidence of the age of the genome.

We analyze four HTRFs from the human genome, which are chromosome 7 (57,937,500 – 58,056,406 bp), chromosome 8 (46,832,500 – 47,458,334 bp), chromosome 22 (16,505,625 – 16,627,187 bp), and chromosome Y (25,000 – 117,031 bp). The method for getting a consensus tandem repeat unit that are proposed in this paper is compared with the three conventional programs or algorithms, which are TRF (Tandem Repeat Finder) [8], SRF (Spectral Repeat Finder) [9], and tandem repeat detection using PT (Period Transform) [10,11]. TRF and SRF are the representative program for finding tandem repeat by using string matching algorithm and signal processing algorithm, respectively. A perfect HTRF is constructed by using the derived tandem repeat unit, and it is compared with the original HTRF. Also, the structure of an HTRF is shown in a diagram representation by using the consensus tandem repeat units.

2   System Modeling and Algorithm

The modeling of HTRF consists of three stages, which are *TR Extractor*, *TR Analyzer*, and *MTR Analyzer*. Figure 1 shows the system structure for analysis of HTRF. TR Extrator gets each tandem repeat from a given HTRF. The individual extracted tandem repeat is analyzed by TR Analyzer. The analysis results of TR Analyzer are the original tandem repeats as well as the properties of the individual tandem repeats such as repeat unit, number of repeat unit, and homology. MTR Analyzer, then, analyzes the relationships among the individual tandem repeats by using the results of TR Analyzer.

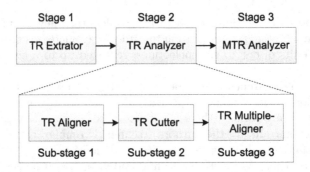

Fig. 1. System structure for analysis of HTRF.

2.1 TR Extractor

There are one or more tandem repeats that are homologous each other in an HTRF. We assume that there are N tandem repeats in an HTRF and each tandem repeat has n tandem repeat units whose length is l. Then, we can express i'th tandem repeat in an HTRF as $TR_i(n, l)$. TR Extractor divides each tandem repeat from an HTRF, which means that it gets all $TR_i(n, l)$ for the HTRF. However, TR Extrator defines tandem repeat when the number and the length of tandem repeat units is greater than δ_n and δ_l, respectively.

2.2 TR Analyzer

TR Analyzer derives a consensus tandem repeat from each tandem repeat in an HTRF. TR Analyzer consists of *TR Aligner*, *TR Cutter*, and *TR Multiple-Aligner*. These three sub-stages are iteratively processed for better performances.

TR Aligner. Two DNA sequence blocks which are a reference sequence and a target sequence are aligned, and the identity of the two sequence blocks are recorded in TR Aligner. If we assume that the length of the sequence block is B and a sequence block that begins from i'th nucleotide base of a tandem repeat is $S(i)$, the reference sequence of the firstly performed TR Aligner is generally $S(1)$. Also, the target sequence is moved from $S(1)$ to $S(x - B + 1)$ where x is the length of the tandem repeat.

The alignment of the two DNA sequences is conducted by dynamic algorithm or greedy algorithm [12]. The identity of the two sequences as a result of the alignment is recorded to $I(i, j)$ where i is the index of the reference sequence and j is the index of the target sequence. Since the reference sequence is fixed in TR Aligner, $I(i, j)$ is a function of variable j. Then, $I(i, j)$ of a perfect tandem repeat becomes 1 when $j = i + l \times k$ and $1 \leq i, j \leq x$ where k is an integer since same sequence blocks are arranged periodically in a perfect tandem repeat. The identity may have its peak point even if the tandem repeat is broken because it still has the attribute of the repetition of the tandem repeat. By using this characteristic of the identity of the tandem repeat, we can get the index of each tandem repeat unit.

TR Cutter. The identity function of a broken tandem repeat is generally fluctuated because its original perfect tandem repeat is randomly broken by biological phenomena such as insertion, deletion, and substitution. Thus, TR Cutter performs two processes to derive the index of each tandem repeat unit. First, TR Cutter makes the identity function be smoothed by averaging it locally. That is, a smoothed version of identity function $M(i,j)$ is defined as follows.

Algorithm 1. Recursive process of three sub-stages of TR Analyzer: TR Aligner, TR Cutter, and TR Multiple-Aligner.

```
 1: procedure TR ALINER(broken_tandem_repeat)
 2:     reference_sequence_index ← 0
 3:     consensus_unit_index ← −1
 4:     while reference_sequence_index ≠ consensus_unit_index do
 5:         Calculate identity function                           ▷ TR Aligner
 6:         Get all tandem repeat units                           ▷ TR Cutter
 7:         Get a consensus unit                          ▷ TR Multiple-Aligner
 8:         Substitute the index of the consensus unit to consensus_unit_index
 9:     end while
10:     return consensus_unit_index
11: end procedure
```

$$M(i,j) = \frac{\sum_{k=j-\lfloor W/2 \rfloor}^{j+\lceil W/2 \rceil -1} I(i,k)}{W} \tag{1}$$

where W is the smoothing window size.

The smoothing process removes the fluctuation of the identity function so that only the start points of real tandem repeat units have their local peak value of identity. The smaller the window size of the smoothing process is, the more peak points that are not the start of tandem repeat unit exist. However, too large window size of the smoothing process may remove the local peak point of a real tandem repeat unit. Therefore, the proper window size is required to leave only the local peak points of the real tandem repeat units in the smoothing process of the identity function.

Then, TR Cutter gets the start index of all the tandem repeat units by differentiating the function $M(i,j)$. The differentiated function $M'(i,j)$ of $M(i,j)$ is as follows.

$$M'(i,j) = M(i,j+1) - M(i,j). \tag{2}$$

After calculating $M'(i,j)$, TR Cutter can find all the tandem repeat units by recording the points when $M'(i,j)$ is 0.

TR Multiple-Aligner. The tandem repeat units that are gotten by TR Cutter are aligned by TR Multiple-Aligner. By using the multiple sequence alignment of the tandem repeat units, TR Multiple-Aligner can get the consensus tandem repeat unit. A direct method of the multiple sequence alignment is the dynamic programming technique to identify the globally optimal alignment solution [13,14]. However, computational complexity of the direct method is basically too high, which takes $O(l^n)$ time where l and n are the average length and the number of tandem repeat units, respectively. Thus, we here use a suboptimal method that utilizes pairwise sequence alignment, which is similar to other suboptimal methods [15,16]. In our method, all pairwise sequence alignments between each pair of tandem repeat units are performed, and the tandem repeat unit that has the highest average alignment score with other tandem repeat units is chosen as the consensus tandem repeat unit. The proposed suboptimal method for finding the consensus tandem repeat unit takes $O((nl)^2)$ time, which reduces many computations compared with the dynamic programming technique particularly when l and n are large.

After the consensus tandem repeat unit is chosen, the three sub-stages of TR Analyzer, which are TR Aligner, TR Cutter, and TR Multiple-Aligner, are re-performed to get a more accurate consensus tandem repeat unit. The recursive process of TR Analyzer is conducted until the reference sequence is not changed in TR Aligner. Algorithm 1 shows the pseudo code of TR Analyzer.

2.3   MTR Analyzer

After TR Extractor divides all tandem repeats from the target HTRF and TR Analyzer derives the consensus tandem repeats of the individual tandem repeats, MTR Analyzer derives a consensus tandem repeat unit among all the tandem repeats. Sine the tandem repeats in an HTRF are highly homologous and are expected to be an identical tandem repeat originally, the consensus tandem repeat unit that is gotten from the multiple tandem repeats increases the reliability of the originality. Also, there are many reverse-complement directional homologous tandem repeats as well as forward directional homologous tandem repeats. Thus, we should not only consider the forward direction but also the reverse-complement direction of homology.

The derivation of the consensus tandem repeat unit is performed by multiple sequence alignment. Thus, we also apply the sub-optimal method of multiple sequence alignment that is used in TR Multiple-Aligner to the derivation of the consensus tandem repeat unit of HTRF for the purpose of reducing the computational complexity.

We can describe an HTRF as a diagram by using the derived consensus tandem repeat units. Figure 2 shows an example of a diagram of an HTRF. The HTRF shown in Fig. 2 has two different tandem repeat units and each tandem repeat appears twice. The tandem repeats made by the first tandem repeat unit are shown twice in forward and reverse directions, and then the tandem repeats made by the second tandem repeat unit are shown twice in only forward direction. Also, a region that is not a tandem repeat exists between the two

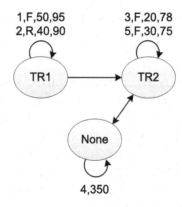

Fig. 2. Example of diagram representation of HTRF.

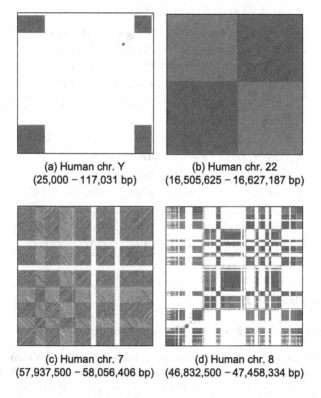

(a) Human chr. Y
(25,000 – 117,031 bp)

(b) Human chr. 22
(16,505,625 – 16,627,187 bp)

(c) Human chr. 7
(57,937,500 – 58,056,406 bp)

(d) Human chr. 8
(46,832,500 – 47,458,334 bp)

Fig. 3. Dot plot pattern of repetitive element arrays of the human genome.

tandem repeats made by the second tandem repeat unit. The four elements that are written above tandem repeat unit in the diagram are order, direction (F is forward direction and R is reverse direction), number of repetitions, and identity in percentage. Also, the two elements below *None* vertex is order and number of nucleotide bases.

Table 1. Consensus tandem repeat units of each MTRA.

Human chromosome	Consensus tandem repeat unit	Average identity
Chr. Y	TR1: *TAGGTCTCATTGAGGACAGATAGAGAGCAGA*	0.95
	CTGTGCAACCTTTAGAGTCTGCATTGGGCC	
Chr. 22	TR1: *GCAGCAGTGTTCTGGAATCCTATG*	0.86
	TGAGGGACAAACACTCAGAACCCA	
Chr. 7	TR1: *TTCAACTCTGTGAGATGAATGCACACATC*	0.88
	ACAAAGAAGTTTCTCAGAATGCTTCTGTC	
	TAGTTTTTATGTGAAGATATTTCCTTTTC	
	CACCATAGGCCTCAAAGTGCTCCAAATG	
	TCCACTTGCAGATTCTACAAAAAGAGTG	
	TTTCAAAACTGCTCAATCAAAAGAAAGG	
Chr. 8	TR1: *CCCACTGAGGCCTATAGTGAAAAACTGAA*	0.76
	TATCCCATGATAAAAACTAGAAAGAAGCT	
	ATCTGTGAAACTGCTTTGTGATGTGTGCA	
	TTCAGCTCACAGAGTTAAACCTTTCTTT	
	TGATTCAGCAGGTTGGAAACACTCTTTT	
	TGTAGAATCTGCAAGGGGATATTTGGAG	
	TR2: *CCAAGGAGGCCTCTCCCATCCCAGAAGCCCC*	
	CAGGGCTGTCCCGGGCGGGCTGTAAAGCCCC	
	AGGCTTTGGAGCAGGGTGCCTGTGTCTCTCG	
	CAGAAGGCCCCCACAAGCGAAAACGGGGCCG	
	CAGGGTGGCGTGGGAGGGCCGCAGGGACTCA	
	GGGGGACGTTGAGGCAGGCAGAGGGGAGAAG	
	CGGCGAGACTGCAGGGAATGCTGGGAGCCTC	

3 Experimental Results

3.1 Analysis of HTRF of the Human Genome

We analyzed four HTRFs of the human genome by using the proposed system modeling. The analyzed HTRFs are chromosome 7 (57,937,500 – 58,056,406 bp), chromosome 8 (46,832,500 – 47,458,334 bp), chromosome 22 (16,505,625 – 16,627, 187 bp), and chromosome Y (25,000 – 117,031 bp). The human genome were obtained from the NCBI (National Center for Biotechnology Information) databases.

We first analyzed the repetitive elements and repetitive element arrays of the DNA sequences by using our analysis program, REMiner and REMiner Viewer [6,7]. The dot plot patterns of repetitive elements and repetitive element arrays of individual DNA sequences are shown in Fig. 3. According to the protocol of

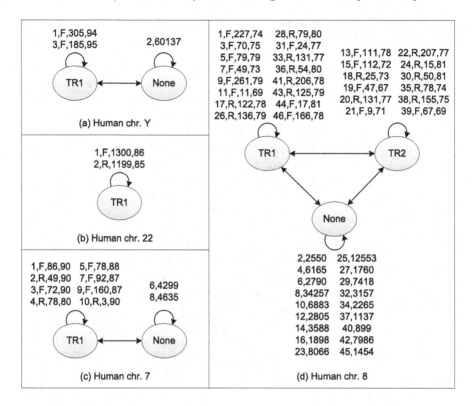

Fig. 4. Diagram representation of HTRF of the human genome.

dot plot, a square is the pattern of a tandem repeat and a rectangle shows the relationship between two tandem repeats [6, 7, 17].

The HTRF of the human chromosome Y in Fig. 3 (a) has two tandem repeats and they are homologous directly, whereas the two tandem repeats of the HTRF of the human chromosome 22 are homologous inversely as shown in Fig. 3 (b). Also, there are many tandem repeats in the HTRF of the human chromosome 7 but they are all homologous directly or inversely as shown in Fig. 3 (c). This means that the tandem repeats all come from a same tandem repeat. In Fig. 3 (d), there are much more tandem repeats and they come from two original tandem repeats.

The consensus tandem repeat units are derived as results of our proposed analysis tool for HTRF. Table 1 shows the consensus tandem repeat units of the four target DNA sequences. The average identity is the mean values of the identity between the perfect tandem repeat that is made by the consensus tandem repeat unit and individual broken tandem repeat. The average identity shows the homology of each tandem repeat in the HTRF and the brokenness level of the HTRF, which can be graphically shown in Fig. 3.

Table 2. Comparison of consensus tandem repeat unit of proposed scheme and conventional schemes (chromosome Y and 22).

Human chromosome	Algorithm	Consensus tandem repeat unit	Average identity	Number of fails
Chr. Y	TR analyzer	*TAGGTCTCATTGAGGACAGAT* *AGAGAGCAGACTGTGCAACC* *TTTAGAGTCTGCATTGGGCC*	0.95	0/2
	TRF	*TAGGTCTCATTGAGGACAGAT* *AGAGAGCAGACTGTGCAACC* *TTTAGAGTCTGCATTGGGCC*	0.95	0/2
	SRF	*TAGGTCTCATTGAGGACAGAT* *AGAGAGCAGACTGTGCAACC* *TTTAGAGTCTGCATTGGGCC*	0.95	0/2
	PTF	*TAGGTCTCATTGAGGACAGAT* *AGAGAGCAGACTGTGCAACC* *TTTAGAGTCTGCATTGGGCC*	0.95	0/2
Chr. 22	TR analyzer	*GCAGCAGTGTTCTGGAATCCTATG* *TGAGGGACAAACACTCAGAACCCA*	0.86	0/2
	TRF	*GCAGCAGTGTTCTGGAATCCTATG* *TGAGGGACAAACACTCAGAACCCA*	0.86	0/2
	SRF	*GCAGCAGTGTTCTGGAATCCTATG* *TGAGGGACAAACACTCAGAACCCA*	0.86	0/2
	PTF	·	·	2/2

Based on the consensus tandem repeat units, we describe each HTRF as a diagram in Fig. 4. By using the new representation of HTRF, we can easily see the overall structure of HTRF and the relationships among individual tandem repeats in HTRF. Furthermore, the original perfect tandem repeat array of the HTRF can be restored and the brokenness level of the HTRF can be calculated by using the consensus tandem repeat units.

3.2 Proposed Algorithm vs. Conventional Algorithm

There are many conventional algorithms that find tandem repeats although they did not consider multiple homologous tandem repeats simultaneously. Most of them can also derive the consensus tandem repeat unit of a tandem repeat. Thus, the conventional algorithms can be used to derive the consensus tandem repeat unit of a tandem repeat that is the function of TR Analyzer in our proposed system for the analysis of HTRF. In this subsection, TR Analyzer is compared with the representative conventional schemes that derive the consensus tandem repeat unit, which are TRF (Tandem Repeat Finder) [8], SRF (Spectral Repeat Finder) [9], and tandem repeat detection using PT (Period Transform) [10,11].

TRF is the representative program of string matching algorithms for finding tandem repeat. It uses pattern recognition criteria that is constructed statistically

Table 3. Comparison of consensus tandem repeat unit of proposed scheme and conventional schemes (chromosome 7).

Human chromosome	Algorithm	Consensus tandem repeat unit	Average identity	Number of fails
Chr. 7	TR analyzer	*TTCAACTCTGTGAGATGAATGC*	0.88	0/8
		ACACATCACAAAGAAGTTTCTC		
		AGAATGCTTCTGTCTAGTTTTT		
		ATGTGAAGATATTTCCTTTTC		
		CACCATAGGCCTCAAAGTGCT		
		CCAAATGTCCACTTGCAGATT		
		CTACAAAAAGAGTGTTTCAAA		
		ACTGCTCAATCAAAAGAAAGG		
	TRF	*TTCAACTCTGTGAGATGAATGC*	0.88	0/8
		ACACATCACAAAGAAGTTTGT		
		CAGAATGCTTCTGTCTAGTTT		
		TTATGTGAAGATATATTCTTT		
		TCCACCATAGGCCTCAAAGTG		
		CTCCAAATGTCCACTGCAGAT		
		TCTACAAAAAGAGTGTTTGAA		
		ATTGCTCAATCAAAAGAAATG		
	SRF	*TTCAACTCTGTGAGATGAATGC*	0.79	2/8
		ACACATCACAAAGAAGTTTCTC		
		AGAATGCTTCTGTCTAGTTTT		
		TATGTGAAGATATTTCCTTTT		
		CCACCATAGGCCTCAAAGCGC		
		TCCAAATGTCCACTTGCAGAT		
		TCTACAAAAAGAGTGTTTAAA		
		ACTGCTCAATCAAAAGAAAGG		
	PTF	*TTCAACTCTGTGAGGTGAATGC*	0.80	6/8
		ACATATCATAAAGAAGTTTGTC		
		AGAATGCTTCTGTCTAGTTTT		
		TATGTGAAGATATATCCTTTT		
		CCACCATAGGCCCCAAAGTGC		
		TCCAAATGTCCACTGCAGATT		
		CTATAAAAATAGTGTTTTAAA		
		ACTGCTCAATTAAAAGTAATG		

Table 4. Comparison of consensus tandem repeat unit of proposed scheme and conventional schemes (chromosome 8 - TR1).

Human chromosome		Consensus tandem repeat unit	Average identity	Number of fails
Chr. 8 (TR1)	TR analyzer	CCCACTGAGGCCTATAGTGAAA	0.77	90/16
		AACTGAATATCCCATGATAAAA		
		ACTAGAAAGAAGCTATCTGTGA		
		AACTGCTTTGTGATGTGTGCA		
		TTCAGCTCACAGAGTTAAACC		
		TTTCTTTTGATTCAGCAGGTT		
		GGAAACACTCTTTTTGTAGAA		
		TCTGCAAGGGGATATTTGGAG		
	TRF	CGCTTTGAGGCCTATGGTGGAA	0.78	0/16
		AAGGAAATATCTTCACATAAAA		
		ACTAGACAGAAGCATTCTCAGA		
		AACTTCTTTGTGATGTGTGCA		
		TTCAACTCACAGAGTTGAACC		
		TTCCTTTTGATAGAGCAGTTT		
		TGAAACACTCTTTTTGTAGAA		
		TCTGCAAGTGGATATTTGGAG		
	SRF	CGCATTGAGGCCTATAGTGTAA	0.73	0/16
		AACTGAATATCCAGTGATAAAA		
		ACAAGAGAGAAGCTATCTGTGA		
		ACCTGCTTAGTGATATGTGGAT		
		TCAGCTCACATAGTTAAACCTT		
		ACTTTTGATTCAGCTGTTTGTG		
		GAAACACTCTTTTTGTAAAAT		
		CTGCCAATAGACATTTCAAAG		
	PTF	CCCCCAAAGGCCAAAAGTCAAA	0.64	15/16
		ATCTGAATATCCCGTGAAAAAA		
		ACTATAAAGAAAATATCTGAGA		
		AAATACTTTGTGGTGTAAAGA		
		GTCATCTCAGAGAGTTAAAAC		
		TTTCTTTTGATAAAACAATTT		
		GAAAAAACTTTTTTGTAAAAT		
		CTCTGAAAGGTAATTTTAGAG		

Table 5. Comparison of consensus tandem repeat unit of proposed scheme and conventional schemes (chromosome 8 - TR2).

Human chromosome	Algorithm	Consensus tandem repeat unit	Average identity	Number of fails
Chr. 8 (TR2)	TR analyzer	CCAAGGAGGCCTCTCCCATCCC	0.76	0/12
		AGAAGCCCCCAGGGCTGTCCCG		
		GGCGGGCTGTAAAGCCCCAGGC		
		TTTGGAGCAGGGTGCCTGTGTC		
		TCTCGCAGAAGGCCCCCACAAG		
		CGAAAACGGGGCCGCAGGGTGG		
		CGTGGGAGGGCCGCAGGGACTC		
		AGGGGGACGTTGAGGCAGGCA		
		GAGGGGAGAAGCGGCGAGACT		
		GCAGGGAATGCTGGGAGCCTC		
	TRF	CCAAGGAGGCCTCTCCCATCCCAG	0.71	3/12
		AAGCCCCAGGGCTGTCCCAGGCAG		
		GCTGTAAAGCCCCAGGCTTTGGAG		
		CAGGGTGCCTGTGTCTCTCGCGGA		
		AGGCCCCACAAGCGAAAACGGGGT		
		CGCAGGGTGGCGTGGGCGGGTCAC		
		AGGGACTCAGGGGACATTGAGGCA		
		GGCAGAGGGGAGAAGCAGCAAGA		
		CAGCAGGGAATGCTGGGAGCCTC		
	SRF	CCAGGAGGCCTCTCCCATCCCCGA	0.69	2/12
		AGCCCTCAGGGCTGTCCCGGACTT		
		GGTGTAAAGCCCCAGGCTTTGGAG		
		CAGGGTGACTGTGTCTCTGGCGGA		
		AGGCCCTGACAAGCGAAAACGGGG		
		TAGCAGGGTGGCGTGGGCGGGTCA		
		TGGGGACTCAGCGGGACGTTGAGG		
		AAGGCCGAGGGGAGAAGCAGCAAG		
		AAAGCAGGGAGTGCTGGGAGCCTC		
	PTF	TCAAGGAGGCCTCTCCCATTCCAG	0.65	11/12
		AAGCCCCCAGGGCTGTTCCTGTTT		
		GATTGTAACTCTTCAGGCTTTGGA		
		TTAGGGTACCTGTGTCTCTGGTGG		
		AAGGGCCCCAAAAGCGAGACCCGG		
		GGGCAAGGTGGAAGGTGGCGGGGG		
		CAGGGACCCAGGGGAAAGCTGAGA		
		CAGGCGGAGGGGAGAAGTGGGAAG		
		ACCTCAGGCAATGCTGGGAGCCTT		

and it is the most widely used tool for identification of tandem repeat for its high accuracy. SRF is the representative program of signal processing algorithms to identify tandem repeat. It finds repetitions by converting the target DNA sequence from time domain to frequency domain using Fourier transform. The tandem repeat detection using PT, which is called *PTF* in this paper, is one of the algorithms for detecting tandem repeat based on signal processing. However, it does not use Fourier transforms but uses period transform to find repetitions.

We performed experiments to derive the consensus tandem repeat unit of each tandem repeat of the human genome by using the conventional schemes. Table 2 through Table 5 compare the results of our proposed scheme with those of three conventional schemes. Among the conventional schemes, TRF finds the most exact consensus tandem repeat unit in all the given tandem repeats, whereas PTF shows poor performance to detect consensus tandem repeat units. The *number of fails* in Table 2 through Table 5 means the number of the cases that a consensus tandem repeat is not detected because the given DNA sequence is not determined to be a tandem repeat. Thus, PTF is only usable to find the consensus tandem repeat units of the tandem repeats of human chromosome Y. This is because PTF does not consider the mutations of insertion and deletion of nucleotide bases. Although the performance of TRF is similar to TR Analyzer, TRF is inadequate to find the consensus tandem repeat unit of tandem repeats that are lengthy and highly broken like *TR2* of human chromosome 8 as shown in Table 5. Therefore, TR Analyzer is the most appropriate tool to derive the consensus tandem repeat unit to date, though there are many other tools that can be substituted (Tables 3 and 4).

4   Conclusions and Further Works

We proposed a system model for analyzing HTRF, which derives the consensus tandem repeat units based on the homology of the multiple tandem repeats and shows the structure of HTRF though a simple diagram representation. The proposed system model was performed on four HTRFs of the human genome, which are chromosome 7 (57,937,500 – 58,056,406 bp), chromosome 8 (46,832,500 – 47,458,334 bp), chromosome 22 (16,505,625 – 16,627,187 bp), and chromosome Y (25,000 – 117,031 bp). The algorithm for deriving a consensus tandem repeat unit of a tandem repeat in the proposed system model can be substituted by a conventional scheme that finds tandem repeat. However, in view of deriving an exact consensus tandem repeat unit, the experimental results showed that the proposed algorithm is the most appropriate for deriving a consensus tandem repeat unit to date.

The analysis of HTRF was performed based on the hypothesis that the homologous tandem repeats of an HTRF are originated from a same tandem repeat and HTRFs are very important to biological phenomenon. This hypothesis is sufficiently plausible considering the high identity of the homologous tandem repeats of an HTRF and their highly structured unique patterns. However, since the hypothesis should be verified biologically, we are going to perform the biological experiments of HTRF with the systematic analysis.

References

1. Kazazian, H.H.: Mobile elements: drivers of genome evolution. Sci. **303**, 1626–2632 (2004)
2. Prak, E.T., Kazazian, H.H.: Mobile elements and the human genome. Nat. Rev. Genet. **1**, 134–144 (2000)
3. Sinden, R.R.: Biological implications of the DNA structures associated with disease-causing triplet repeats. Am. J. Hum. Genet. **64**, 346–353 (1999)
4. Christian, M., Dennis, J., John, M.: STRBase: a short tandem repeat DNA database for the human identity testing community. Nucleic Acids Res. **29**, 320–322 (2001)
5. Hauth, A.M., Joseph, D.A.: Beyond tandem repeats: complex pattern structures and distant regions of similarity. Bioinform. **18**, S31–S37 (2002)
6. Chung, B.I., Lee, K.H., Shin, K.S., Kim, W.C., Kwon, D.N., You, R.N., Lee, Y.K., Cho, K., Cho, D.H.: REMiner: a tool for unbiased mining and analysis of repetitive elements and their arrangement structures of large chromosomes. Genomics **98**, 381–389 (2011)
7. Kim, W.C., Lee, K.H., Shin, K.S., You, R.N., Lee, Y.K., Cho, K., Cho, D.H.: REMiner-II: A tool for rapid identification and configuration of repetitive element arrays from large mammalian chromosomes as a single query. Genomics **100**, 131–140 (2012)
8. Benson, G.: Tandem repeats finder: a program to analyze DNA sequences. Nucleic Acids Res. **27**, 573–580 (1999)
9. Sharma, D., Issac, B., Raghava, G.P., Ramaswamy, R.: Spectral repeat finder (SRF): identification of repetitive sequences using fourier transformation. Bioinform. **20**, 1405–1412 (2004)
10. Buchner, M., Janjarasjitt, S.: Detection and visualization of tandem repeats in DNA sequences. IEEE Trans. Signal Process. **51**, 2280–2287 (2003)
11. Brodzik, A.K.: Quaternionic periodicity transform: an algebraic solution to the tandem repeat detection problem. Bioinform. **23**, 694–700 (2007)
12. Zhang, Z., Schwartz, S., Wagner, L., Miller, W.: A greedy algorithm for aligning DNA sequences. J. Comput. Biol. **7**, 203–214 (2000)
13. Wang, L., Jiang, T.: On the complexity of multiple sequence alignment. J. Comput. Biol. **1**, 337–348 (1994)
14. Just, W.: Computational complexity of multiple sequence alignment with SP-score. J. Comput. Biol. **8**, 615–623 (2001)
15. Humberto, C., David, L.: The multiple sequence alignment problem in biology. SIAM J. Appl. Math. **48**, 1073–1082 (1998)
16. Lipman, D.J., Altschul, S.F., Kececioglu, J.D.: A tool for multiple sequence alignment. Proc. Natl. Acad. Sci. U.S.A. **86**, 4412–4415 (1989)
17. Edgar, R.C., Myers, E.W.: PILER: identification and classification of genomic repeats. Bioinform. **21**, i152–i158 (2005)

A Preliminary Analysis on HEp-2 Pattern Classification: Evaluating Strategies Based on Support Vector Machines and Subclass Discriminant Analysis

Ihtesham Ul Islam, Santa Di Cataldo, Andrea Bottino$^{(\boxtimes)}$, Enrico Macii,
and Elisa Ficarra

Department of Control and Computer Engineering, Politecnico di Torino,
Corso Duca degli Abruzzi 24, 10129 Torino, Italy
{ihtesham.ulislam,santa.dicataldo,andrea.bottino,
enrico.macii,elisa.ficarra}@polito.it

Abstract. The categorization of different staining patterns in HEp-2 cell slides by means of indirect immunofluorescence (IIF) is important for the differential diagnosis of autoimmune diseases. The clinical practice usually relies on the visual evaluation of the slides, which is time-consuming and subject to the specialist's experience. Thus, there is a growing demand for computer-aided solutions capable of automatically classifying HEp-2 staining patterns. In the attempt to identify the most suited strategy for this task, in this work we compare two approaches based on Support Vector Machines and Subclass Discriminant Analysis. These techniques classify the available samples, characterized through a limited set of optimal textural attributes that are identified with a feature selection scheme. Our experimental results show that both strategies have a good concordance with the diagnosis of the human specialist and show the better performance of the Subclass Discriminant Analysis (91 % accuracy) compared to Support Vector Machines (87 % accuracy).

Keywords: Indirect immunofluorescence imaging · HEp-2 staining pattern classification · Support vector machines · Subclass discriminant analysis · Pattern recognition

1 Introduction

Indirect immunofluorescence (IIF) is an imaging modality detecting abundance of those molecules that induce an immune response in the sample tissue. This technique uses the specificity of antibodies to their antigen in order to bind fluorescent dyes to specific biomolecule targets within a cell. The screening for anti-nuclear antibodies by IIF is a standard method in the current diagnostic approach to a number of important autoimmune pathologies such as systemic rheumatic diseases as well as multiple sclerosis and diabetes [1]. This screening,

© Springer-Verlag Berlin Heidelberg 2014
M. Fernández-Chimeno et al. (Eds.): BIOSTEC 2013, CCIS 452, pp. 176–190, 2014.
DOI: 10.1007/978-3-662-44485-6_13

which makes use of a fluorescence microscope, is typically done by visual inspection on cultured cells of the HEp-2 cell line: the specialist observes the IIF slide at the microscope (see Fig. 1 for an example), and makes a diagnosis based on the perceived intensity of the fluorescence signal and on the type of the staining pattern. Fluorescence intensity evaluation is needed for classifying between positive, intermediate and negative (*i.e.* absence of fluorescence) samples. Then, specific staining patterns on positive and intermediate samples reveal the presence of different antibodies and, thus, different types of autoimmune diseases. Therefore, a correct description of staining patterns is fundamental for the differential diagnosis of the pathologies. Examples of the six main staining patterns described by literature (homogeneous, fine speckled, coarse speckled, nucleolar, cytoplasmic or centromere) are reported in Fig. 2. They are distinguished as follows:

- **Homogeneous:** diffuse staining of the entire nucleus, with or without apparent masking of the nucleoli.
- **Nucleolar:** fluorescent staining of the nucleoli within the nucleus, sharply separated from the unstained nucleoplasm.
- **Coarse/Fine Speckled:** fluorescent aggregates throughout the nucleus which can be very fine to very coarse depending on the type of antibody present.
- **Centromere:** discrete uniform speckles throughout the nucleus, the number corresponds to a multiple of the normal chromosome number.
- **Cytoplasmic Fluorescence:** granular or fibrous fluorescence in the cytoplasm.

The manual classification of HEp-2 staining pattern suffers from usual problems in medical imaging, that is (i) the reliability of the results is subject to the specialist's experience and expertise, and (ii) the analysis of large volume of images is a tedious and time-consuming operation, translating into higher costs for the health system. Studies report very high inter- and intra-laboratory variability for this type of screening (up-to 10 %), that can be even higher in case of non-specialized structures [1]. This variability impacts on the reliability of the obtained results and, most of all, on their reproducibility.

Thus, in the last few years, reliable automatic systems for automating the whole IIF process have been in great demand and several tools have been proposed [2–6]. Nevertheless, the accurate classification of the staining patterns still remains a challenge.

Several classification schemes have been applied: among the others, learning vector quantization [3], decision tree induction algorithms [4,5], and multi-expert systems [6]. Unfortunately, direct comparison of the results presented by different works is not possible, since they were obtained on different datasets and on different classes. However, it is worth noting that textural features are generally acknowledged for being the most appropriate for staining pattern classification.

In this work, we compare two techniques we implemented that classify the cells into one of the six staining patterns addressed by literature. The first is based on Support Vector Machines (SVM). This approach was already introduced in our previous work [7], and it is described again here for the sake of completeness. The second technique is a novel procedure based on Subclass Discriminant Analysis (SDA), a recent dimensionality reduction method that has been proven successful in different problems. SDA aims at improving the classification of a large number of different data distributions, whether they are composed by compact sets or not, by describing the underlying distribution of each class using a mixture of Gaussians. Since some of the staining patterns are characterized by a relevant within-class variance, SDA appears to be a promising method to improve their classification accuracy.

In our approach, each cell is initially characterized with a set of features based on statistical measurements of the grey-level distributions and on frequency-domain transformations. The dimension of this feature vector is then reduced applying different procedures, aiming at selecting the subsets of feature variables that are best suited to the classification with both SVM and SDA.

After a description of the dataset employed for training and testing our methods, Sect. 2, and a description of the two classification techniques, Sect. 3, this work presents in Sect. 4 the results of our experiments, aimed at identifying the best IIF classification technique. Discussion and conclusions are presented in Sects. 4.3 and 5, respectively.

2   Materials

The dataset used in our experiments contains IIF images that are publicly available at [8]. It is composed of 14 annotated IIF images acquired using slides of HEp-2 substrate at the fixed dilution of 1:80, as recommended by the guidelines in [9]. The images were acquired with a resolution of 1388×1038 pixels and a color depth of 24 bits. The acquisition unit consisted of a fluorescence microscope (40-fold magnification) coupled with a 50 W mercury vapour lamp and with a

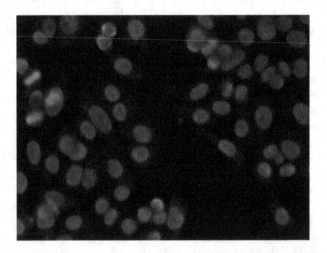

Fig. 1. HEp-2 IIF image.

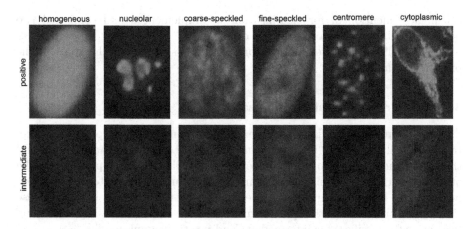

Fig. 2. Examples of staining patterns that are considered relevant to diagnostic purposes, either with intermediate and positive fluorescence intensity.

Table 1. HEp-2 cell dataset characterization.

Pattern	N. of samples	Int.	Pos.
Homogeneous	150	47	103
Nucleolar	102	46	56
Coarse speckled	109	41	68
Fine speckled	94	48	46
Centromere	208	119	89
Cytoplasmic	58	24	34
Tot.	721	325	396

digital camera (SLIM system by Das srl) having a CCD with square pixel of 6.45 μm side. An example of the available images can be seen in Fig. 1.

From these images, a set of samples of HeP-2 cells have been extracted. Specialists manually segmented each cell at a workstation monitor, labelling it with the corresponding fluorescence intensity level (either intermediate or positive) and staining pattern. The latter can be distinguished in the six classes described in the introduction.

The dataset contains a total of 721 cells, 325 of which with intermediate and 396 with positive fluorescence intensity. A full characterization of dataset is reported in Table 1.

3   Methods

Our approach combines texture analysis and feature selection techniques in order to obtain a limited set of image features that is optimal for the classification task.

As already mentioned, for classification we implemented and compared two different methods, based on SVM and on SDA.

In the following subsections we provide details about all the steps of the proposed techniques.

3.1 Size and Contrast Normalization

Size and intensity normalisation of the samples is a necessary preprocessing step. In fact, small differences in the dimensions of the cell images are normal, and these differences are completely independent from their staining pattern. On the other hand, there are considerable variations of fluorescence intensity between intermediate and positive samples. Reducing such variability helps to decrease the noise in the classification process and avoids as well the necessity of training two separate classifiers for intermediate and positive samples.

Size normalization was obtained by re-sampling all the cell images to 64×64 pixels dimension. Contrast normalization consisted in linearly remapping the intensity values so that 1 % of data is saturated at low and high intensities.

3.2 Feature Extraction

Textural analysis techniques have already been proven successful in HEp-2 cell staining characterization [7]. In fact, they are able to describe the most relevant image variations occurring in the cell allowing to differentiate between the staining patterns.

The two major approaches for textural analysis are either based on statistical methods describing the distribution of grey-levels in the image or on frequency-domain measurements of image variations. In our work we propose a combination of both of them in order to extract a comprehensive set of features able to fully characterize the staining pattern of the cell.

A first set of features was computed based on Gray-Level Co-occurrence Matrices (GLCM), a well established technique that extracts textural information from the spatial relationship between intensity values at specified offsets in the image. More specifically, textural features are computed from a set of grey-tone spacial dependence matrices reporting the distribution of co-occurring values between neighbouring pixels according to different angles and distances [10]. In practice, the $(i, j)_{d,\theta}$ element of a GLCM contains the probability for a pair of pixels located at a neighbourhood distance d and direction θ to have gray levels i and j, respectively.

In our work, we extracted 44 GLCM textural features reported in Table 2, based on four 16×16 GLCMs computed for a fixed unitarian neighbourhood distance and a varying angle $\theta = 0^o, 45^o, 90^o, 135^o$ (see [7] for details). The features are based on well-established statistical measurements whose characterization can be found in [10–12]. The use of 4 different directions is aimed at making the method less sensitive to rotations in the images.

Table 2. GLCM features.

N.	Feature	Ref.
1	Uniformity	[10]
2	Entropy	[11]
3	Dissimilarity	[11]
4	Contrast	[10]
5	Inverse difference	[12]
6	Correlation	[10]
7	Homogeneity	[11]
8	Autocorrelation	[11]
9	Cluster Shade	[11]
10	Cluster Prominence	[11]
11	Maximum probability	[11]
12	Sum of Squares	[10]
13	Sum Average	[10]
14	Sum Variance	[10]
15	Sum Entropy	[10]
16	Difference variance	[10]
17	Difference entropy	[10]
18	Information measures of correlation (1)	[10]
19	Information measures of correlation (2)	[10]
20	Maximal correlation coefficient	[10]
21	Inverse difference normalized (INN)	[12]
22	Inverse difference moment normalized (IDN)	[12]

Besides statistical methods, a largely used approach to extract relevant textural information for image compression and classification is based on frequency-domain transformations [13]. The underlying concept is the transformation of the image spatial information into a different space whose coordinate system has an interpretation that is closely related to the description of image texture.

In our work, we computed the two-dimensional Discrete Cosine Transform (DCT) [14] of the normalized images and then extracted 328 DCT coefficients (described in details in our previous work [7]) representing different patterns of image variation and directional information of the texture. The same approach was already successfully applied for texture classification and pattern recognition [13].

Combining GLCM and DCT sets, we obtained a total number of $44 + 328 = 372$ features to characterize each sample image.

3.3   Classification Based on Support Vector Machines

The first classification method we implemented was already introduced in our previous work [7]. It is based on Support Vector Machines (SVM), a well-established machine learning technique that has been proven successful for classification and regression purposes in many applications [15].

The classification is based on the implicit mapping of data to a high-dimensional space via a kernel function, and on the identification of the maximum-margin hyperplane that separates the given training instances in this space.

In our work we used SVM with radial basis kernel, optimizing the kernel parameters by means of ten-fold cross-validation technique and a grid search, as suggested in [15].

Feature Selection. Feature selection (FS) strategies were applied in order to select a limited set of optimal features able to improve the accuracy of the staining pattern classifier.

SVM are widely acknowledged for their built-in feature selection capability, as they implicitly map data in a transformed domain where the features that are crucial to the classification purpose are emphasized [16]. Nevertheless, the combination of SVM with feature selection strategies, besides improving training efficiency, can further enhance the accuracy of classification. In fact, although the presence of irrelevant features does not change the hyperplane margin of SVM, it may increase the radius of the training data points, impacting on SVM's generalization capability and also increasing the probability of over-fitting [17].

In our work we applied feature selection in two sequential steps. The first is based on minimum-Redundancy-Maximum-Relevance (mRMR) algorithm, whose better performance over the conventional top-ranking methods has been widely demonstrated in the literature [18]. The mRMR algorithm sorts the features that are most relevant for characterizing the classification variable, pointing at the contemporaneous minimization of their mutual similarity and maximization of their correlation with the classification variable. The number of the candidate features selected by mRMR was arbitrarily set to 50.

As for mRMR to work at its best the classification variables have to be categorical and not continuous, we preventively performed features discretization on the input data. For this purpose, we applied CAIM (class-attribute interdependence maximization) algorithm [19], which is best suited to work with supervised data, as it generates a minimal number of discrete intervals by maximizing the class-attribute interdependence.

The output of mRMR is a generic candidate feature set, which is independent from the classification algorithm [18] and not necessarily optimal for SVM. Therefore, we applied as second FS step a Sequential Forward Selection (SFS) scheme in order to iteratively construct the subset of optimal features that is best suited for SVM classification.

Classical SFS [20] works towards the minimization of the misclassification error: starting from an initial empty set, at each iteration the feature providing the greatest classification accuracy improvement is added, until no more

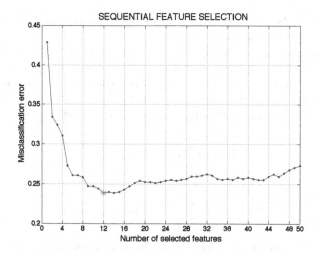

Fig. 3. Sequential Feature Selection strategy: misclassification error vs. number of selected features at each iteration. The optimal feature set size is 12.

improvement is obtained. As this implementation tends to be trapped in local minima, in our approach we proceeded with the iterations until all the available features were added, and then we selected the feature set with the best classification accuracy. The final dimension of this optimal set was found to be 12 (see Fig. 3).

3.4   Classification Based on Subclass Discriminant Analysis

SDA belongs to family of Discriminant analysis (DA) algorithms, which have been used for dimensionality reduction and feature extraction in many applications of computer vision and pattern recognition [21–24]. These algorithms map a set of samples $X = (x_1, x_2, \ldots, x_n)$, associated to a class label $\in [1, C]$ and belonging to a high-dimensional feature space $\in \Re^D$, to a low-dimensional subspace $\in \Re^d$, with $d \ll D$, where the data can be more easily separated according to their class-labels. Therefore, DA problem can be generally stated as finding the transformation matrix $V = (v_1, v_2, \ldots, v_d)$, with $v_i \in \Re^D$, for mapping a sample x into the final d-dimensional subspace.

In most DA algorithms, the transformation matrix V is found by maximizing the so-called Fisher-Rao's criterion:

$$J(V) = \frac{|V^T A V|}{|V^T B V|} \tag{1}$$

where A and B are symmetric and positive-defined matrices. The solution to this problem is given by the generalized eigenvalue decomposition:

$$AV = BV\Lambda \tag{2}$$

Where V is (as above) the desired transformation matrix, and Λ is a diagonal matrix of its corresponding eigenvalues.

Linear Discriminant Analysis (LDA) is probably the most well-known DA technique. This method assumes that the C classes the data belong to are homoscedastic, that is their underlying distributions are Gaussian with common variance and different means. In (1), LDA uses $A = S_B$, the between-class matrix, and $B = S_W$, the within-class scatter matrix, defined as:

$$S_B = \sum_{i=1}^{C} (\mu_i - \mu)(\mu_i - \mu)^T \tag{3}$$

$$S_W = \frac{1}{n} \sum_{i=1}^{C} \sum_{j=1}^{n_i} (x_{ij} - \mu_i)(x_{ij} - \mu_i)^T \tag{4}$$

where C is the number of classes, μ_i is the sample mean for class i, μ is the global mean, x_{ij} is the jth sample of class i and n_i the number of samples in class i.

LDA provides the $(C\text{-}1)$-dimensional subspace that maximizes the between-class variance and minimizes the within-class variance in any particular data set. In other words, it guarantees maximal class separability and, possibly, optimizes the accuracy in later classification.

However, the assumption of having C homoscedastic classes is the very limitation of this method. LDA works well for linear problems and fails to provide optimal subspaces for inherently non-linear structures in data. Several extensions of LDA have been introduced in literature to effectively classify data with non-linearities [25].

To this end, one of the most effective approaches is the Subclass Discriminant Analysis (SDA), proposed in [26]. The main idea of SDA it is to find a way to describe a large number of different data distributions, whether they are composed by compact sets or not, by describing the underlying distribution of each class using a mixture of Gaussians. This is achieved by dividing the classes into subclasses. Therefore, the problem to be solved is to find the optimal number of subclasses maximizing the classification accuracy in the reduced space. In SDA, the transformation matrix V is found by defining the between-subclass scatter matrix S_B in Eq. (1) as:

$$S_B = \sum_{i=1}^{C-1} \sum_{j=1}^{H_i} \sum_{k=i+1}^{C} \sum_{l=1}^{H_k} p_{ij}p_{kl}(\mu_{ij} - \mu_{kl})(\mu_{ij} - \mu_{kl})^T \tag{5}$$

where H_i is the number of subclasses of class i, μ_{ij} and p_{ij} are the mean and prior probability of the jth subclass of class i, respectively. The priors are estimated as $p_{ij} = n_{ij}/n$, where n_{ij} is the number of samples in the jth subclass of class i. In the simplest case of SDA with no class subdivisions, this equation reduces to that of LDA.

In order to select the optimal number of subclasses, in [26], the authors propose two different methods. The first is based on a stability criterion described in [27]. However, as pointed out in [28], when data have a Gaussian homoscedastic subclass structure, the minimization of the metric used in this criterion is not guaranteed. Authors in [28] hypothesize that this is likely to happen also for heteroscedastic classes.

The second selection criterion is based on a Leave-one-object test. In practice, for each subdivision, a leave-one-out cross validation (LOOCV) is applied, and the optimal subdivision is the one giving the maximal recognition rate. The problem with this strategy is that it has very high computational costs, especially when the dataset to classify is large and the number of classes is high. This is exactly what is happening in our case, where the initial classes are 6 and the samples are 721.

Therefore, to overcome these problems, we used a different formulation of the optimality criterion, similar to the leave-one-object test, but based on a stratified 5-fold cross validation, which optimizes the accuracies obtained with a k-Nearest Neighbour (kNN) classifier. A value of 8 for k has been heuristically found to provide good classification results.

Our implementation differs from the original SDA formulation for two other details. The first concerns the clustering methods used to divide classes into subclasses. In [26] data are assigned to subclasses by first sorting the class samples with a Nearest-Neighbour based algorithm and then dividing the obtained list into a set of clusters of the same size. However, this method does not allow to model efficiently the non-linearity present in the data, as in the case of staining patterns under analysis. Therefore, we used the K-means algorithm, which partitions the samples into k clusters by minimizing iteratively the sum, over all clusters, of the within-cluster sums of sample-to-cluster-centroid distances. Since, in this method, the centroids are initially set at random, different initialization results in different divisions. Hence, we repeated the clustering 20 times and kept the solution providing the minimal sum of all within-cluster distances.

The second difference is that, instead of increasing the number of subclasses for each class of the same amount at each iteration, all the possible permutation of class subdivisions are created by iteratively incrementing by one the number of subclasses of a single class in a set of nested loops. For a specific class r, the subdivision process is stopped when the minimal number of samples in the H_r clusters obtained with K-means drops below a predefined threshold. In order to reduce the computational times, the clusters created in inner loops are computed only once and cached for further use.

The classification accuracy of our method is computed as the average accuracy of the different CV rounds. It should be underlined that, given the differences between training and test sets, different optimal subclasses subdivision are likely to be obtained at each CV iteration.

Feature Selection. As well as for the SVM classifier, we applied FS strategies to SDA too. In this case, we used only the reduced feature set obtained with mRMR. This has been done for two reasons.

First, while mRMR is independent from the classification method, SFS relies on the classifier output, which makes it unfeasible with the computational cost of SDA.

Second, it can be easily shown that the rank of matrix S_B, and therefore of the dimensionality d of the reduced subspace obtained from Eq. (2), is given by $min(H-1, rank(S_X))$, where H is the total number of subclasses and $rank(S_X)$ is equal (or minor) to the number of features characterizing each sample. While the number of features selected with mRMR (50) is a reasonable upper bound for d, reducing it further might hamper the possibility to obtain a good classification accuracy in problems, like the one tackled in this paper, in which the data present high non-linearities.

4   Results

The two classification methods presented in Sect. 3 were tested on the same annotated IIF images, using the staining pattern information provided by the specialists as ground truth for cross-validation.

4.1   SVM Classification

We recall here the experimental results on SVM classification, already reported in our previous paper [7], for comparison with SDA approach.

As for SVM classification, experiments were run on the following datasets:

dataset I, the initial 372 elements feature set;
dataset II, the 50 elements candidate set selected by mRMR feature selection;
dataset III, the final 12 elements feature vector obtained with combination of
   mRMR + SFS.

The accuracy results of the 10-fold cross-validation, grouped by staining pattern, are reported in Table 3. The last row of the table shows the overall accuracy obtained by SVM in each dataset.

The following initial remarks can be drawn from the analysis of the results of Table 3:

- SVM classifier obtained an average accuracy of 86.96 % in the six staining patterns. The maximum and minimum per-class accuracy were 98.17 % (coarse speckled pattern) and 71.28 % (fine speckled pattern).
- FS significantly improved the classification performances (+9.01 % on the overall average accuracy). This confirms the considerations drawn in Sect. 3.3 about the weakness of the implicit feature selection ability claimed by SVM. As it can be seen, mRMR improved the per-class accuracy of all the staining patterns (see results on dataset II compared to those on dataset I). The combination of mRMR+SFS (dataset III) further improved the average accuracy of SVM. While per-class accuracies of centromere and cytoplasmic patterns were slightly decreased (of, respectively, −1.44 % and −3.45 % w.r.t. dataset II), the fine speckle pattern, that had lowest per-class accuracy, is the class that obtained the best improvement (+9.58 % w.r.t. dataset II and +25.53 % w.r.t. dataset I). This non-uniform behaviour is not surprising, since SFS optimized the average classification accuracy in the overall dataset and not the accuracies of the single classes.

Table 3. SVM Classification results: accuracy rate (%).

Pattern	Dataset		
	I	II	III
Homogeneous	78.66	84.00	86.00
Nucleolar	89.22	93.14	93.14
Coarse speckled	92.66	95.41	98.17
Fine speckled	45.75	61.70	71.28
Centromere	84.13	88.46	87.02
Cytoplasmic	58.62	86.21	82.76
Overall	77.95	85.58	86.96

4.2 SDA Classification

Table 4, which is, again, organized by staining class, summarizes the classification results obtained with SDA. As already explained in Sect. 3.4, SFS strategy was not applied in combination with SDA classifier. Therefore, the table contains only results on dataset I (the initial 372 feature set) and dataset II (50 feature set obtained after mRMR).

LDA results (which are those obtained with SDA with no class subdivisions) are also provided for comparison, in order to demonstrate the effective capabilities of SDA to better classify datasets with high non-linearities. Finally, in the last row we show the overall accuracies obtained in the four cases.

Analysing the results, some considerations can be drawn:

- as expected, the overall accuracy of SDA outperforms that of LDA (+7.29 % on dataset I and +7.13 % on dataset II). Concerning the per-class results, better results are obtained in most of the cases (except for centromere class for dataset I, and coarse speckled class for dataset II), with, as best improvements, a +17.34 % in dataset I for homogeneous class and + 23.51 % in dataset II for fine speckled class;
- as in the SVM experiments, FS effectively improves the SDA accuracies of all classes (the best improvement being the +9.71 % of fine speckled). The overall improvement is +4.75 %
- the best average accuracy obtained is 90.79 %, with dataset II, which outperforms the best accuracy obtained by SVM with mRMR+SFS feature selection (86.96 %, dataset III). The best per-class improvements have been obtained for fine speckled (+18.14 %) and cytoplasmic class (+17.24 %), while coarse speckled and centromere class obtained slightly lower accuracies (respectively, −6.48 % and −4.85 %).

4.3 Discussion

The results presented in Tables 3 and 4 suggest that the proposed algorithm is a good solution for the automated classification of immunofluorescence cell patterns.

Table 4. SDA Classification results: accuracy rate (%).

Pattern	Dataset			
	I		II	
	LDA	**SDA**	LDA	**SDA**
Homogeneous	63.33	80.67	80.00	85.33
Nucleolar	90.29	94.29	91.19	96.14
Coarse speckled	87.19	88.05	92.55	91.69
Fine speckled	62.87	79.71	65.91	89.42
Centromere	75.96	73.53	80.78	82.17
Cytoplasmic	92.88	100.00	91.52	100.00
Overall	78.75	86.04	83.66	90.79

As a matter of facts, the accuracy rate is comparable to the one obtained by the specialists, whose inter-laboratory variability is generally assessed around 10 % or even higher [1]. Besides that, differently from human operators, our technique provides fully-repeatable results that are based on objective and quantitative features of the images.

As for the classification techniques, the same results show that SDA technique, in combination with a proper selection of the most relevant features, outperforms the best accuracy achievable with SVM on the same dataset (II) and even with those obtained by SVM on dataset III, specifically optimized for that technique with a two-step FS process. Therefore, our experiments show the capabilities of SDA to describe in a more suitable way the underlying distributions of each of the staining pattern class, improving their classification accuracies.

5   Conclusions

In this paper we proposed the comparison of two approaches, based on SVM and SDA, for the automatic classification of staining patterns in HEp-2 cell IIF images. Texture descriptors based on GLCM and DCT coefficients are first exploited to extract a 372-size characteristic vector for each cell. Then, a feature selection algorithm is applied to obtain a reduced candidate feature set that improves the classification accuracies of the two methods.

Feature selection is based on the mRMR algorithm, which sorts the features that are most relevant for characterizing the classification variable. The 50 top-ranked features were selected. In the case of SVM-based method, a two-steps feature selection procedure, coupling mRMR with SFS algorithm, is implemented in order to further improve classification accuracies.

The two approaches provide average classification accuracies of about 87 % and 91 %, respectively. These results are comparable with those of human specialists. Conversely, they are completely repeatable since our automated technique does not depend on the subjectivity of the operator. Moreover, our experiments show the effectiveness of SDA into describing more precisely, compared to SVM, the underlying distributions of each of the staining pattern class.

As future steps, we plan to work on:

(1) a better characterization of cell patterns, which can be insensitive to changes in size, rotation and intensity;
(2) an improvement of the SDA classifier in terms of computational efficiency. For this purpose, methods selecting a priori the classes that effectively needs to be partitioned, like the one described in [29], will be investigated;

Moreover, we plan to develop a pipeline for automatic cells segmentation in IIF images and to combine it with our pattern classification algorithm in order to obtain a complete automated approach for the computer-aided diagnosis (CAD) of autoimmune diseases.

References

1. Egerer, K., Roggenbuck, D., Hiemann, R., Weyer, M.G., Buttner, T., Radau, B., Krause, R., Lehmann, B., Feist, E., Burmester, G.R.: Automated evaluation of autoantibodies on human epithelial-2 cells as an approach to standardize cell-based immunofluorescence tests. Arthritis Res. Ther. **12**(2), 1–9 (2010)
2. Creemers, C., Guerti, K., Geerts, S., Van Cotthem, K., Ledda, A., Spruyt, V.: HEp-2 cell pattern segmentation for the support of autoimmune disease diagnosis. In: Proceedings of ISABEL 2011, vol. 28, pp. 1–5 (2011)
3. Hsieh, R.Y., Huang, Y.C., Chung, C.W., Huang, Y.L.: HEp-2 cell classification in indirect immunofuorescence images. In: Proceedings of ICICS 2009, vol. 26, pp. 211–214 (2009)
4. Perner, P., Perner, H., Muller, B.: Mining knowledge for HEp-2 cell image classification. Artif. Intell. Med. **26**, 161–173 (2002)
5. Sack, U., Knoechner, S., Warschkau, H., Pigla, U., Emmrich, F., Kamprad, M.: Computer-assisted classification of HEp-2 immunofluorescence patterns in autoimmune diagnostics. Autoimmun. Rev. **2**(5), 298–304 (2003)
6. Soda, P., Iannello, G.: A hybrid multi-expert systems for HEp-2 staining pattern classification. In: Proceedings of ICIAP 2007, pp. 685–690 (2007)
7. Di Cataldo, S., Bottino, A., Ficarra, E., Macii, E.: Applying textural features to the classification of HEp-2 cell patterns in IIF images. In: 21st International Conference on Pattern Recognition (ICPR 2012), Tsukuba, Japan, 11–15 November (2012)
8. MIVIA Lab. http://nerone.diiie.unisa.it/zope/mivia/databases/db_database/biomedical/. Accessed: September 2012
9. Tozzoli, R., Bizzaro, N., Tonutti, E., Villalta, D., Bassetti, D., Manoni, F., Piazza, A., Pradella, M., Rizzotti, P.: Guidelines for the laboratory use of autoantibody tests in the diagnosis and monitoring of autoimmune rheumatic diseases. Am. J. Clin. Pathol. **117**(2), 316–324 (2002)
10. Haralick, R.M., Shanmugam, K., Dinstein, I.: Textural features for image classification. IEEE Trans. Syst. Man Cybern. **3**(6), 610–621 (1973)
11. Soh, L., Tsatsoulis, C.: Texture analysis of SAR sea ice imagery using gray level co-occurrence matrices. IEEE Trans. Geosci. Remote Sens. **37**(2), 780–795 (1999)

12. Clausi, D.A.: An analysis of co-occurrence texture statistics as a function of grey level quantization. Can. J. Remote Sens. **28**(1), 45–62 (2002)
13. Sorwar, G., Abraham, A., Dooley, L.S.: Texture classification based on DCT and soft computing. In: Proceedings of FUZZ-IEEE 01, 2–5 December 2001
14. Ahmed, N., Natarajan, T., Rao, K.R.: Discrete cosine transform. IEEE Trans. Comput. **23**, 90–93 (1974)
15. Chang, C.-C., Lin, C.-J.: Libsvm: a library for support vector machines. ACM Trans. Intell. Syst. Technol. **2**(3), 27:1–27:27 (2011)
16. Temko, A., Camprubi, C.N.: Classification of acoustic events using SVM-based clustering schemes. Pattern Recognit. **39**(4), 682–694 (2006)
17. Weston, J., Mukherjee, S., Chapelle, O., Pontil, M., Poggio, T., Vapnik, V.: Feature selection for SVMs. Adv. Neural Inf. Process. Syst. **13**, 668–674 (2000)
18. Peng, H., Long, F., Ding, C.: Feature selection based on mutual information: criteria of max-dependency, max-relevance, and min-redundancy. IEEE Trans. PAMI **27**, 1226–1238 (2005)
19. Kurgan, L.A., Cios, K.J.: Caim discretization algorithm. IEEE Trans. Knowl. Data Eng. **16**(2), 145–153 (2004)
20. Ververidis, D., Kotropoulos, C.: Fast and accurate feature subset selection applied into speech emotion recognition. Els. Sig. Process. **88**(12), 2956–2970 (2008)
21. Fukunaga, K.: Introduction to Statistical Pattern Recognition, 2nd edn. Academic Press, San Diego (1990)
22. Swets, D.L., Weng, J.J.: Using discriminant eigenfeatures for image retrieval. IEEE Trans. Pattern Anal. Mach. Intell. **18**(8), 831–836 (1996)
23. Etemad, K., Chellapa, R.: Discriminant analysis for recognition of human face images. J. Optical Soc. Am. A **14**(8), 1724–1733 (1997)
24. Belhumeur, P.N., Hespanha, J.P., Kriegman, D.J.: Eigenfaces vs. Fisherfaces: recognition using class specific linear projection. IEEE Trans. Pattern Anal. Mach. Intell. **19**(7), 711–720 (1997)
25. Boulgouris, N.V., Plataniotis, K.N., Micheli-Tzanakou, E.: Discriminant analysis for dimensionality reduction: an overview of recent developments. In: Biometrics: Theory, Methods, and Applications. Wiley, New York (2009)
26. Zhu, M., Martinez, A.M.: Subclass discriminant analysis. IEEE Trans. PAMI **28**(8), 1274–1286 (2006)
27. Martinez, A.M., Zhu, M.: Where are linear feature extraction methods applicable? IEEE Trans. Pattern Anal. Mach. Intell. **27**(12), 1934–1944 (2005)
28. Gkalelis, N., Mezaris, V., Kompatsiaris, I.: Mixture subclass discriminant analysis. IEEE Sig. Process. Lett. **18**(5), 319–322 (2011)
29. Kim, S.-W.: A pre-clustering technique for optimizing subclass discriminant analysis. Pattern Recogn. Lett. **31**(6), 462–468 (2010)

Predicting Protein Localization Using a Domain Adaptation Approach

Nic Herndon$^{(\boxtimes)}$ and Doina Caragea

Kansas State University, Manhattan, KS 66506, USA
{nherndon,dcaragea}@ksu.edu

Abstract. A challenge arising from the ever-increasing volume of biological data generated by next generation sequencing technologies is the
annotation of this data, e.g. identification of gene structure from the location of splice sites, or prediction of protein function/localization. The
annotation can be achieved by using automated classification algorithms.
Supervised classification requires large amounts of labeled data for the
problem at hand. For many problems, labeled data is not available. However, labeled data might be available for a similar, related problem. To
leverage the labeled data available for the related problem, we propose an
algorithm that builds a naïve Bayes classifier for biological sequences in a
domain adaptation setting. Specifically, it uses the existing large corpus
of labeled data from a source organism, in conjunction with any available labeled data and lots of unlabeled data from a target organism, thus
alleviating the need to manually label a large number of sequences for a
supervised classifier. When tested on the task of predicting protein localization from the composition of the protein, this algorithm performed
better than the multinomial naïve Bayes classifier. However, on a more
difficult task, of splice site prediction, the results were not satisfactory.

Keywords: Naïve Bayes · Domain adaptation · Supervised learning ·
Semi-supervised learning · Self-training · Biological sequences · Protein
localization

1 Introduction

The widespread use of next generation sequencing (NGS) technologies in the
recent years has resulted in an increase in the volume of biological data generated,
including both DNA sequences and also derived protein sequences. A challenge
arising from the increased volume of data consists of organizing, analyzing, and
interpreting this data, in order to create or improve genome assemblies or genome
annotation, or to predict protein function, structure and localization, among
others. Some of these problems can be framed as biological sequence classification
problems, i.e., assigning one of several labels to a DNA or protein sequence based
on its content (e.g., predicting the presence or absence of an acceptor or donor
splice site in DNA sequences centered around GT or AG dimers; or determining

© Springer-Verlag Berlin Heidelberg 2014
M. Fernández-Chimeno et al. (Eds.): BIOSTEC 2013, CCIS 452, pp. 191–206, 2014.
DOI: 10.1007/978-3-662-44485-6_14

where a protein is localized, such as in cytoplasm, inner membrane, periplasm, outer membrane, or extracellular space, a.k.a., protein localization).

Using machine learning or statistical inference methods allows labeling of biological data several orders of magnitude faster than it can be done manually, and with high accuracy. For example, hidden Markov models are currently used in gene prediction algorithms, and support vector machines have shown promising results with handwritten digit classification [29], optical character recognition [16,23] and translation initiation sites classification based on proximity to start codon within sequence window [16] or based on positional nucleotide incidences [32], classification into malign or benign of gene expression profiles [18], *ab initio* gene prediction [2], classification of DNA sequences into sequences with splice site at a determined location or not [1,5,10,11,13,20,25,26,28,31], and classifying the function of genes based on gene expression data [3].

However, using a supervised classifier trained on a source domain to predict data on a different target domain usually results in reduced classification accuracy. Instead of using the supervised classifier, an algorithm developed in the domain adaptation framework can be employed to transfer knowledge from the source domain to the target domain. Such an algorithm has to take into consideration the fact that some, if not all, of the features have different probabilities in the target and source domains [12]. In other words, some of the features that are correlated to a label in the source domain might not be correlated to the same or any label in the target domain, while, some of the features have the same label correlations between the source and target domains. The former ones are known as domain specific features and the latter ones are generalizable features [12].

Domain adaptation algorithms are particularly applicable to many biological problems for which there is a large corpus of labeled data for some well studied organisms and much less labeled data for an organism of interest. Thus, when studying a new organism, it would be preferable if the knowledge from other, more extensively studied organism(s), could be applied to a lesser studied organism. This would alleviate the need to manually generate enough labeled data to use a machine learning algorithm to make predictions on the biological sequences from the target domain. Instead, we could filter out the domain specific features from the source domain and use only the generalizable features between the source and target domains, together with the target specific features, to classify the data.

Towards this goal, we modified the Adapted Naïve Bayes (ANB) algorithm to make it suitable for the biological data. We chose this algorithm because Naïve Bayes based algorithms are faster and require no tuning. In addition, this algorithm was successfully used by Tan *et al.* [27] on text classification for sentiment analysis, discussed in Sect. 2. It combines a weighted version of the multinomial Naïve Bayes classifier with the Expectation-Maximization algorithm. In the maximization step, the class probabilities and the conditional feature probabilities given the class are calculated using a weighted combination between the labeled data from the source domain and the unlabeled data from the target domain. In the expectation step, the conditional class probabilities given the instance are calculated with the probability values from the maximization step using Bayes theorem. The two steps are repeated until the probabilities in the expectation step converge. With each iteration, the weight is shifted from the source data to the target data. The key modifications we made to this algorithm are the use of labeled data from the target domain, and the incorporation of self-training [14,22,30] to make it feasible for biological data, as presented in more detail in Sect. 3.

We tested the ANB classifier on three biological datasets, as described in the Sect. 3.4, for classifying localization of proteins, and predicting splice site locations. The experimental results, Sect. 3.6, show that this classifier achieves better classification accuracy than a Naïve Bayes classifier trained on the source domain and tested on the target domain, especially when the two domains are less related.

2 Related Work

Up to now, most of the work in domain adaptation has been on non-biological problems. For instance, text classification has received a lot of attention in the domain classification framework. One example, the Naïve Bayes Transfer Classification algorithm [4], assumes that the source and target data have different distributions. It trains a classifier on source data and then applies the Expectation-Maximization (EM) algorithm to fit the classifier for the target data, using the Kullback-Liebler divergence to determine the trade-off parameters in the EM algorithm. When tested on datasets from Newsgroups, SRAA and Reuters for the task of top-category classification of text documents this algorithm performed better than support vector machine and Naïve Bayes classifiers.

Another algorithm derived from the Naïve Bayes classifier that uses domain adaptation is the Adapted Naïve Bayes classifier [27], which identifies and uses only the generalizable features from the source domain, and the unlabeled data with all the features from the target domain to build a classifier for the target domain. This algorithm was evaluated on transferring the sentiment analysis classifier from a source domain to several target domains. The prediction rate was promising, with Micro F1 values between 0.69 and 0.90, and Macro F1 values between 0.59 and 0.91. However, the classifier did not use any labeled data from the target domain.

Nigam et al. [17] showed empirically that combining a small labeled dataset with a large unlabeled dataset from the same or different domains can reduce the classification error of text documents by up to 30 %. Their algorithm also uses a combination of Expectation Maximization and the Naïve Bayes classifier by first learning a classifier on the labeled data which is then used to classify the unlabeled data. The combination of these datasets trains a new classifier and iterates until convergence. By augmenting the labeled data with unlabeled data the number of labeled instances was smaller compared to using only labeled data.

For biological sequences, most domain adaptation algorithms employed support vector machines. For example, Sonnenburg et al. [26] used a Support Vector

Machine with weighted degree kernel to classify DNA sequences into sequences that have or not have a splice site at the location of interest. Even though the training data was highly skewed towards the negative class, their classifier achieved good accuracy.

For more work on domain adaptation and transfer learning, see the survey by Pan and Yang [19].

3   Methodology

3.1   Identifying and Selecting Generalizable Features from the Source Domain

To successfully adapt a classifier from the source domain to the target domain, the classifier has to identify in the source domain the subset of the features that generalize well and are highly correlated with the label. Then, use a combination of only these features from the source domain and all the features from the target domain to predict the labels in the target domain.

We used the feature selection method proposed by Tan *et al.* [27]. Theoretically, the set of features in each domain can be split into four categories, based on two selection criteria. The first selection criterion is the level of correlation. The features have varying degrees of correlation with the label assigned to a sequence. Based on the correlation between the feature and the label, the features can be divided into features that are highly related to the labels, and features that are less related to the labels. The second selection criterion is the specificity of the features. Based on this criterion, the features can be divided into features that are very specific to a domain, and features that generalize well across related domains.

To select these features from the source domain we rank all the features from the source domain based on their probabilities. The features that are generalizable would most likely occur frequently in both domains, and should be ranked higher, as shown in Fig. 1. Moreover, the features that are correlated to the labels should be ranked higher. Therefore, we use the following measure to rank the features in the source domain:

$$f(w) = \log \frac{P_s(w) \cdot P_t(w)}{|P_s(w) - P_t(w)| + \alpha} \tag{1}$$

where P_s and P_t are the probability of the feature w in the source and target domain, respectively. The numerator ranks higher the features that occur frequently in both domains, since the larger both probabilities are the larger the numerator is, and thus the higher the rank of the feature is. The denominator ranks higher the features that have similar probabilities (i.e., the generalizable features), since the closer the probabilities are for a feature in both domains, the smaller the denominator value is, and thus the higher the rank. The additional value in the denominator, α, is used to prevent division by zero. The higher its value is the more influence the numerator has in ranking the features, and vice

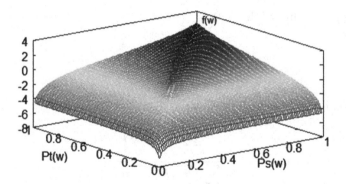

Fig. 1. Ranking of features in the source domain using Eq. (1). The rank of a feature is higher if it has a high probability or occurs with similar probability in the target domain. Note: This graph was drawn using Octave [6].

versa. To limit its influence on ranking the features, we chose a small value for this parameter, 0.0001. The probability of a feature in either domain is

$$P(w) = \frac{N(w) + \beta}{|D| + 2 \cdot \beta} \tag{2}$$

where N is the number of instances in the domain in which the feature w occurs, D is the total number of instances in the domain and β is a smoothing factor, which is used to prevent the probability of a feature to be 0 (which would make the numerator in (1) equal to 0, and the logarithm function is undefined for 0). We chose a small value for β as well, 0.0001, to limit its influence on the ranking of features. Note that the values for α and β do not have to be the same, but they can be, as used by Tan *et al.* [27] and in our case.

3.2   Multinomial Naïve Bayes (MNB) Classifier

The multinomial naïve Bayes classifier [15] assumes that the sample data used to train the classifier is representative of the population data on which the classifier will be used. In addition, it assumes that the frequency of the features determines the label assigned to an instance, and that the position of a feature is irrelevant (the naïve Bayes assumption). Thus, using Bayes' property a classifier can approximate the posterior probability, i.e., the probability of a class given an unclassified instance, as being proportional to the product of the prior probability of the class, and the feature conditional probabilities given an instance from the sample data:

$$P(c_k \mid d_i) \propto P(c_k) \prod_{t \in |V|} [P(w_t \mid c_k)]^{N_{t,i}} \tag{3}$$

where the probability of the class is

$$P(c_k) = \frac{\sum_{i \in |D|} P(c_k \mid d_i)}{|D|} \tag{4}$$

and the conditional probability is

$$P(w_t \mid c_k) = \frac{\sum_{i \in |D|} N_{t,i} \cdot P(c_k \mid d_i) + 1}{\sum_{t \in |V|} \sum_{i \in |D|} N_{t,i} \cdot P(c_k \mid d_i) + |V|} \tag{5}$$

Here, $N_{t,i}$ is the number of times feature w_t occurs in instance d_i, $|V|$ is the number of features, and $|D|$ is the number of instances.

3.3 Adapted Naïve Bayes Classifier for Biological Sequences

One limitation of the MNB classifier is that it can only be trained on one domain, and when the trained classifier is used on a different domain, in most cases, its classification accuracy decreases. To address this, we used the Adapted Naïve Bayes (ANB) classifier proposed by Tan *et al.* [27], with two modifications: we used the labeled data from the target domain, and employed the self-training technique. These will be described in more detail shortly.

The ANB algorithm is a combination of the expectation-maximization (*EM*) algorithm and a weighted multinomial Naïve Bayes algorithm. Similar to the *EM* algorithm, it has two steps that are iterated until convergence. In the first step, the *M*-step, we simultaneously calculate the class probability and the class conditional probability of a feature. However, unlike the *EM* algorithm that uses the data from one domain to calculate these values, this algorithm uses a weighted combination of the data from the source domain and the target domain.

$$P(c_k) = \frac{(1 - \lambda) \sum_{i \in D_s} P(c_k \mid d_i) + \lambda \sum_{i \in D_t} P(c_k \mid d_i)}{(1 - \lambda)|D_s| + \lambda|D_t|} \tag{6}$$

$$P(w_t \mid c_k) = \frac{(1 - \lambda)(\eta_t N_{t,k}^s) + \lambda N_{t,k}^t + 1}{(1 - \lambda) \sum_{t \in |V|} \eta_t N_{t,k}^s + \lambda \sum_{t \in |V|} N_{t,k}^t + 1} \tag{7}$$

where $N_{t,k}$ is the number of times feature w_t occurs in a domain in instances labeled with class k:

$$N_{t,k} = \sum_{i \in D} N_{t,i} P(c_k \mid d_i) \tag{8}$$

λ is the weight factor between the source and target domains:

$$\lambda = \min\{\delta \cdot \tau, 1\} \tag{9}$$

and τ is the iteration number. $\delta \in (0,1)$ is a constant that determines how fast the weight shifts from the source domain to the target domain, and η_t is 1 if feature t in the source domain is a generalizable feature, 0 otherwise.

Unlike the algorithm proposed by Tan et al. [27], which considers that all the instances from the target domain are unlabeled and does not use them during the first iteration (i.e., $\lambda = 0$), it is reasonable to assume that there is a small number of labeled instances in the target domain, and our algorithm uses any labeled data from the target domain in the first and subsequent iterations. In the first iteration we use only labeled instances from the source and target domains to calculate the probability distributions for the class conditional probabilities given the instance. In subsequent iterations we use the class of the instance for the labeled data from the source and target domains and the probability distribution of the class for the unlabeled data from the target domain.

In the second step, the E-step, we estimate the probability of the class for each instance with the values obtained from the M-step.

$$P(c_k \mid d_i) \propto P(c_k) \prod_{t \in |V|} [P(w_t \mid c_k)]^{N_{t,i}} \tag{10}$$

The second modification we made to the ANB classifier [27], is our use of self-training, i.e., at each iteration, we select, proportional to the class distribution, the instances with the top class probability, and consider these to be labeled in the subsequent iterations. This improves the prediction accuracy of our classifier because it does not allow the unlabeled data to alter the class distribution from the target labeled data.

The two steps, E and M, are repeated until the instance conditional probabilities values in (10) converge (or a given number of iterations is reached). The algorithm is summarized in Algorithm 1.

3.4   Data Sets

We used three data sets to evaluate our classifier, two for the task of protein localization and one for the task of splice site prediction. The first data set, PSORTb v2.0[1] [8], was first introduced in [9], and contains proteins from gram-negative and gram-positive bacteria and their primary localization information: cytoplasm, inner membrane, periplasm, outer membrane, and extracellular space. For our experiments, we identified classes that appear in both datasets, and used 480 proteins from gram-positive bacteria (194 from cytoplasm, 103 from inner membrane, and 183 from extracellular space) and 777 proteins from gram-negative bacteria (278 from cytoplasm, 309 from inner membrane, and 190 from extracellular space). The second data set, TargetP[2], was first introduced in [7], and contains plant and non-plant proteins and their subcellular localization: mitochondrial, chloroplast, secretory pathway, and "other." From this data set

---

[1] Downloaded from http://www.psort.org/dataset/datasetv2.html
[2] Downloaded from http://www.cbs.dtu.dk/services/TargetP/datasets/datasets.php

---

Algorithm 1. Outline of the Adapted Naïve Bayes algorithm for biological sequences.

1: Select generalizable features from the source domain, i.e., the top ranked features using Equation (1).
2: For each class simultaneously calculate the class probability and the class conditional probability of each feature using Equations (6-7). For the source domain use all labeled instances, and only the generalizable features. For the target domain use only labeled instances, and all features.
3: Select, proportional to the class distribution, the target instances with the top class probability, and consider these to be labeled in the subsequent iterations.
4: **while** labels assigned to unlabeled data change **do**
5:    **M-step**: Same as step 2 but use the class for labeled and self-trained instances from the target domain, and the class distribution for unlabeled instances.
6:    Same as step 3.
7:    **E-step**: Calculate the class distribution for unlabeled training instances from the target domain using Equation (10).
8: **end while**
9: Use classifier to label new target data.

---

we used 799 plant proteins (368 mitochondrial, 269 secretory pathway and 162 "other") and 2,738 non-plant proteins (371 mitochondrial, 715 secretory pathway and 1652 "other"). Predicting protein localization is an important biological problem because the function of the proteins is related to their localization. The third data set[3], first introduced in [24], contains DNA sequences of 141 base pairs centered around the donor splice site dimer AG and the label of whether or not that AG dimer is a true splice site. The sequences are from five organisms, *C.elegans* as the source domain, and *C.remanei, P.pacificus, D.melanogaster,* and *A.thaliana* as target domains. In each dataset there are about 1% positive instances. Accurately predicting splice sites is important for genome annotation [2,21].

3.5   Data Preparation and Experimental Setup

Protein Localization. We represent each sequence as a count of occurrences of k-mers. We use a sliding window approach to count the k-mer frequencies. For example, the protein sequence LLRSYRS would be transformed when using 2-mers into 1, 1, 2, 1, 1 which are the counts corresponding to the occurrences of features LL, LR, RS, SY, YR.

In order to obtain unbiased estimates for classifier performance we use five-fold cross validation. We use all labeled data from the source domain for training (tSL) and randomly split the target domain data into 3 sets: 20% used as labeled data for training (tTL), 60% used as unlabeled data for training (tTU), and 20% used as test data (TTL), as shown in Fig. 2(a). So, we train our classifier on tSL + tTL + tTU and test it on TTL.

---

[3] Downloaded from ftp://ftp.tuebingen.mpg.de/fml/cwidmer/

We wanted to answer several questions – specifically, how does the performance of the classifier vary with:

Q1. Features used (i.e., 3-mers, 2-mers, or 1-mers)?
Q2. Number of features used in the target domain (i.e., keep all features, remove at most 50 % of the least occurring features)?
Q3. Number of features retained in the source domain after selecting the generalizable features?
Q4. Variation with the size of the target labeled/unlabeled data set (i.e., train on $100 \% \, tSL + x\% \, tTL + y\% \, tTU$, where $x \in \{5, 10, 20\}$ and $y \in \{20, 40, 60\}$)?
Q5. The distance between the source and target domains?
Q6. The choice of the source and target domains?

As baselines, we compared our classifier (**ANB**) with the multinomial naïve Bayes classifier trained on all source data (**MNB s**), the multinomial naïve Bayes classifier trained on 5 % target data (**MNB 5t**), and the multinomial naïve Bayes classifier trained on 80 % target data (**MNB 80t**). Each classifier was tested on 20 % of target data. The expectation is that the prediction accuracy of our classifier will be lower bounded by **MNB 5t**, upper bounded by **MNB 80t**, and be better than **MNB s**.

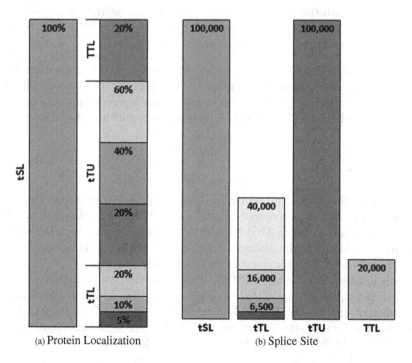

(a) Protein Localization   (b) Splice Site

Fig. 2. Experimental setup. We use 3 datasets to train our algorithm – source domain labeled (tSL), target domain labeled (tTL), and target domain unlabeled (tTU) – and 1 to test it – target domain labeled (TTL).

To evaluate our classifier we used the area under the receiver operating characteristic (auROC), as the class distributions are relatively balanced.

Splice Site Prediction. Similar to the protein localization task, we represent each sequence as a count of occurrences of 8-mers. We use a sliding window approach to count the 8-mer frequencies.

For the splice site prediction there are 3 folds for each organism. From each fold we use the dataset with 100,000 instances for the source domain (tSL), the datasets with 2,500, 6,500, 16,000, and 40,000 instances for the target domain as labeled (tTL), the 100,000 datasets for the target domain as unlabeled (tTU), and 20,000 instances to test our algorithm on the target domain (TTL), as shown in Fig. 2(b). Then we averaged the results over the 3 folds to obtain unbiased estimates. Just like for protein localization, we train our classifier on tSL + tTL + tTU and test it on TTL.

We wanted to answer similar questions to the protein localization task:

Q1. Number of features used in the target domain (i.e., keep all features, remove at most 50 % of the least occurring features)?
Q2. Number of features retained in the source domain after selecting the generalizable features?
Q3. Variation with the size of the target labeled data set (i.e., train on 100,000 tSL + z tTL + 100,000 tTU, where $z \in \{2, 500, 6, 500, 16, 000, 40, 000\}$)?
Q4. The distance between the source and target domains?

As baseline, we compared our classifier with the best overall algorithm in [24].

To evaluate our classifier we used the area under precision-recall curve (auPRC), a metric that is preferred over area under a receiver operating characteristic curve when the class distribution is skewed, which is the case with this dataset.

3.6   Results

Protein Localization. This section provides empirical evidence that augmenting the labeled data from a source domain with labeled and unlabeled data from the target domain with the ANB algorithm improves the classification accuracy compared to using only the limited labeled data from the target domain or using only the data from a source domain with the MNB classifier.

Table 1 shows the average auROC values over the five-fold cross validation trials for our algorithm and for the baseline algorithms. For our algorithm, we used different amounts of labeled and unlabeled data from the target domain. For example, the top-left value is the auROC for our algorithm trained on 5 % labeled data and 20 % unlabeled data. For each dataset and the features used the largest auROC value for the ANB is highlighted.

We noted that the performance of the ANB classifier varies, as follows:

Table 1. A comparison, on the protein localization task, between the Adapted Naïve Bayes classifier (ANB), the multinomial naïve Bayes classifier trained on all source data (MNB s), the multinomial naïve Bayes classifier trained on 5 % target data (MNB 5t), and the multinomial naïve Bayes classifier trained on 80 % target data (MNB 80t). The results are reported as average auROC values over five-fold cross validation trials. For the ANB classifier, the row headings indicated how much target unlabeled data was used in training the classifier and the column headings indicate how much target labeled data was used. The best values for the ANB are highlighted. Note that ANB is bounded by MNB 5t and MNB 80t, and that ANB predicts more accurately as the length of k-mers increases.

PSORTb dataset — Gram^{+} as source and gram^{-} as target

tTU	ANB 5%	ANB 10%	ANB 20%	MNB s	MNB 5t	MNB 80t
Features are 1-mers						
20%	91.42	91.70	**92.08**			
40%	90.68	90.82	91.68	92.74	92.18	93.52
60%	89.00	90.20	91.90			
Features are 2-mers						
20%	93.58	93.66	**93.94**			
40%	92.84	92.68	93.90	93.30	91.90	94.24
60%	92.92	93.58	93.50			
Features are 3-mers						
20%	93.80	93.80	**94.24**			
40%	92.62	92.78	93.14	91.94	85.80	95.52
60%	91.34	92.40	93.08			

TargetP dataset — Plant as source and non-plant as target

tTU	ANB 5%	ANB 10%	ANB 20%	MNB s	MNB 5t	MNB 80t
Features are 1-mers						
20%	65.26	69.84	**73.98**			
40%	62.90	66.24	69.16	76.38	79.90	81.28
60%	60.88	64.52	70.40			
Features are 2-mers						
20%	65.78	71.84	**79.38**			
40%	62.12	67.02	69.34	78.62	82.60	83.96
60%	60.28	63.08	67.14			
Features are 3-mers						
20%	75.82	81.44	**85.66**			
40%	74.04	79.72	83.46	66.82	63.86	88.36
60%	76.18	76.36	77.96			

Gram^{-} as source and gram^{+} as target

tTU	ANB 5%	ANB 10%	ANB 20%	MNB s	MNB 5t	MNB 80t
Features are 1-mers						
20%	92.78	93.20	**93.46**			
40%	89.78	93.26	91.18	93.60	91.42	95.56
60%	89.12	87.28	93.02			
Features are 2-mers						
20%	90.90	94.52	94.66			
40%	91.80	92.06	**95.02**	94.42	88.52	96.16
60%	94.26	94.28	94.28			
Features are 3-mers						
20%	95.90	95.20	**96.14**			
40%	92.80	94.40	94.60	95.78	81.18	95.44
60%	92.78	92.82	94.60			

Non-plant as source and plant as target

tTU	ANB 5%	ANB 10%	ANB 20%	MNB s	MNB 5t	MNB 80t
Features are 1-mers						
20%	72.96	71.90	**77.04**			
40%	69.22	71.96	76.96	76.18	73.66	85.14
60%	67.16	73.40	75.48			
Features are 2-mers						
20%	78.24	78.10	**78.68**			
40%	72.72	75.14	78.62	78.36	75.08	88.52
60%	73.80	73.62	75.92			
Features are 3-mers						
20%	82.00	80.92	**85.96**			
40%	73.82	74.42	79.90	89.68	68.60	86.28
60%	69.04	72.56	78.48			

A1. The best results were obtained when using 3-mers as features. This makes sense since longer k-mers capture more information associated with the relative position of each amino-acid. When using 3-mers, our algorithm provides between 9.84 % and 34.14 % better classification accuracy when compared to multinomial naïve Bayes classifier trained on 5 % of the labeled data from the target domain, and between 0.37 % and 28.2 % when compared to the multinomial naïve Bayes classifier trained on labeled data from the source domain, except when the plant proteins are the target domain.

A2. When trying to establish how many features from the target domain should be used we determined that removing any features does not improve the performance of our algorithm.

A3. When trying to ascertain how many features from the source domain should be kept after ranking them with Eq. 1, we determined that the best results were obtained when at least 50 % of the features were kept, i.e., the 50 % top-ranked features and any other features with the same rank as the last feature kept.

A4. For most cases, the largest auROC values for our algorithm were obtained when using the least amount of target unlabeled data. This would suggest that even though using unlabeled data is beneficial, using too much unlabeled data is detrimental because the unlabeled instances act as noise and corrupt the prediction from the target labeled data. In addition, intuitively, using more labeled data from the target domain should lead to better prediction accuracy. This was indeed the case with our classifier.

A5. When the source and target domains are close the classifier learned is better. For example, the auROC is higher for the PSORTb datasets than for the TargetP datasets. Therefore, the closer the target domain is to the source domain the better the classifier learned.

A6. For the PSORTb dataset, the ANB classifier had better prediction accuracy when the gram-negative proteins were used as the source domain than when the gram-positive proteins were used as the source domain. Similarly, for the TargetP dataset, we obtained better predictions when using non-plant proteins as the source domain than when using plant proteins as the source domain. This is because in both cases there were more gram-negative instances and more non-plant instances, respectively, than gram-positive instances and plant instances, respectively.

It is interesting to note that in some instances the multinomial naïve Bayes classifier trained on the source domain performed better than our algorithm. This occurred mainly when our algorithm used 5 % or 10 % of the target labeled data and when the features were 1-mers or 2-mers. However, this is somewhat expected, as using very little labeled data from the target domain does not provide a representative sample for the population, and using short k-mers does not capture the relative position of the amino-acids.

Splice Site Prediction. Although our algorithm works well for the protein localization task, the results for the third dataset, on the splice site prediction

Table 2. auPRC values for the 4 target organisms based on the number of labeled target instances used for training: 2,500, 6,500, 16,000, and 40,000. We show for comparison with our algorithm the values for the best overall algorithm in [24], $SVM_{S,T}$.

	2,500	6,500	16,000	40,000
ANB	1.13	1.13	1.13	1.10
SVM	77.06	77.80	77.89	79.02

(a) *C.remanei*

	2,500	6,500	16,000	40,000
ANB	1.00	0.97	1.07	1.10
SVM	64.72	66.39	68.44	71.00

(b) *P.pacificus*

	2,500	6,500	16,000	40,000
ANB	1.07	1.13	1.07	1.03
SVM	40.80	37.87	52.33	58.17

(c) *D.melanogaster*

	2,500	6,500	16,000	40,000
ANB	1.20	1.17	1.20	1.17
SVM	24.21	27.30	38.49	49.75

(d) *A.thaliana*

task, were very poor, as shown in Table 2. Our algorithm always gravitated towards classifying each instance as not containing a splice site. We believe that this is due mainly because the 8-mers indicating a splice site occur with low frequency and their relative position to the splice site is important. We will discuss in Sect. 4 how we propose to address this issue in future work.

We noted that the performance of the ANB classifier varies, as follows:

A1. Similar to protein localization task, removing any features does not improve the performance of our algorithm.
A2. In terms of the number of features from the source domain to keep after ranking them with Eq. 1, we determined that the best results were obtained when at least 50 % of the features were kept, i.e., the 50 % top-ranked features and any other features with the same rank as the last feature kept.
A3. The auPRC values for our algorithm were very similar regardless of the amount of target labeled data.
A4. The classification performance of our algorithm did not decrease as the distance between the source and target domains increased, as we would have expected.

The last two observations lead us to believe that our features need to take into consideration the locations of the 8-mers to improve the classification accuracy of our classifier on the splice site prediction task.

4   Conclusions and Future Work

In this paper, we proposed a domain adaptation classifier for biological sequences. This algorithm showed promising classification performance in our experiments. Our analysis indicates that the closer the target domain is to the source domain the better is the classifier learned. Other conclusions drawn from our observations: using 2-mers or 3-mers results in better prediction, with small differences between them; removing features from the target domain reduces the accuracy of classifier; having more target labeled data increases the accuracy of classifier; and adding too much target unlabeled data decreases the accuracy of classifier.

In future work we would like to investigate how would assigning different weights to the data used for training influence the accuracy prediction of the algorithm. We would like to assign higher weight to the labeled data from the target domain since this is more likely to correctly predict the class of the target test data than the labeled data from the source domain or the unlabeled data from the target domain.

We would also like to evaluate other methods for selecting the generalizable features. For example, we would like to investigate if selecting generalizable features using the mutual information of the features instead of their probabilities, in Eq. (1), leads to better classification accuracy.

Another aspect we would like to improve is the accuracy of this classifier on the splice site dataset, to get accuracy that is similar to state of the art splice site classifiers, e.g., SVM classifiers. We would like to reduce the number of motifs with different clustering strategies, and identify more discriminative motifs using Gibbs sampling or MEME. In addition, we would like to run experiments on smaller splice site datasets to better understand the characteristics of this problem.

Acknowledgements. The computing for this project was performed on the Beocat Research Cluster at Kansas State University, which is funded in part by NSF grants CNS-1006860, EPS-1006860, EPS-0919443, and MRI-1126709.

References

1. Baten, A., Chang, B., Halgamuge, S., Li, J.: Splice site identification using probabilistic parameters and SVM classification. BMC Bioinform. **7**(Suppl. 5), S15 (2006)
2. Bernal, A., Crammer, K., Hatzigeorgiou, A., Pereira, F.: Global discriminative learning for higher-accuracy computational gene prediction. PLoS Comput. Biol. **3**(3), e54 (2007)
3. Brown, M.P.S., Grundy, W.N., Lin, D., Cristianini, N., Sugnet, C., Furey, T.S., Ares Jr., M., Haussler, D.: Knowledge-based analysis of microarray gene expression data using support vector machines. PNAS **97**(1), 262–267 (2000)
4. Dai, W., Xue, G., Yang, Q., Yu, Y.: Transferring naïve bayes classifiers for text classification. In: Proceedings of the 22nd AAAI Conference on Artificial Intelligence (2007)
5. Degroeve, S., Saeys, Y., De Baets, B., Rouzé, P., Van De Peer, Y.: Splicemachine: predicting splice sites from high-dimensional local context representations. Bioinformatics **21**(8), 1332–1338 (2005)
6. Eaton, J.W., Bateman, D., Hauberg, S.: GNU Octave Manual Version 3. Network Theory Ltd., Bristol (2008)
7. Emanuelsson, O., Nielsen, H., Brunak, S., von Heijne, G.: Predicting subcellular localization of proteins based on their N-terminal amino acid sequence. J. Mol. Biol. **300**(4), 1005–1016 (2000)

8. Gardy, J.L., Laird, M.R., Chen, F., Rey, S., Walsh, C.J., Ester, M., Brinkman, F.S.L.: Psortb v. 2.0: Expanded prediction of bacterial protein subcellular localization and insights gained from comparative proteome analysis. Bioinformatics **21**(5), 617–623 (2005)

9. Gardy, J.L., Spencer, C., Wang, K., Ester, M., Tusnády, G.E., Simon, I., Hua, S., deFays, K., Lambert, C., Nakai, K., Brinkman, F.S.: Psort-b: improving protein subcellular localization prediction for gram-negative bacteria. Nucleic Acids Res. **31**(13), 3613–3617 (2003)

10. Huang, J., Li, T., Chen, K., Wu, J.: An approach of encoding for prediction of splice sites using svm. Biochimie **88**, 923–929 (2006)

11. Jaakkola, T.S., Haussler, D.: Exploiting generative models in discriminative classifiers. In: Proceedings of the 1998 Conference on Advances in Neural Information Processing Systems II, pp. 487–493. MIT Press, Cambridge (1999)

12. Jiang, J., Zhai, C.: A two-stage approach to domain adaptation for statistical classifiers. In: Proceedings of the Sixteenth ACM Conference on Information and Knowledge Management, CIKM '07, pp. 401–410. ACM, New York (2007)

13. Lorena, A.C., de Carvalho, A.C.P.L.F.: Human splice site identification with multiclass support vector machines and bagging. In: Kaynak, O., Alpaydın, E., Oja, E., Xu, L. (eds.) ICANN/ICONIP 2003. LNCS, vol. 2714, pp. 234–241. Springer, Heidelberg (2003)

14. Maeireizo, B., Litman, D., Hwa, R.: Co-training for predicting emotions with spoken dialogue data. In: Proceedings of the ACL 2004 on Interactive Poster and Demonstration Sessions, ACLdemo '04. Association for Computational Linguistics, Stroudsburg (2004)

15. Mccallum, A., Nigam, K.: A comparison of event models for naive bayes text classification. In: AAAI-98 Workshop on 'Learning for Text Categorization' (1998)

16. Müller, K.-R., Mika, S., Rätsch, G., Tsuda, S., Schölkopf, B.: An introduction to kernel-based learning algorithms. IEEE Trans. Neural Networks **12**(2), 181–202 (2001)

17. Nigam, K., Mccallum, A., Thrun, S., Mitchell, T.: Text classification from labeled and unlabeled documents using EM. Mach. Learn. **39**(2–3), 103–134 (1999)

18. Noble, W.S.: What is a support vector machine? Nat Biotechnol. **24**(12), 1565–1567 (2006)

19. Pan, S.J., Yang, Q.: A survey on transfer learning. IEEE Trans. Knowl. Data Eng. **22**(10), 1345–1359 (2010)

20. Rätsch, G., Sonnenburg, S.: Accurate splice site detection for caenorhabditis elegans. In: Schölkopf, B., Tsuda, K., Vert, J.-P. (eds.) Kernel Methods in Computational Biology, pp. 277–298. MIT Press, Cambridge (2004)

21. Rätsch, G., Sonnenburg, S., Srinivasan, J., Witte, H., Müller, K.-R., Sommer, R., Schölkopf, B.: Improving the c. elegans genome annotation using machine learning. PLoS Comput. Biol. **3**, e20 (2007)

22. Riloff, E., Wiebe, J., Wilson, T.: Learning subjective nouns using extraction pattern bootstrapping. In: Proceedings of the Seventh Conference on Natural Language Learning at HLT-NAACL 2003, CONLL '03, vol. 4, pp. 25–32. Association for Computational Linguistics, Stroudsburg (2003)

23. Schölkopf, B., Smola, A.J.: Learning with Kernels: Support Vector Machines, Regularization, Optimization, and Beyond. MIT Press, Cambridge (2001)

24. Schweikert, G., Widmer, C., Schölkopf, B., Rätsch, G.: An empirical analysis of domain adaptation algorithms for genomic sequence analysis. In: NIPS'08, pp. 1433–1440 (2008)

25. Sonnenburg, S., Rätsch, G., Jagota, A., Müller, K.-R.: New methods for splice site recognition. In: Dorronsoro, J.R. (ed.) ICANN 2002. LNCS, vol. 2415, pp. 329–336. Springer, Heidelberg (2002)
26. Sonnenburg, S., Schweikert, G., Philips, P., Behr, J., Rätsch, G.: Accurate splice site prediction using support vector machines. BMC Bioinf. 8(Suppl. 10), 1–16 (2007)
27. Tan, S., Cheng, X., Wang, Y., Xu, H.: Adapting Naive Bayes to domain adaptation for sentiment analysis. In: Boughanem, M., Berrut, C., Mothe, J., Soule-Dupuy, C. (eds.) ECIR 2009. LNCS, vol. 5478, pp. 337–349. Springer, Heidelberg (2009)
28. Tsuda, K., Kawanabe, M., Rätsch, G., Sonnenburg, S., Müller, K.-R.: A new discriminative kernel from probabilistic models. Neural Comput. 14(10), 2397–2414 (2002)
29. Vapnik, V.N.: The Nature of Statistical Learning Theory. Springer-Verlag New York Inc., New York (1995)
30. Yarowsky, D.: Unsupervised word sense disambiguation rivaling supervised methods. In: Proceedings of the 33rd Annual Meeting on Association for Computational Linguistics, ACL '95, pp. 189–196. Association for Computational Linguistics, Stroudsburg (1995)
31. Zhang, Y., Chu, C.-H., Chen, Y., Zha, H., Ji, X.: Splice site prediction using support vector machines with a bayes kernel. Expert Syst. Appl. 30(1), 73–81 (2006)
32. Zien, A., Rätsch, G., Mika, S., Schölkopf, B., Lengauer, T., Müller, K.-R.: Engineering support vector machine kernels that recognize translation initiation sites. Bioinformatics 16(9), 799–807 (2000)

Optimization of Cost Sensitive Models to Improve Prediction of Molecular Functions

Sebastián García-López[1], Jorge Alberto Jaramillo-Garzón[1,2(✉)], and German Castellanos-Dominguez[1]

[1] Signal Processing and Recognition Group, Universidad Nacional de Colombia, Campus la Nubia, Km 7 vía al Magdalena, Manizales, Colombia
[2] Grupo de Automática y Electrónica, Instituto Tecnológico Metropolitano, Cll 54A No 30-01, Medellín, Colombia
{sgarcialop,jajaramillog,cgcastellanosd}@unal.edu.co

Abstract. The prediction of unknown protein functions is one of the main concerns at field of computational biology. This fact is reflected specifically in the prediction of molecular functions such as catalytic and binding activities. This, along with the massive amount of information has made that tools based on machine learning techniques have increase their popularity in the last years. However, these tools are confronted to several problems associated to the treated data, one of them is the learning with large imbalance between their categories. There exist several techniques to overcomes the class imbalance, but most of them present many weakness that difficult the obtaining of reliable results. Moreover, models based on cost sensitive learning seems to be a good choice to deal with imbalance data, yet, the obtaining of a optimal cost matrix still remains an open issue. In this paper, a methodology to calculate a optimal cost matrix for models based on cost sensitive learning is proposed. The results show the superiority of this approach compared with several techniques in the state of the art regarding to class imbalance. Tests were applied to prediction of molecular functions in Embryophyta plants.

Keywords: Molecular functions prediction · Proteins · Cuckoo search · Cost sensitive learning · Class imbalance

1 Introduction

Nowadays, modern biology has seen an increasing use of computational techniques for large scale and complex biological data analysis. Several computational machine learning techniques are applied [16]. For example, to classify different types of samples in gene expression of microarrays data [2] or mass spectrometry based proteomics data [1]. In this context, there is a vast number of problems associated with nature of data. In particular, given that same protein can be associated to several functional classes, classification problem with

© Springer-Verlag Berlin Heidelberg 2014
M. Fernández-Chimeno et al. (Eds.): BIOSTEC 2013, CCIS 452, pp. 207–222, 2014.
DOI: 10.1007/978-3-662-44485-6_15

multiple labels arises. A straightforward way to cope with this issue is the "one-against-all" strategy, by which a binary classifier is trained per class to take independent decisions about protein membership. Yet, this approach leads to a high imbalance between sample number per each class, magnifying an already present disparity of their sizes, and thereby producing a large bias towards category having more information [21].

There are several ways to address class imbalance problems, being the most representative sampling, boosting, and cost sensitive strategies. Sampling techniques can be divided into oversampling and subsampling. Former techniques reproduces samples of minority class until they reach the same size as the majority class, but it induces two major problems: (i) over-training and (ii) noise addition in training set, affecting reliability of protein localization [12]. Although, subsampling eliminates samples of majority class until reaching the same size of category having fewer samples (i.e., minority class), this technique may eliminate useful information if sample selection criteria are not properly selected [12]. Boosting strategies, in turn, are designed to train a set of individually trained classifiers in an iterative way, such that every new classifier emphasizes on incorrectly learned instances by previous trained classifier [7]. Boosting methods, however, are prone to fail if there is not enough data [20] or if training data holds too much noise [6]. Finally, models based on cost sensitive learning assume different costs (or penalties) whenever examples are misclassified. This process is modelled by a cost matrix that is a numerical representation of penalty of misclassifying examples from one category to another. Conventionally, such models assume that costs are fixed; but since this condition is far for being matched in real-world applications it posses as an open problem [18].

In this paper, an efficient methodology of obtaining the optimal cost matrix for a cost sensitive learning model is proposed. This methodology is applied to the prediction of molecular functions in *Embryophyta* plants and is compared with a broad spectrum of class-balance strategies to obtain a comprehensive analysis of the problem. Results show that cost sensitive models are highly reliable and can outperform many commonly used balance strategies in the prediction of molecular functions.

2   Class-Balance Strategies

Generally, class-balance strategies are divided into following explained below strategies: sampling, boosting, and cost sensitive.

2.1   Sampling Strategies

Synthetic Minority Oversampling Technique (SMOTE). In this case, new synthetic samples are generated that are addressed to minority class [5]. Further, these samples are computed by interpolation among several closely spaced real samples. In this way, the decision boundary of the minority class becomes more general [11]. Synthetic samples are generated as follows: for each real sample under consideration, represented as a feature vector, distance between it and its nearest neighbors is taken. The result is multiplied by a random number ranging within interval $(0, 1)$ with uniform probability, and this output is added to original feature vector. This procedure causes selection of a random point along the line segment between two neighboring samples.

Subsampling Based on Particle Swarm Optimization (PSO). This technique is based on search of an optimal sample subset from a given majority class that maximizes generalization capability of classifier. To this end, Metaheuristic optimization strategy is used. PSO algorithm creates several subsets of majority class and evaluates its classification performance. When completion criterion is accomplished, samples are ranked by their frequency selection. After the frequency listing of selected samples is obtained, a balanced dataset is constructed by combination of samples, which belong to majority class with major frequency indexes, and minority class [24], as summarized in Fig. 1.

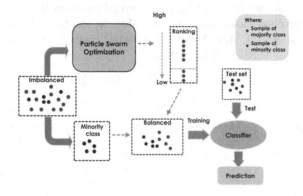

Fig. 1. Schematic representation of PSO subsampling based on algorithm [24].

2.2   Boosting Strategies

Boosting Algorithm (AdaBoost). Boosting algorithms place iteratively different weights on training data at each iteration, in such a way that boosting increases those weights associated to incorrectly classified examples, but decreases weights related to correctly classified examples. Thus, training system is forced to focus on rare items. AdaBoost is the most representative boosting technique, which generates a set of classifiers to be further combined using the weighted majority voting [19]. Basically, a weak classifier is trained using instances drawn from an iteratively updated distribution of training data. Introduced distribution update ensures that instances misclassified by previous classifier are more likely to be included in training data of the next classifier. To make final decision, each classifier has a different power of decision depending on its performance during training procedure.

2.3   Cost Sensitive Strategies

Cost Sensitive Learning. This strategy attempts to minimize costs associated to their decisions rather than simply reaching high precision. Given a cost specification for either correct or incorrect predictions, sample can be assigned to that class that leads to lower expected cost, where the expected value is computed using conditional probability of each class for a given sample. So, assuming c_{ij} as inputs associated to a cost matrix C holding cost to predict a class i when the true class is j. If $i = j$, prediction is assumed as correct, whereas if $i \neq j$, prediction is incorrect. Given a sample x, optimal prediction for each i is a class that minimizes:

$$L(x, i) = \sum_j P(j|x)c_{ij} \tag{1}$$

where $L(x, i)$ is the sum over alternative possibilities for true class of sample x. In this framework, the goal of algorithm is to produce a classifier estimating probability $P(j|x)$. So, for a given x, prediction can be carried out as if i were the true class. As quoted in [9], the main idea behind decision-making based on cost sensitivity learning is that it may be optimal to act as if one class were true even when other class looks like more probable. In the biclass case, the optimal prediction is the class 1, if and only if, expected cost of prediction is less than or equal to the one predicting class 0, i.e.:

$$P(j = 0|x)c_{10} + P(j = 1|x)c_{11} < P(j = 0|x)c_{00} + P(j = 1|x)c_{01} \tag{2a}$$

$$(1 - p)c_{10} + pc_{11} < (1 - p)c_{00} + pc_{01}, \tag{2b}$$

where $p = P(j = 1|x)$. Therefore, threshold for optimal decision making is as follows:

$$p^* = \frac{c_{10} - c_{00}}{c_{10} - c_{00} + c_{01} - c_{11}} \tag{3}$$

MetaCost. This technique assumes an unaltered classifier, but adjusts learning according to a given cost matrix. Initially, training set is taken to constitute multiple subsets via bootstrap, where each obtained subset is used to build an classifier ensemble making the final decision [8]. Classifiers are combined through a majority vote rule to determine the probability of each data object x belonging to each class label. Next, each data object of training data is relabeled based on evaluation provided for an introduced conditional risk function, as follows:

$$R(i|x) = \sum_j P(j|x)c_{ij} \tag{4}$$

Lastly, the classification algorithm makes the decision over relabeled training data.

3   Proposed Optimal Cost Matrix Based on CuckooCost Search

The *Cuckoo Search* is based on parasitic behavior exposed by some species of Cuckoo birds that has as natural strategy to leave eggs in host nest created by

Table 1. Scaled cost matrix.

	Actual negative	Actual positive
Predicted negative	$c_{00} = 0$	$c_{01} = k^+/k^-$
Predicted positive	$c_{10} = 1$	$c_{11} = 0$

other birds. This eggs presents the particularity to have a big similitude with host eggs, so, the more similar they are, the greater your chance of survival.

Based on this statement, Cuckoo Search uses three hypothesized rules:

- Each cuckoo lays one egg at a time, but dumps it in a randomly chosen nest.
- The best nests with high quality of eggs (solutions) should carry over to the next generations.
- The number of available host nests is fixed, and a host can discover an alien egg with a probability $p_a \in [0,1]$. In this case, the host bird can either throw the egg away or abandon the nest so as to build a completely new nest in a new location.

For simplicity, the last assumption can be approximated by a fraction p_a of the n nests being replaced by new nests (with new random solutions at new locations). For a maximization problem, the quality or fitness of a solution can simply be proportional to the objective function. Given a solution at time t, noted as $\boldsymbol{x}_i^t$, thus, generation of new solutions along the time t is defined as:

$$\boldsymbol{x}_i^{(t+1)} = \boldsymbol{x}_i^t + \alpha \oplus \mathrm{L\acute{e}vy}(\lambda) \tag{5}$$

where α is a scale parameter while λ is the step size of the Cuckoo Search optimization. Notation $\oplus$ stands for a direct summation operator.

In bi-class problems, the minority class (i.e., the category with less samples) is assumed to have higher misclassification cost k^+ (usually, this samples relate to category of interest). Due to big amount of data, likewise, category with more samples must have lower misclassification cost k^-. From the above, if a cost matrix is given, the optimal decisions remain unchanged if their cost (in this case the inputs of the cost matrix) are multiplied by a scaling factor [17]. This normalization allows to change the baseline in which costs are measured. Therefore, if each element of the cost matrix that is multiplied by $1/k^-$, can be expressed as shown in Table 1.

Since costs are normalized to the unchanged optimal decision, value k^- can always be set to 1. Therefore k^+/k^- must be bigger than 1 [9]. This relationship is termed *cost-sensitive rescale ratio* or cost ratio [17]. In order to deal with class-imbalance using rescaling, different costs are to be incurred for different classes. So, the optimal rescale ratio (*imbalance rescale ratio*) of positive class to negative class, r, is defined as $r_I = n^-/n^+$. Therefore, to handle unequal misclassification and class-imbalance at the same time, both the cost-sensitive rescale ratio r_C and the imbalance rescale ratio r_I should be taken under consideration [17]. Merging both scale factors, cost ratio of the cost matrix can obtained as $\varphi = r_C r_I$, where $\varphi \geq r_I$.

Suggested cost using Cuckoo Search, termed *CuckooCost*, accomplishes optimal parameter values to achieve the best possible classification performance. Each nest represents a solution set in the searching space, i.e., each egg on the nest represents a parameter to be used in the model optimization. In this case, the cost ratio and classifier parameters are handled to improve the performance of cost sensitive learning.

It is worth noting that in Cuckoo Search, both parameters p_a and α are to explore efficiently over searching space and allow to find together globally and locally improved solutions. Additionally, these parameters directly influence the convergence rate of used optimization algorithm. For instance, if value p_a tends to be small and α value is large, the algorithm tends to increment the iteration number to converge to an optimal value. On the other hand, if p_a is large but α is small, the algorithm convergence speed tends to be very high but it is more likely to converge to a local optimum. In this work, an improvement to Cuckoo Search proposed in [23] is used that consists in restraining range of p_a and α; within this values search may change dynamically during each iteration, through the following equations:

$$c = \frac{1}{N_T} \ln\left(\frac{\alpha_{\min}}{\alpha_{\max}}\right), \, l\alpha \in [\alpha_{\min}, \alpha_{\max}] \tag{6a}$$

$$p_a = p_{\max} - \frac{N_I}{N_T}(p_{\max} - p_{\min}), \, p_a \in [p_{\min}, p_{\max}] \tag{6b}$$

$$\alpha = \alpha_{\max} \exp(c\,N_I) \tag{6c}$$

where N_T is the total number of iterations present in the optimization, N_I is the current iteration in the algorithm,

The best nest is the one holding the optimal parameters to induce a dependable cost sensitive model.

4 Experimental Setup

4.1 Database

Used database that is a subset of the one constructed in [15] holds 1098 proteins belonging to *Embryophyta* taxonomy of the Uniprot [14], with at least one annotation in the molecular function ontology of the Gene Ontology Annotation project [3]. Sequences predicted by computational tools and with no real experimental evidence are discarded. Proteins are associated to one or more of the seven categories that are shown in Table 2. The dataset does not contain protein sequences with a sequence identity superior to 40 % in order to avoid neither bias nor overtraining in the training dataset. All the proteins are mapped into feature vectors enclosing several statistical and physical-chemical attributes (see Table 3).

Table 2. Dataset description.

Class	Biological name	Samples	Imbalance ratio
GO 0003677	DNA binding	143	1 : 7.68
GO 0003700	Sequence-specific DNA binding transcription factor activity	102	1 : 10.76
GO 0003824	Catalytic activity	401	1 : 2.74
GO 0005215	Transporter activity	133	1 : 8.26
GO 0016787	Hydrolase activity	237	1 : 4.63
GO 0030234	Enzyme regulator activity	46	1 : 23.87
GO 0030528	Transcription regulator activity	152	1 : 7.22

Table 3. Description of the feature space (taken from [15]).

Feature	Description	Number
Chemical-physical	Length of the sequences	1
	Molecular weight	1
	Percentage of positively charged residues (%)	1
	Percentage of negatively charged residues (%)	1
	Isoelectric point	1
	GRAVY - hydropathic index	1
Primary structure	Frequency of each aminoacids	20
	Frequency of each dimers	400
Secundary structure	Frequency of structures	3
	Frequency of dimers in structures	9
	TOTAL	**438**

4.2 Fitness Function Approach

Performance of CuckooCost depends largely on a function that can properly guide searching process of the optimal hyperparameters. For this purpose, we propose the following fitness function that combines two variables that directly influence the classification process: area under ROC curve, $\mho$, and the total cost, ς, as follows:

$$\Theta(\mho, \varsigma) = \lambda(\mho) + (1 - \lambda)(\varsigma) \tag{7}$$

The aim of proposed fitness function is to maximize the overall classification performance as well as to minimize the cost associated with the wrong classified samples. Optimal value of free parameter of the fitness function is searched within a range interval $[0, 1.]$ Heuristically, the best value is fixed at $\lambda = [0.1]$.

4.3   Class Imbalance and Classification Schemes

To mitigate the effect generated by multi-label samples in the dataset, as well as to reduce classification complexity and to obtain a better interpretation of results, *against-vs-all* learning strategy is used. Nevertheless, the usage of this strategy leads to additional problems such as highly class imbalance on data space. To overcomes this issue, the following class balance strategies are considered: AdaBoost (Ada), SMOTE, Subsampling based on particle swarm optimization (SPSO), and cost sensitive learning (CS). Also, the proposed Meta-Cost (MC) is considered in two versions: (i) without matrix cost optimization via CuckooCost (MC), (*ii*) Cost sensitive learning and MetaCost within Cuck-ooCost (MCCu). During classification testing, support vector machines (SVM) with Gaussian Kernel is used, except the test with AdaBoost, for which Naive Bayes is employed as weak classifier and twenty iterations for Boosting technique. Parameter tuning needed in SVM and Gaussian Kernel (penalty constant C and dispersion γ) are carried out by using particle swarm optimization. However, PSO is not accomplished in cost sensitive learning strategies (CS and Meta-Cost), mainly, since the optimization based on Cuckoo Search turns to be more effective that PSO. So, CuckooCost takes γ and penalty constant C as hyperparameters in the optimization problem. To evaluate the performance of molecular function classification, cross-validation is used over ten folds. Besides, chosen a priori search parameter ranges of CuckooCost are the following:

$$1 \leq \varphi \leq 1.5 R_d$$
$$0.00030518 \leq C \leq 4096$$
$$0.000030518 \leq \gamma \leq 32$$

where φ is the cost ratio extracted from cost matrix, and R_d is the imbalance ratio.

4.4   Evaluation Metrics

Performance measures non-susceptible to unbalance data phenomena are used to obtain a reliably evaluation of accomplished classification. Measures such as sensitivity, specificity, geometric mean, and ROC area (AUC) are used, which are defined as:

$$\text{Sensitivity}: \quad S_e = \frac{T_P}{T_P + F_N} \tag{8a}$$

$$\text{Specificity}: \quad S_p = \frac{T_N}{T_N + F_P} \tag{8b}$$

$$\text{Geometric mean}: \quad \mu_G = \sqrt{S_e S_p} \tag{8c}$$

$$\text{ROC area}: \quad \mho = \frac{1 + T_P' - F_P'}{2} \tag{8d}$$

where T_P, T_N, F_N, and F_P are the true positive, true negative, false negative, and false positive values obtained from confusion matrix, respectively; T_P' is the true positive rate, F_P' is the false positive rate.

Additionally, a metric measuring the classification bias degree, termed relative sensitivity, is used that is defined as $r_S = S_e/S_p$, as given in [22].

Data Complexity Measures. Degree of data imbalance is not the only factor leading to a biased learning. Elements associated with data complexity may also influence learning models. Particularly, data complexity can be related to difficulties inherent in data, shortcomings in classification algorithms, and the low representation present in the data space [4,12]. The following measures are used to quantify data complexity:

(i) **Overlap Measures:** They explore both range and distribution of values in each category, and verify the overlap between them. The measures following overlap measures are used:
   - *Volume of Overlap Region* (VOR): For a given feature set, $\{f_i\}$, VOR measures the amount of overlap in boundary region between two categories, $c_i : i = (1,2)$, and is defined as [4,13]:

$$VOR = \prod_i \frac{\min(\max(f_i, c_1), \max(f_i, c_2)) - \max(\min(f_i, c_1), \min(f_i, c_2))}{\max(\max(f_i, c_1), \max(f_i, c_2)) - \min(\min(f_i, c_1), \min(f_i, c_2))} \tag{9}$$

   - *Fisher's Discriminant Ratio:* For a multidimensional problem, all features not necessarily have to contribute to class discrimination. As long as there exists one discriminating feature, the problem is suitable. Therefore, we use the maximum f over all the feature dimensions to describe a problem [4,13]. This measure also serves as indicator of quality in the dataset representation, i.e., if its value tends to be low, there is little contribution in the overall discrimination of the dataset, which may indicate a weak representation of the data. Fisher's discriminant ratio is defined as:

$$\kappa = \max \frac{(\mu_1 - \mu_2)^2}{\sigma_1^2 + \sigma_2^2} \tag{10}$$

   where μ_1 and μ_2 are the feature mean of the classes 1 and 2, while σ_1 and σ_2 are the feature variance of same classes, respectively.
   - **Scatter Matrix of Difference between inter/intra Classes:** It measures the distance between the class distribution and indicates on improved separability as its value is greater [10]. Being complementary to **VOR** and κ, this metric is described as:

$$J_4 = \text{tr}\{S_b - S_w\} \tag{11}$$

where, $S_w = \sum_{i=1}^{C} \frac{n_i}{n} \widehat{\Sigma}_i$ and $S_b = \sum_{i=1}^{C} \frac{n_i}{n}(m_i - m)(m_i - m)^\top$, being $\widehat{\Sigma}_i$ covariance matrix of i-th class, m_i is the sample mean of i-th class and

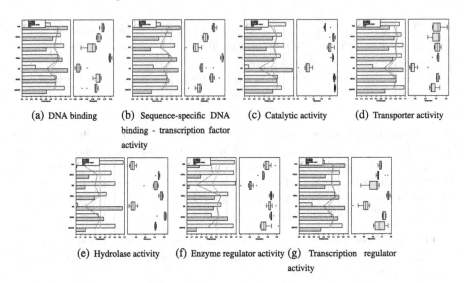

(a) DNA binding   (b) Sequence-specific DNA   (c) Catalytic activity   (d) Transporter activity
binding - transcription factor
activity

(e) Hydrolase activity   (f) Enzyme regulator activity (g) Transcription regulator
activity

Fig. 2. Molecular function prediction results.

m the sample mean of the whole dataset. Notation tr stands for matrix trace, C is the number of classes, and n is the number of samples in whole dataset.

(ii) **Measures of Geometry, Topology, and Density of Manifolds:** These metrics give indirect information about separation between categories. It is assumed that a category is composed by a collection of one or more manifolds, forming the support of the probability distribution of a given class. The shape, position and interconnectivity of manifolds give a hint of its overlap [4,13]. To evaluate the complexity of manifolds, the leave-one-out error for a one-nearest-neighbour classifier, Δ, is used.

5   Results and Discussion

Figure 2 summarizes obtained classification results that are displayed by bars and lines at different color scales. Each subfigure holds information about behavior of the geometric mean (drawn in red color), area under ROC curve (AUC) (green), sensitivity (light blue), and specificity (light cyan). Each row depicts each one considered class-balance strategies, which are ranked in ascending order according to used balance strategy, that is, oversampling (**SMOTE**), subsampling (**SPSO**), cost-sensitive learning without any optimized parameters and the same strategy using CuckooCost as well (**CS, CSCu, MC, MCCU**), and Boosting (**AdaBoost**). On the right side of the graph, mean values of boxplots are also shown regarding classifier dispersions obtained by each balance technique.

Table 4 that includes information concerning data complexity involved in categories describes measurements determining overlap and separability between

Table 4. Table of data complexity measurements in the datasets.

Categories	κ	VOR	$J4$	Δ (%)	Imbalance
GO 0003677	1,162564308	1,518414e-45	366,65	42	1:7,68
GO 0003700	1,258898151	8,325292e-43	153,09	54,7	1:10,76
GO 0003824	0,095424389	1,503915e-39	114,67	41,3	1:2,74
GO 0005215	1,275657636	1,974715e-67	3045,37	19,4	1:8,26
GO 0016787	0,004254501	6,654359e-07	-0,472	53,9	1:4,63
GO 0030234	0,265168845	1,247835e-26	14,712	79,3	1:23,87
GO 0030528	0,954652410	1,125151e-37	255,43	37,6	1:7,22

classes (VOR, $J4$, κ). Also included measurements of nonlinearity in the classifiers (Δ) are compared to information of imbalance level for each used dataset. This comparison is intended to provide information about difficulty to induce reliable learning models in each biclass problem.

Measures such as $J4$ and κ tend to be favorable as they increase in value, indicating a greater separability, otherwise VOR tends to be better as its value approaches zero, indicating a smaller area of overlap. According to the values given in Table 4, the most complex space is the set belonging to Hydrolase activity (GO 0016787), showing a low value at $J4$ and VOR highest compared to values achieved by other classes. This fact is proved by the results obtained for this class, as seen in Fig. 2(a), where all techniques show poor performance balance. This suggests that a very poor data representation is present in this class.

Also, as seen from values listed in Table 4, level of imbalance is not as significant as compared with the values of overlap between the data. Then, one can infer that data complexity may deteriorate more severely the learning process in protein prediction compared to the class imbalance. But this happens only when level of overlap and separability is to big compared with imbalance ratio itself. Therefore, it is convenient to use complexity measures as a complement to the imbalance degree to be certain about problem complexity.

Despite observed complexity, the best behavior for Hydrolase activity is obtained by SPSO that provides a value of geometric mean (GM) and ROC area (AUC) just over 50 %, but with very low dispersion in the prediction. Yet, obtained difference is not representative if compared to the performance of CSCu method. In datasets with higher imbalance between categories (such as Enzyme regulator activity and Sequence-specific DNA binding transcription factor activity (GO 0030234 and GO 0003700)), it is worth noting that CSCu performs considerable superiority over other compared techniques. In fact, its performance overcomes in five of the seven categories (GO 0003677, GO 0003700, GO 0003824, GO 0030234 and GO 0030528), while for the remaining 2 sets (GO 0005215 and GO 0016787), it is one of the highest performing prediction techniques, as seen in Table 5.

Table 5. Prediction performances with several balancing strategies.

Categories	SMOTE		SPSO		CS		CSCu		MC		MCCu		Ada	
	AUC	GM	AUC	GM	AUC	GM	AUC	GM	AUC	GM	AUC	GM	AUC	GM
GO 0003677	0,693	0,668	0,708	0,707	0,615	0,519	**0,797**	**0,796**	0,684	0,659	0,718	0,713	0,766	0,747
GO 0003700	0,654	0,599	0,721	0,721	0,679	0,629	**0,815**	**0,810**	0,617	0,566	0,668	0,655	0,773	0,744
GO 0003824	0,664	0,658	0,667	0,667	0,53	0,292	**0,671**	**0,671**	0,618	0,592	0,661	0,654	0,599	0,536
GO 0005215	0,778	0,752	0,811	0,81	0,643	0,562	0.815	0.810	0,803	0,788	**0,839**	**0,835**	0,812	0,766
GO 0016787	0,505	0,405	**0,516**	**0,513**	0,497	0,188	0.491	0.473	0,499	0,395	0,499	0,443	0,485	0,128
GO 0030234	0,568	0,429	0,663	0,642	0,618	0,613	**0,696**	**0,683**	0,515	0,205	0,617	0,518	0,675	0,502
GO 0030528	0,659	0,621	0,717	0,714	0,595	0,493	**0,784**	**0,783**	0,68	0,662	0,676	0,66	0,723	0,691
Total	0,646	0,59	0,686	0,682	0,596	0,47	**0,724**	**0,718**	0,63	0,552	0,668	0,64	0,69	0,588

On the other hand, both AdaBoost and SMOTE techniques obtain the worst prediction results, especially, in Hidrolase activity, Enzyme regulator activity, and Transcription regulator activity (GO 0016787, GO 0030234, and GO 0030528). Since there is a high probability of inducing extra noise during training set when synthetic samples are added, therefore, we may infer that in the presence of sets with high overlap, oversampling technique is not an option making unreliable the model for prediction of molecular functions. In case of AdaBoost, a high overlap may decrease considerably the generalization capability of used classifier, which mostly is forced to have complex decision boundaries.

As shown in Fig. 2 and Table 5, the use of CuckooCost improves the performance of considered methods based on cost sensitive learning (CS, MC, CSCu, MCCu). Moreover, it clearly shows a substantial improvement in Meta-Cost and cost sensitive learning in overall performance (increased GM and AUC), as well as the reliability of the results by decreasing the classification dispersion in every category. Although MetaCost tends to improve when using CuckooCost strategy in transporter activity (GO 0005215), still there is a slight increase in terms of the variability of the results. MetaCost accomplishes the resampling procedure using Bootstrap strategy, when taking a portion of the training set to create a subset in each iteration. Further, each subset is taken by a number of base classifiers equal to the number of iterations for the algorithm selected and the final classification decision is made in committee by a vote of each classifier. So, if the number of iterations in MetaCost is not adequate and additionally the dataset has a substantial degree of imbalance, as it is in this case, the number of samples of interest, i.e., the samples belonging to this category used for each base classifier might not be enough. As a result, the variability of performance increases.

For all considered categories, generally, there exist cases where some balance techniques show very similar values of μ_G compared with their $\mho$ values, mainly, in SPSO and CSCu. It happens, particularly, when the numeric difference between sensitivity and specificity becomes too close.

Assuming $\xi = S_e - S_p$, then, it holds that:

$$T_P' = \frac{T_P}{T_P + F_N} = S_e$$

$$F_P' = \frac{F_P}{T_N + F_P}$$

However, $F_P/(T_N + F_P) = 1 - T_N/(T_N + F_P) = 1 - S_p$. Moreover, taking into account that the ROC curves shows a comparison between T_P' vs F_P', then, $\mho(T_P', F_P') = (1 + T_P' - F_P')/2$. So, $\mho$ (AUC) can be expressed in terms of S_e and S_p, as follows:

$$\mho(S_e, S_p) = \frac{1 + S_e - (1 - S_p)}{2}$$

$$= \frac{S_e + S_p}{2}$$

Therefore expressing the sensitivity in terms of specificity, i.e., $S_e = S_p + \xi$, one can infer that if the numeric distance between sensitivity and specificity is shortened, that is, $\xi \to 0$, then:

$$\lim_{\xi \to 0} \mho(S_e, S_p) = \lim_{\xi \to 0} \frac{S_e + S_p}{2}$$

$$= \lim_{\xi \to 0} \frac{S_p + S_p + \xi}{2} = S_p$$

Now, if considering the geometric mean, the following holds:

$$\lim_{\xi \to 0} \mu_G(S_e, S_p) = \lim_{\xi \to 0} \sqrt{S_e\, S_p}$$

$$= \lim_{\xi \to 0} \sqrt{(S_p + \xi)S_p} = S_p$$

So, the above expressions indicate that if ξ tends to be much smaller, both $\mho$ and μ_G tend increasingly to the same value.

The case when $\mho = \mu_G$ also takes place when the balancing techniques have very little classification bias. This fact can be corroborated using the relative sensitivity (R_S) [22]: If $R_S \to 1$, then, $S_e/S_p \to 1$, in turn, $S_e = S_p$.

The R_S values for each technique are shown in Table 6. As seen, SPSO and CSCu techniques show less bias in their classification performance having values more close to one. That is, both classifiers tend to get a trade-off between sensibility and specificity values, as shown with the points in Fig. 2, for which $\mho = \mu_G$. On contrast, SMOTE tends to get more specificity, although sampling techniques try to become more sensitive to increase distribution of samples in those categories having lower representation. Lastly, it must be quoted that both CS and MC carry out quite a substantial improvement when they use CuckooCost to optimize their parameters. Initially, CS is to sensitive but it has a small specificity, contrary case to MC, which has a big specificity. When CuckooCost is used, both strategies accomplish similar performance, specially, in terms of CS.

Table 6. Table of relative sensitivity.

Categories	SMOTE	SPSO	CS	CSCu	MC	MCCu	Ada
GO 0003677	0,582	0,996	3,301	1.129	0,582	0,796	1,572
GO 0003700	0,427	1,067	2,185	1.239	0,433	0,676	1,701
GO 0003824	0,766	0,973	11,057	1.045	0,555	0,755	0,406
GO 0005215	0,593	1,002	2,885	0.794	0,688	0,842	0,762
GO 0016787	0,251	1,234	25,892	0.576	0,241	0,371	0,017
GO 0030234	0,208	1,652	0,783	0.684	0,044	0,297	0,269
GO 0030528	0,503	1,203	3,541	1.120	0,631	0,651	1,824

6   Conclusions and Future Work

A method to optimize free parameters associated to cost sensitive learning, which is applied to prediction of molecular functions in embryophita plants, is proposed. The method is devoted to rule directly sensitivity and specificity on classifier performance (related to the costs involved misclassifying samples belonging to each category). The optimization is carried out over cost matrix elements, which are tuned by adapting those elements outside the main diagonal, in order to build the cost ratio. The variation of the cost ratio, along with the classification parameters are used as hyperparameters in the optimization problem, since the metric intrinsically modifies the fitness function. To this purpose, a metaheuristic optimization technique called Cuckoo Search is suggested. The methodology takes as fitness function both the maximization of ROC area (AUC) and minimization of total cost; being both variables important in the classification model. This work shows that the use of models based on cost sensitivity learning are competitive, reliable, and even superior to other balance techniques in the state of the art, specially, in applications related to bioinformatics. As future work, the approach of new fitness functions that lead to better classification results is to be considered.

Acknowledgements. This work is partially funded by the Research office (DIMA) at the Universidad Nacional de Colombia at Manizales and the Colombian National Research Centre (COLCIENCIAS) through grant No.111952128388 and the *"jovenes investigadores e innovadores - 2010 Virginia Gutierrez de Pineda"* fellowship.

References

1. Aebersold, R., Mann, M., et al.: Mass spectrometry-based proteomics. Nat. **422**(6928), 198–207 (2003)
2. Allison, D.B., Cui, X., Page, G.P., Sabripour, M.: Microarray data analysis: from disarray to consolidation and consensus. Nat. Rev. Genet. **7**(1), 55–65 (2006)

3. Ashburner, M., Ball, C.A., Blake, J.A., Botstein, D., Butler, H., Cherry, J.M., Davis, A.P., Dolinski, K., Dwight, S.S., Eppig, J.T., et al.: Gene ontology: tool for the unification of biology. Nat. Genet. **25**(1), 25 (2000)
4. Basu, M.: Data Complexity in Pattern Recognition. Springer, New York (2006)
5. Chawla, N.V., Bowyer, K.W., Hall, L.O., Kegelmeyer, W.P.: Smote: synthetic minority over-sampling technique. J. Artif. Intell. Res. **16**, 321–357 (2002)
6. Dietterich, T.G.: An experimental comparison of three methods for constructing ensembles of decision trees: bagging, boosting, and randomization. Mach. Learn. **40**(2), 139–157 (2000)
7. Ding, Z.: Diversified ensemble classifiers for highly imbalanced data learning and its application in bioinformatics. Ph.D thesis, Georgia State University (2011)
8. Domingos, P.: Metacost: a general method for making classifiers cost-sensitive. In: Proceedings of the Fifth ACM SIGKDD International Conference on Knowledge Discovery and Data Mining, pp. 155–164. ACM (1999)
9. Elkan, C.: The foundations of cost-sensitive learning. In: International Joint Conference on Artificial Intelligence, vol. 17, pp. 973–978. Lawrence Erlbaum Associates Ltd (2001)
10. García-López, S., Jaramillo-Garzón, J.A., Higuita-Vásquez, J.C., Castellanos-Domínguez, C.G.: Wrapper and filter metrics for PSO-based class balance applied to protein subcellular localization. In: 2012 Biostec-Bioinformatics (2012)
11. Grzymala-Busse, J.W., Stefanowski, J., Wilk, S.: A comparison of two approaches to data mining from imbalanced data. J. Intell. Manuf. **16**(6), 565–573 (2005)
12. He, H., Garcia, E.A.: Learning from imbalanced data. IEEE Trans. Knowl. Data Eng. **21**(9), 1263–1284 (2009)
13. Ho, T.K., Basu, M.: Complexity measures of supervised classification problems. IEEE Trans. Pattern Anal. Mach. Intell. **24**(3), 289–300 (2002)
14. Jain, E., Bairoch, A., Duvaud, S., Phan, I., Redaschi, N., Suzek, B., Martin, M., McGarvey, P., Gasteiger, E.: Infrastructure for the life sciences: design and implementation of the uniprot website. BMC Bioinform. **10**(1), 136 (2009)
15. Jaramillo-Garzón, J.A., Gallardo-Chacón, J.J., Castellanos-Domínguez, C.G., Perera-Lluna, A.: Predictability of gene ontology slim-terms from primary structure information in embryophyta plant proteins. BMC Bioinform. **14**(1), 68 (2013)
16. Larrañaga, P., Calvo, B., Santana, R., Bielza, C., Galdiano, J., Inza, I., Lozano, J.A., Armañanzas, R., Santafé, G., Pérez, A., et al.: Machine learning in bioinformatics. Briefings Bioinform. **7**(1), 86–112 (2006)
17. Liu, X.Y., Zhou, Z.H.: The influence of class imbalance on cost-sensitive learning: an empirical study. In: 2006 Sixth International Conference on Data Mining, ICDM'06, pp. 970–974. IEEE (2006)
18. Liu, X.-Y., Zhou, Z.-H.: Towards cost-sensitive learning for real-world applications. In: Cao, L., Huang, J.Z., Bailey, J., Koh, Y.S., Luo, J. (eds.) PAKDD Workshops 2011. LNCS, vol. 7104, pp. 494–505. Springer, Heidelberg (2012)
19. Polikar, R.: Ensemble based systems in decision making. IEEE Circuits Syst. Mag. **6**(3), 21–45 (2006)
20. Schapire, R.E.: A brief introduction to boosting. In: International Joint Conference on Artificial Intelligence, vol. 16, pp. 1401–1406. Lawrence Erlbaum Associates Ltd (1999)
21. Sonnenburg, S., Schweikert, G., Philips, P., Behr, J., Rätsch, G.: Accurate splice site prediction using support vector machines. BMC Bioinform. **8**(Suppl 10), S7 (2007)
22. Su, C.T., Hsiao, Y.H.: An evaluation of the robustness of MTS for imbalanced data. IEEE Trans. Knowl. Data Eng. **19**(10), 1321–1332 (2007)

23. Mohanna, E., Valian, E., Tavakoli, S.: Improved cuckoo search algorithm for global optimization. Int. J. Commun. Inf. Technol. **1**(1), 31–44 (2011)
24. Yang, P., Xu, L., Zhou, B.B., Zhang, Z., Zomaya, A.Y.: A particle swarm based hybrid system for imbalanced medical data sampling. BMC Genomics **10** (Suppl 3), S34 (2009)

Bio-inspired Systems
and Signal Processing

Dynamic Changes of Nasal Airflow Resistance During Provocation with Birch Pollen Allergen

Tiina M. Seppänen[1](✉), Olli-Pekka Alho[2], and Tapio Seppänen[1]

[1] Department of Computer Science and Engineering,
University of Oulu, Oulu, Finland
{tiina.seppanen, tapio.seppanen}@ee.oulu.fi
[2] Department of Otorhinolaryngology, University of Oulu, Oulu, Finland
olli-pekka.alho@oulu.fi

Abstract. Over 500 million people suffer from allergic rhinitis around the world. This huge problem causes, in addition to individual impacts, a substantial economic burden to societies. There is a lack of an objective measurement method producing a reliable, accurate and continuous measurement data about the dynamic changes in nasal function. Here, a method to assess the nasal airflow resistance as a continuous signal is proposed and used to compute resistance values during the birch pollen provocation test. The required pressure recording is measured using a nasopharyngeal catheter and the flow recording is measured using respiratory effort belts calibrated with the new method. Ten birch pollen allergic and eleven non-allergic volunteers were challenged with control solution and allergen solution. Continuous nasal airflow resistance signals were computed and analyzed for the dynamic changes in the nasal airflow resistance. The derived signals show in great detail the intensity and timing differences in subjects' reactions. Quantitative results of resistance changes indicate that allergic and non-allergic subjects can be differentiated in a statistically significant degree using the proposed method. The method opens entirely new possibilities to research accurately the dynamic changes in non-stationary nasal function and could increase the reliability and accuracy of diagnostics and assessment of the effect of nasal treatments.

Keywords: Allergy · Challenge · Nasal resistance · Provocation test · Respiration · Polygraphic recorder · Spirometer

1 Introduction

Allergic rhinitis is a major global health problem due to its prevalence, impact on quality of life, along with the impact on work/school performance and productivity. It is also a substantial economic burden to societies. Allergic rhinitis is a systemic inflammatory condition which is inheritable and links with other illnesses like asthma [1–3]. Patients from all countries, ethnic groups and ages suffer from allergic rhinitis, and the prevalence is increasing in most countries of the world to the extent that in some countries over 50 % of adolescents are reporting such symptoms [4]. Using a conservative estimate, over 500 million people suffer from allergic rhinitis around the

© Springer-Verlag Berlin Heidelberg 2014
M. Fernández-Chimeno et al. (Eds.): BIOSTEC 2013, CCIS 452, pp. 225–239, 2014.
DOI: 10.1007/978-3-662-44485-6_16

world. Therefore, specific guidelines and programs on the problem have been released, for example by European Union and World Health Organization [5–7].

Allergic rhinitis is diagnosed when the patient has allergic symptoms and specific antigens are detected in the blood. Typical symptoms include nasal obstruction, rhinorrhea, nasal itching, sneezing and eye irritation [6]. The presence of the allergy can also be verified by nasal provocation tests in which subjects are challenged with the suspected allergen. After provocation, changes in their subjective feelings of symptoms are recorded and the amount of secretions and the respiratory function of the nose are measured. Visual Analogue Scale (VAS) is commonly used as a method to measure subjective feelings of nasal obstruction [6]. Nasal provocation tests are done for example in the diagnosis of chronic rhinitis, at the beginning of desensitization and in the diagnosis of work-related respiratory diseases (occupational asthma, occupational rhinitis). To rule out non-specific nasal hyper-reactivity, the nasal mucosa is usually challenged with a control solution before the actual allergen solution.

Nasal function is difficult to quantify directly by clinical examination, which calls for objective measurement methods. Examples of these include peak inspiratory flow measurement (PNIF), acoustic rhinometry and rhinomanometry [8, 9]. PNIF is a noninvasive method that measures the nasal airflow during maximal forced nasal inspiration. Acoustic rhinometry, in its turn, assesses nasal geometry by measuring cross-sectional area of the nose as a function of the distance from the nostril. Rhinomanometer involves the simultaneous measurement of pressure and airflow from the values of which nasal airflow resistance is determined [10]. The resistance is characteristically described as a number that derives from one or more breathing cycles of data. In nasal provocation tests, the major response to measure is the rise in nasal airflow resistance. The rise is rapid (seconds or minutes) and the timing differs in different individuals. This makes it difficult to be detected with a rhinomanometer. One possibility is to determine the resistance with the rhinomanometer in certain time-intervals, but this has been indicated to give inconsistent and variable results with low reproducibility [11–13].

There is thus clearly a demand for a measurement method giving a reliable, accurate and continuous measurement data about the nasal airflow resistance. This kind of measurement could provide much more information about the fast changes in nasal function for instance during provocation tests.

In this paper, a new method to assess nasal function is proposed that produces nasal airflow resistance as a continuous signal at any sampling frequency allowing for analysis of dynamic changes in the resistance. The method is used to study nasal responses of test subjects from birch pollen allergic group and control group.

2 Methods

2.1 Study Subjects

The study protocol was approved by the institutional Ethics Committee of Oulu University Hospital. In Finland, the birch pollen is a common cause of the allergic symptoms such as intermittent seasonal allergic rhinitis for which reason it was chosen as a substance for provocation tests. Ten (seven males) birch pollen allergic and eleven (eight males) non birch pollen allergic adult volunteers were recruited. They gave written informed consent and their background information was gathered using a questionnaire. The mean (SD) age of the allergic and non-allergic subjects was 24 (1) and 24 (3) years, respectively. The subjects had to be free of heart diseases, brain circulatory disorders and surgical operations of nose. Pregnant ones were rejected as well. The subjects were not allowed to be under medication that affects the function of their nose during a specific time period before the measurement. Additionally, they had to be free of any acute respiratory symptoms during the prior two weeks to the measurements. Before measurement, they were not allowed to have a smoke for four hours and heavy meal, caffeine or other stimulants for two hours.

Measurements were carried out in the spring time before the birch pollen season. An ear, nose and throat specialist examined all the subjects. Before measurements, the total IgE and the specific IgE for birch pollen were determined from blood to verify their allergy or non-allergy status (Table 1). Allergen specific IgE value under 0.35 kU/l means negative result for allergen. Values 0.35–0.69 kU/l, 0.70–3.49 kU/l, 3.50–17.4 kU/l and values above 17.5 kU/l represent the low level, the moderate level, the high level and the very high level of allergen specific IgE, respectively. As can be seen from Table 1, allergic subjects had different levels of allergy.

Table 1. Total IgE and specific IgE for birch pollen values [kU/l] for both groups.

Allergic subject	Total IgE	Specific IgE	Non-allergic subject	Total IgE	Specific IgE
1	119	1.58	1	12	0.00
2	143	55.70	2	6	0.00
3	504	66.00	3	9	0.00
4	59	26.20	4	9	0.02
5	68	51.20	5	27	0.00
6	128	13.20	6	13	0.00
7	95	16.50	7	19	0.18
8	24	3.23	8	3	0.00
9	60	2.18	9	8	0.00
10	165	57.40	10	24	0.00
			11	9	0.00

2.2   Measurement Devices

The pressure and respiratory effort belt signals were recorded with a polygraphic recorder (TrackIt, Lifelines Ltd, Hampshire, UK) with the sampling frequency of 100 Hz. The pressure recording was measured with a 1-mm diameter nasopharyngeal catheter (CH 06, Unomedical A/S, Denmark). The differential pressure sensor (Braebon Ultima Dual Airflow Pressure Transducer) referenced to the atmospheric pressure was connected to the catheter. A sterile filter (Minisart, Sartorius Ltd, Epsom, Uk) was used for protection in between the catheter and the pressure sensor. The pressure data of the recorder was calibrated to physical units (Pascal). Respiratory effort belts (Ultima SmartBelt, Braebon Medical Corp., Ogdensburg, NY, USA) were attached to the subjects' rib cage and abdomen. For calibrating the signals from respiratory effort belts, simultaneous respiratory airflow signal was recorded with a spirometer (SpiroStar USB, Medikro Oy, Kuopio, Finland).

2.3   Challenge Protocol

An immunologically standardized, water-based commercial 1:10 000 SQU/ml extract of birch (Allergologisk Laboratorium A/S, Copenhagen, Denmark) was used in the nasal provocation test. The diluent solution of the allergen extract was used as a control solution (ALK, A/S, Copenhagen, Denmark). Both solutions were administered into the nasal cavities (bilateral challenge) by pump spray.

At first, the rib cage belt was placed on the xyphoid process and the abdominal belt near the umbilicus. Then, the subjects sat peacefully for a period of 30 min prior to the measurement to adapt themselves to the environment. They were instructed to sit in back upright position avoiding movements and speaking during all measurements. First, flow and respiratory effort belt signals were recorded for one minute with the spirometer and polygraphic recorder, respectively (Fig. 1). The data was used for calibrating the respiratory effort belt signals to flow signal as described in Sect. 2.4. The respiratory effort belts were kept on during the whole measurement protocol.

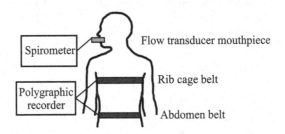

Fig. 1. Measurement of airflow and respiratory effort belt signals for belt calibration

Next, the spirometer was removed from the subject. A nasopharyngeal catheter was inserted 8 cm deep along the floor of nasal cavity into the nasopharynx, the tip of the catheter lying 1 cm anterior from the back wall of the nasopharynx. Air was blown with the syringe through the catheter to inhibit the nasal secrete blocking it. This was done before each protocol phase and every time that the catheter blocking was detected. Measurement setup of the signals needed for the computation of the nasal airflow resistance is depicted in Fig. 2.

At the first protocol phase, the baseline was recorded for 10 min. At the second protocol phase, to rule out nonspecific nasal hyper-reactivity, the nasal mucosa was challenged with a control solution sprayed carefully on the anterior nasal mucosa of both nasal cavities. After that, pressure and airflow were recorded for 5 min. At the third protocol phase, the allergen solution was inserted carefully on the anterior nasal mucosa of both nasal cavities, after which pressure and airflow were recorded for 20 min. After inserting the solution, the recording was started as soon as possible but first waiting for the reactions such as sneezing to settle. After every phase, the subjects were asked about their worst sensation of obstruction in VAS scale during the phase. The VAS scale was from zero (totally open) to seven (totally obstructed). Finally, the nasopharyngeal catheter was removed and the calibration data collection was repeated with the spirometer.

After recording, all the signals were validated manually by using specially-made visualization software. All detected disturbances, originated for example from sneezing, snuffling, mouth opening and moving, were deleted from signals before analysis. Specific care was taken to maintain the correct synchrony between the signals.

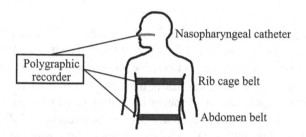

Fig. 2. Measurement of the nasal pressure and respiratory effort belt signals for the nasal airflow resistance computation.

2.4   Calibration Method of the Respiratory Effort Belts

A prediction of the respiratory airflow $\hat{y}$ is commonly calculated from the dimensional changes of the respiratory effort belt signals by applying the method of multiple linear regression [14]. The conventional model is established by fitting the following model to the time-synchronized signals:

$$y = \beta_1 x_1 + \beta_2 x_2 + \varepsilon, \tag{1}$$

where the respiratory effort belt signals x_1 and x_2 from the rib cage and abdomen, respectively, are the predictor variables, parameters β_1 and β_2 are regression coefficients and ε is a zero-mean Gaussian error. In this model, one sample of each predictor variable is used at a time to predict the response variable y.

In this study, we use our previously published calibration method [15], which is based on the MISO (Multiple-Input Single-Output) system model consisting of a polynomial FIR (Finite Impulse Response) filter bank and a delay element, see Fig. 3.

This method extends the conventional one in an important way: it uses a number of N consecutive signal samples and linear filtering for each prediction. In the model representation, vector notation (bold letter type) is used below to denote that N consecutive signal samples of each predictor variable are included as components, and that the parameters are now vectors of dimension N. The calibration model can be established as follows:

$$y = \boldsymbol{\beta}_1^T \boldsymbol{x}_1 + \boldsymbol{\beta}_2^T \boldsymbol{x}_2 + \varepsilon, \tag{2}$$

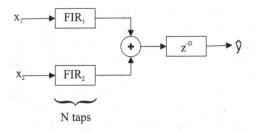

Fig. 3. Extended respiratory effort belt calibration method as a MISO system.

where $\boldsymbol{\beta}_1^T$ and $\boldsymbol{\beta}_2^T$ denote the N tap coefficients of filters FIR_1 and FIR_2 on Fig. 3, respectively. $\boldsymbol{x}_1$ and $\boldsymbol{x}_2$ are vectors including N consecutive samples from the rib cage signal and abdomen signal, respectively: $\boldsymbol{x}_1 = [x_{11}, x_{12}, \ldots, x_{1N}]^T$ and $\boldsymbol{x}_2 = [x_{21}, x_{22}, \ldots, x_{2N}]^T$. Superscript T denotes matrix transpose in the formula.

The spirometer signal and simultaneous respiratory effort belt signals are input to the regression analysis which yields optimal tap coefficients ($\boldsymbol{\beta}_1^T$ and $\boldsymbol{\beta}_2^T$) and minimal prediction error for both filters. During the calibration, tap coefficients are estimated with the method of least-squares.

There is always a delay between the spirometer signal and the respiratory effort belt signals due to two reasons. Firstly, spirometer measures airflow from mouth and respiratory effort belts signals are measured from the rib cage and abdomen. A delay occurs due to the time it takes for the airflow to propagate from the mouth to the lungs and vice versa. Secondly, each measuring device has internal delays. For these reason, the delay element z^{-D} is included at the output, see Fig. 3. The filter tap coefficients were solved for each feasible delay candidate as described above. The minimum error ε in the respiratory airflow prediction determined the optimal delay value.

In our previous study [15], the 0.3 s time window of FIR filters was found to produce the best respiratory airflow prediction. Thus, we used the same window size for FIR filters in this study as well.

2.5   Calculation of Continuous Nasal Airway Resistance Values

Our present measurement system acquires the pressure signal by using a small naso-pharyngeal catheter and the flow signal from the calibrated respiratory effort belts [16–19]. As the pressure signal is measured from inside the nose and the flow estimate signal from the rib cage and abdomen, a small lag between the signals occurs. The cross-correlation function between the signals with a predefined range of lag values is calculated and the maximum peak is found for correcting the misalignment.

In principle, the resistance R is defined by Ohm's law as $R = P / Y$, where P and Y denote pressure difference and flow, respectively. In rhinomanometry, the resistance is read out at a certain reference pressure value such as 75 Pa or 150 Pa. Mathematical models, such as Broms [20], offer parametric means to describe the nonlinear pressure/flow relationship. In this work, we adopted the model of Broms, see Fig. 4.

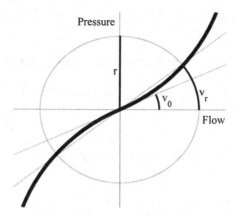

Fig. 4. A diagram of Broms model.

In the model, the pressure/flow relationship is considered to follow the equation

$$v_r = v_0 + cr, \tag{3}$$

where v_r is the angle when radius is r, v_0 is the angle in the origin and c is a constant describing the curvature of the trace. The resistance in radius r, indicated by R_r, is given by

$$R_r = x \tan v_r, \tag{4}$$

where x is a normalization factor depending on the data in hand. It was set to 10 by Broms, because it was best suited for their data [20]. In our study, we adapt x to signal variability (as explained below) and set

$$x = \frac{\sigma_P}{\sigma_Y}. \tag{5}$$

The pressure values can be several orders of magnitude smaller than the flow values. This often makes calculations unstable with noisy data, because the constant c and the angle v_r vary in a relative small range. Therefore, before calculations, we first normalize the pressure and flow values by dividing them with the corresponding standard deviations σ_P and σ_Y of the signals. The normalized data are used in the calculations and the original units are finally restored by using (5) in (4).

The Broms model is applied such that all the measurement data are used for identifying the model parameters in order to calculate a resistance value. Thus, the model expects the data to be stationary, while the nasal system is not stationary. We propose an extension to the Broms model such that it can be used for calculating a continuous nasal airflow resistance value through model adaptation to varying signal statistics.

We use the least-mean-square (LMS) algorithm to adaptively adjust the parameters v_0 and c in time. The normalized pressure signal P' and the flow estimate signal Y' are the filter inputs. The filter length of only one time sample was found to be sufficient. The update formulas of v_0 and c are

$$v_0(k + 1) = v_0(k) + \mu(k)\{v_r(k) - [v_0(k) + c(k)r(k)]\} \tag{6}$$

$$c(k + 1) = c(k) + \mu(k)r(k)\{v_r(k) - [v_0(k) + c(k)r(k)]\}, \tag{7}$$

where

$$r(k) = \sqrt{P'^2(k) + Y'^2(k)} \tag{8}$$

$$v_r(k) = \tan^{-1}\frac{P'(k)}{Y'(k)} \tag{9}$$

The initial values for $v_0(0)$ and $c(0)$ were set as $\pi/4$ and 0.1, respectively. Then, the LMS filter was run with data samples in reversed order to initialize filter coefficients properly. The learning rate parameter $\mu(k)$ was defined as

$$\mu(k) = \frac{10^{-3}}{1 + e^{4 - 10r(k) + 2r^2(k)}}. \tag{10}$$

The learning rate parameter was formulated to dampen the parts of signals, which could potentially produce noisy results. More details can be found in our previous paper [16].

Instantaneous resistance values are calculated over the whole measurement data and shown as dynamic plots over time.

2.6 Statistics

Statistical significance of resistance changes in the test subjects was assessed by Wilcoxon signed-rank test. Statistical significance between the subject groups, in its turn, was assessed by Wilcoxon rank-sum test (Mann-Whitney test). The null-hypothesis for statistical tests was that there are no differences in the medians of given data sets. Statistical dependence between variables of the subject groups was assessed by calculating the Spearman's correlation coefficient. It is a measure of statistical dependence between two variables.

3   Results and Discussion

3.1   Dynamic Changes in Nasal Airflow Resistance

Pressure and respiratory effort belt signals were recorded 10 min in baseline, 5 min after the control challenge and 20 min after the allergen challenge. At first, the respiratory effort belts were calibrated from the first 1 min calibration recording. Then, continuous nasal airflow resistance signals were computed from the pressure and respiratory effort belt signals. Small gaps in the signals can be seen in the figures due to removing of the artifacts during manual validation.

An example case of the continuous resistance signals of birch pollen allergic subject is shown in Fig. 5. The subject shows a non-specific response after having been challenged with the control solution. After the reaction in the beginning, the resistance curve returns approximately to the same level as it was at the *baseline* (about 60 Pa/dm^3/s). After having been challenged with the allergen, a significant allergic reaction occurs. The resistance curve rises immediately and continues rising still about 8 minutes before settling down at about the level of 550 Pa/dm^3/s.

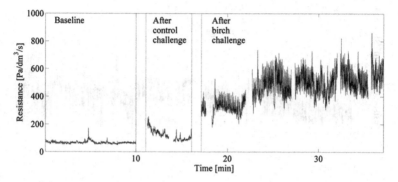

Fig. 5. Resistance curve for allergic subject with a short non-specific response to the control solution and fast and significant reaction to the birch challenge. Each pair of vertical bars marks the period of a challenge.

An example of continuous resistance signal from another birch pollen allergic subject is shown in Fig. 6. After the birch challenge, there is a much slower rise in the resistance than in the previous case (Fig. 5). The resistance rises about 11 min from the level of 150 Pa/dm^3/s in the *baseline* and *after control challenge* to the level of about 500 Pa/dm^3/s.

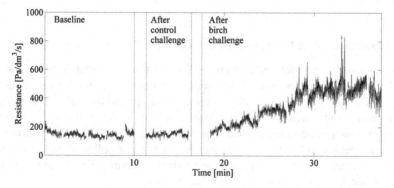

Fig. 6. Resistance curve for allergic subject with a slower significant reaction to the birch challenge.

In this study, we had a unique opportunity to assess also the rise times of resistance and relative rise of resistance after the challenge because we had accurate continuous resistance curves at our disposal. In the group of birch pollen allergic subjects, the median rise time of resistance before it settled to the stable level after the allergen challenge was 9 min. Median of relative rise of resistance after the birch challenge was 76 % and the rise varied from 17 % to 511 %. There was only weak correlation between the rise time and relative rise of resistance after the birch challenge (R = 0.318) confirming the fact that subjects react differently in intensity and in timing to the challenge.

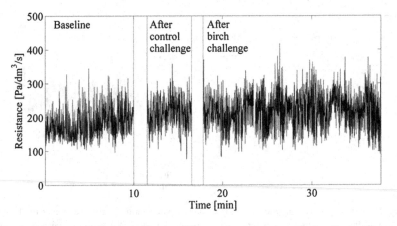

Fig. 7. Resistance curve for non-allergic subject with no reactions during the measurement.

Next, two cases of resistance curves from non-allergic subjects is presented. In Fig. 7, no reaction to either of the challenge solution is detected, but the resistance stays in about 200 $Pa/dm^3/s$ during the whole measurement protocol. In Fig. 8, a short initial reaction in the resistance curve is obvious immediately after the control challenge and the allergen challenge. After the reactions, stable resistance curves follow. These reactions can be non-specific reactions: the nose reacts immediately to any manipulation.

3.2 Resistance Level Changes Between Protocol Phases

First, the respiratory effort belts were calibrated from the first 1 min calibration recording. The continuous nasal airflow resistance was then computed for the last two minutes of all phases: *baseline, after control challenge* and *after allergen challenge*. The last two minutes was chosen to get stable parts of the data and also to exclude the short-term non-specific reactions from the analysis. Tables 2 and 3 present the VAS values and mean nasal airflow resistances for each birch pollen allergic subject and non-allergic subject in the three phases, respectively. Group medians instead of means are given because data size is small and non-normal.

In the group of birch pollen allergic subjects, there was a statistically significant change in the resistance values between the *baseline* and *after allergen challenge*

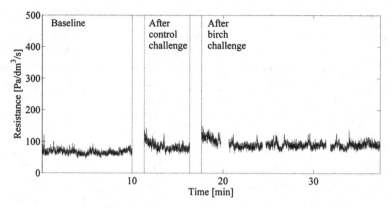

Fig. 8. Resistance curve for non-allergic subject with short non-specific responses after control challenge and after allergen challenge.

Table 2. VAS and nasal airflow resistance values for allergic subjects.

Subject	Baseline		After control challenge		After allergen challenge	
	VAS	Resistance [Pa/dm³/s]	VAS	Resistance [Pa/dm³/s]	VAS	Resistance [Pa/dm³/s]
1	1	151	1	131	3	194
2	2	68	2	60	5	177
3	1	138	1	154	3.5	511
4	2	163	2	205	4	407
5	1	63	1	88	3	545
6	1	146	1	156	4	478
7	2	337	2	535	5	653
8	1	87	1	103	2	168
9	2	131	3	111	3	142
10	1	172	1	179	3	279
Median	1	142	1	143	3.3	343

($p = 0.005$) and in the resistance values between the *after control challenge* and *after allergen challenge* phases ($p = 0.005$). There was no statistically significant change in the resistance values between the *baseline* and *after control challenge* phases ($p = 0.202$). Respectively, in the group of non-allergic subjects, there was no statistically significant change in the resistance values between the *baseline* and *after allergen challenge* ($p = 0.068$) and in the resistance values between the *after control challenge* and *after allergen challenge* ($p = 0.248$). However, there was a statistically significant change in the resistance values between the *baseline* and *after control challenge* phases ($p = 0.005$). Both groups included subjects for whom resistance values increased transiently after control challenge: 50 % in the allergic group and 70 % in the control

Table 3. VAS and nasal airflow resistance values for non-allergic subjects.

Subject	Baseline		After control challenge		After allergen challenge	
	VAS	Resistance [Pa/dm^3/s]	VAS	Resistance [Pa/dm^3/s]	VAS	Resistance [Pa/dm^3/s]
1	1	402	1	462	1	374
2	1	70	1	80	1	92
3	1	131	2	150	1	130
4	1	128	1.5	134	1	124
5	1	178	1	314	1	299
6	1	212	1	214	1	235
7	1	76	1	89	1	108
8	1	114	1	176	1	150
9	1	187	1	211	1	193
10	1	54	3	54	4	71
11	2	216	2	277	1.5	217
Median	1	131	1	176	1	150

group, respectively. These individuals may have non-specific reactions to any nasal manipulations. With this small group size, a statistically significant resistance change in the group mean can easily occur due to random variation which may explain the result with the control group. To summarize the most important finding, the allergic group showed a large change in absolute resistance values during the provocation test, while the non-allergic group did not.

In the *baseline*, the median resistance was 142 Pa/dm^3/s and 131 Pa/dm^3/s for the allergic and non-allergic group, respectively. There was no statistically significant difference in the resistance between the two groups (p = 0.756). *After control challenge*, the median resistance was 143 Pa/dm^3/s and 176 Pa/dm^3/s for the allergic and non-allergic group, respectively. There was no statistically significant difference in the resistance between the two groups (p = 0.512). *After allergen challenge*, the median resistance was 343 Pa/dm^3/s and 150 Pa/dm^3/s for the allergic and non-allergic group, respectively. In this case, there was a statistically significant difference in the resistance between the two groups (p = 0.020). To summarize, the two study groups differ only in the responses to the allergen challenge.

Next, resistance changes of individual subjects are considered. The median change in the subjects' resistance between the *baseline* and *after allergen challenge* was 177 Pa/dm^3/s and 17 Pa/dm^3/s for the allergic and non-allergic group, respectively. There was a statistically significant difference in the resistance change between the two groups (p = 0.000). The median change in the subjects' resistance between the *after control challenge* and *after allergen challenge* was 118 Pa/dm^3/s and −15 Pa/dm^3/s for the allergic and non-allergic group, respectively. There was a statistically significant difference in the resistance change between the two groups (p = 0.001). The median change in the subjects' resistance between the *baseline* and *after control challenge* was 13 Pa/dm^3/s and 19 Pa/dm^3/s for the allergic and non-allergic group, respectively.

There was no statistically significant difference in the resistance change between the two groups (p = 0.349). To summarize, the individual subjects of the allergic group showed large changes in the absolute resistance values during allergen challenge, while individuals from the non-allergic group did not.

The median of relative change in the subjects' resistance between the *baseline* and *after allergen challenge* was 122 % and 11 % for the allergic and non-allergic group, respectively. There was a statistically significant difference in the relative resistance change between the two groups (p = 0.002). The median of relative change in the subjects' resistance between the *after control challenge* and *after allergen challenge* was 81 % and −7 % for the allergic and non-allergic group, respectively. There was a statistically significant difference in the relative resistance change between the two groups (p = 0.000). The median of relative change in the subjects' resistance between the *baseline* and *after control challenge* was 9 % and 15 % for the allergic and non-allergic group, respectively. There was no statistically significant difference in the relative resistance change between the two groups (p = 0.426). To summarize, large relative changes in the individual resistance values occurred only with the allergic group and in the allergen challenge phase.

There was no statistically significant difference in the VAS values of the *baseline* and *after control challenge* between the two groups (p = 0.251 and p = 0.809, respectively). Instead, there was a statistically significant difference between the groups in the VAS values of the *after allergen challenge* (p = 0.0003). When the change in VAS values between the two phases were studied, no statistically significant difference was found between the two groups in VAS change from *baseline* to *after control challenge* (p = 0.468) and statistically significant difference in VAS change from *baseline* to *after allergen challenge* (p = 0.0006) and from *after control solution* to *after allergen solution* (p = 0.00007). In the group of allergic subjects, a strong correlation was found between the S-IgE value and VAS change from *baseline* to *after allergen challenge* (R = 0.735, p = 0.016) and also between the S-IgE value and VAS change from *after control challenge* to *after allergen challenge* (R = 0.735, p = 0.016).

4   Conclusions

Here, a method to assess nasal function was proposed that produces nasal airflow resistance as a continuous signal at any sampling frequency allowing for analysis of dynamic changes in the resistance. The method uses pressure signal from nasopharyngeal catheter and calibrated respiratory effort signals from rib cage and abdomen. An LMS filter extension to the Broms model was presented that computes continuous resistance and adapts to the time-varying characteristics of the non-stationary nasal functioning.

Continuous nasal airflow resistance curves were presented from selected subjects of two subject groups – birch pollen allergic and non-allergic subjects. These curves demonstrate the dynamic changes in the subjects' nasal airflow resistance during the challenge. From the figures, the timing and intensity of the reactions can be seen in great detail. To our knowledge, this is the first time that it is possible to estimate accurately from the nasal airflow resistance: (1) how fast and strong the allergic response occurs, (2) how long it takes the reaction to settle, and (3) whether short non-specific hyper-reactive responses occur with test subjects.

Quantitative results of nasal airflow resistance changes were presented for two subject groups to demonstrate their reactivity to the birch challenge. The allergic group showed a large change in absolute resistance values during the provocation test, while the non-allergic group did not. The two study groups differ only in the responses to the allergen challenge. The individual subjects of the allergic group showed large changes in the absolute resistance values during allergen challenge, whereas individuals from the non-allergic group did not. It should be noted that large relative changes in the individual resistance values occurred only with the allergic group and in the allergen challenge phase. As a conclusion, allergic and non-allergic subjects can be differentiated with statistically significant difference using the presented method.

The proposed method opens entirely new opportunities to research accurately dynamic changes in non-stationary nasal function. It could increase the reliability and accuracy of diagnostics and assessment of the effect of nasal treatments.

Acknowledgements. Allergy Research Foundation and Finnish Cultural Foundation, North Ostrobothnia Regional Fund are gratefully acknowledged for financial support.

References

1. Bousquet, J., et al.: Allergic ahinitis and its impact on asthma (ARIA): achievements in 10 years and future needs. J. Allergy Clin. Immunol. **130**(5), 1049–1062 (2012)
2. Bousquet, J., Van Cauwenberge, P., Khaltaev, N.: Allergic rhinitis and its impact on asthma. J. Allergy Clin. Immunol. **108**, S147–S334 (2001)
3. Shaaban, R., Zureik, M., Soussan, D., Neukirch, C., Heinrich, J., Sunyer, J., Wjst, M., Cerveri, I., Pin, I., Bousquet, J., Jarvis, D., Burney, P.G., Neukirch, F., Leynaert, B.: Rhinitis and onset of asthma: a longitudinal population-based study. Lancet **372**, 1049–1057 (2008)
4. Sears, M.R., Burrows, B., Herbison, G.P., Holdaway, M.D., Flannery, E.M.: Atopy in childhood. II. Relationship to airway responsiveness, hay fewer and asthma. Clin. Exp. Allergy **23**(11), 949–956 (1993)
5. Van Cauwenberge, P., Watelet, J.B., Van Zele, T., Bousquet, J., Burney, P., Zuberbier, T.: the GA²LEN partners: spreading excellence in allergy and asthma: the GA²LEN (Global allergy and asthma European network) project. Allergy **60**, 858–864 (2005)
6. Bousquet, J., et al.: Allergic rhinitis and its impact on asthma (ARIA) 2008 update (in collaboration with the world health organization, GA²LEN and AllerGen). Allergy **63** (Suppl. 86), 8–160 (2008)
7. Bousquet, J., Dahl, R., Khaltaev, N.: Global Alliance against chronic respiratory diseases. Allergy **62**, 216–223 (2007)
8. Clement, P.A., Gordts, F.: Consensus report on acoustic rhinometry and rhinomanometry. Rhinology **43**, 169–179 (2005)
9. Starling-Schwaz, R., Peake, H.L., Salome, C.M., Toelle, B.G., Ng, K.W., Marks, G.B., Lean, M.L., Rimmer, S.J.: Repeatability of peak nasal inspiratory flow measurements and utility for assessing the severity of rhinitis. Allergy **60**, 795–800 (2005)
10. Chaaban, M., Corey, J.P.: Assessing nasal air flow: options and utility. Proc. Am. Thorac. Soc. **8**(1), 70–78 (2011)

11. Pirilä, T., Talvisara, A., Alho, O.-P., Oja, H.: Physiological fluctuations in nasal resistance may interfere with nasal monitoring in the nasal provocation test. Acta Otolaryngol. **117**, 596–600 (1997)
12. Pirilä, T., Nuutinen, J.: Acoustic rhinometry, rhinomanometry and the amount of nasal secretion in the clinical monitoring of the nasal provocation test. Clin. Exp. Allergy **28**(4), 468–477 (1998)
13. Hohlfield, J.M., Holland-Letz, T., Larbig, M., Lavae-Mokhtari, M., Wierenga, E., Kapsenberg, M., van Ree, R., Krug, N., Bufe, A.: Diagnostic value of outcome measures following allergen exposure in an environmental challenge chamber compared with natural conditions. Clin. Exp. Allergy **40**(7), 998–1006 (2010)
14. Tobin, M.: Breathing pattern analysis. Intensive Care Med. **18**, 193–201 (1992)
15. Seppänen, T.M., Alho, O.-P., Koskinen, M., Seppänen, T.: Improved calibration method of respiratory belts by extension of multiple linear regression. In: Proceedings of the 5th European Conference of the International Federation for Medical and Biological Engineering, vol. 37, pp. 161–164 (2012)
16. Seppänen, T., Koskinen, M., Seppänen, T.M., Raappana, A., Alho, O.-P.: Continuous assessment of nasal airflow resistance by adaptive modeling. Physiol. Meas. **30**, 1197–1209 (2009)
17. Seppänen, T., Koskinen, M., Seppänen, T.M., Alho, O.-P.: Addendum to 'continuous assessment of nasal airflow resistance by adaptive modeling' – technical repeatability. Physiol. Meas. **31**, 1547–1551 (2010)
18. Patent 120132 B Finland
19. Patent 0602755-1 Sweden
20. Broms, P., Jonson, B., Lamm, C.J.: Rhinomanometry. II. A system for numerical description of nasal airway resistance. Acta Otolaryngol. **94**, 157–168 (1982)

Identification of Athletes During Walking and Jogging Based on Gait and Electrocardiographic Patterns

Peter Christ[✉] and Ulrich Rückert

Cognitronics and Sensor Systems Group, CITEC, Bielefeld University,
Universitätsstr. 21-23, 33615 Bielefeld, Germany
{pchrist,rueckert}@cit-ec.uni-bielefeld.de
http://www.ks.cit-ec.uni-bielefeld.de

Abstract. We propose a biometric method for identifying athletes based on information extracted from the gait style and the electrocardiographic (ECG) waveform. The required signals are recorded within a non-clinical acquisition setup using a wireless body sensor attached to a chest strap with integrated textile electrodes. Our method combines both sources of information to allow identification despite severe intra-subjects variations in the gait patterns (walking and jogging) and motion related artefacts in the ECG patterns. For identification we use features extracted in time and frequency domain and a standard classifier. Within a treadmill experiment with 22 subjects we obtained an accuracy of 98.1 % for velocities from 3 to 9 km/h. On a second data set consisting of 9 subjects and two sessions of recording, our method achieved 93.8 % despite variations in the patterns due to reapplying the body sensor and an increased velocity (up to 11 km/h).

Keywords: Human identification · Accelerometer · Electrocardiograph (ECG) · Wireless body sensor (WBS) · Pattern recognition

1 Introduction

The identification of humans is important for various applications such as surveillance systems, authorization checks at doors or electronic devices (e.g. computer, smartphone). A variety of biometric characteristics have been investigated such as information from fingerprint, iris and retina, human face, voice, gait or electrocardiograph.

Previous work has shown that discerning, reproducible information on the human is found in the ECG waveform, especially around the QRS complex [7,11]. Moreover, biomechanical differences between the gait style of humans have been investigated and used for identification within video and acceleration sensor based applications [12,18].

We propose a biometric measure combining both sources of information: characteristics in the electrocardiograph (ECG) waveform and the gait style.

© Springer-Verlag Berlin Heidelberg 2014
M. Fernández-Chimeno et al. (Eds.): BIOSTEC 2013, CCIS 452, pp. 240–257, 2014.
DOI: 10.1007/978-3-662-44485-6_17

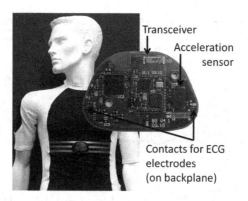

Fig. 1. Our self-made wireless body sensor (WBS) and its integration into a chest strap. The WBS can measure a person's electrocardiograph (ECG) and accelerations of the body along three orthogonal axes.

Unlike other applications, our approach focuses on the identification of athletes during physical exercise using a compact wireless body sensor (WBS) which is worn around the chest (see Fig. 1). The WBS is typically used to measure the heart-rate and the body accelerations of athletes. Our identification method additionally utilizes the sensor measurements to identify the athlete, enabling an automatic annotation of sensor data with the subject's identity. Our goal is to overcome the drawbacks of a manual annotation of measurements for applications in sports medicine and athlete training research. Furthermore, recognizing the subject allows to automatically load personal settings on the WBS or the sport equipment for a customized training. Our identification method is in particular interesting for a WBS which is used with several athletes of a mid-sized group.

Our identification method uses features in time and frequency domain to extract characteristics on the subject which are used as input to a classifier for identification. By combining information from gait and ECG we can successfully identify subjects despite of artefacts in the ECG caused by a slipping of the ECG electrodes and severe variations in the gait patterns between walking and jogging.

Previous work in this field focused on the identification of humans from either gait or ECG waveform characteristics. Mainly ECGs were used which were recorded at rest or with a clinical acquisition setup. The gait based identification was carried out for walking velocities.

Rong et al. proposed a method which uses measurements recorded during walking with an accelerometer located at the subject's waist [19]. The method utilises a segmentation into gait cycles to extract gait patterns. Dynamic time warping is applied to compensate natural changes in walking speed. The actual gait segment is then compared with a reference pattern of the subject and a 1-nearest neighbour classifier is used to recognize the subject. Ailisto et al. evaluated an accelerometer based identification based on similarities between gait segments to protect portable devices [2]. Mäntyjärvi et al. evaluated a gait based

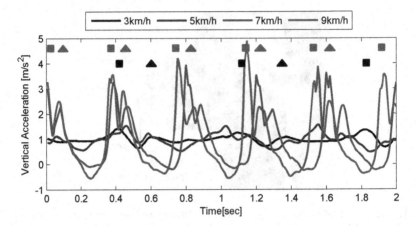

Fig. 2. Vertical acceleration data of a subject walking and jogging at velocities from 3 to 9 km/h. Each stride is represented by two consecutive peaks which correspond to the heel strike (square) and the toe strike (triangle). These peaks are marked for 9 km/h (red) and for 3 km/h (black). Velocity can be increased with either longer strides (increase in signal amplitude) or a higher step frequency (color figure online).

identification for different walking velocities using correlation coefficients derived from a template comparison, frequency coefficients and a histogram based comparison [16]. Gafurov et al. proposed two methods based on histogram similarity and gait cycle length to distinguish acceleration measurements recorded at the lower leg [12].

Several methods have been proposed to identify a human based on ECG measurements. Biel et al. used data from a standard 12-lead ECG recorded during rest to identify subjects using multivariate analysis [4]. Furthermore, the study showed that identification is possible with even one-lead ECGs. Shen et al. also utilises data from one-lead ECGs to distinguish subjects using a template matching and a decision-based neural network [21]. Chan et al. identifies subjects based on ECGs recorded within a non-clinical acquisition setup where the subjects were holding two electrodes on the pads of their thumbs [7]. For classification, three qualitative measures were used: percent residual difference, correlation coefficient, and a novel distance measure based on wavelet transform.

This paper is organized as follows: Sect. 2 describes the identification of a subject based on acceleration and ECG measurements. Information on preprocessing, feature extraction and used classifiers is given. Section 3 explains the conducted experiments for data collection. Section 4 presents the experimental results of our identification method. The results are summarised and discussed in Sect. 5 and a prospect on our future work is given.

2 Identification of a Subject

This section describes the identification of a subject based on gait style and ECG waveform characteristics. We describe the preprocessing of the signals, the feature extraction and the classifiers used for identification.

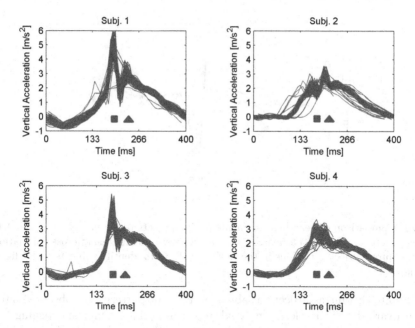

Fig. 3. Alignment of 100 consecutive strides of four subjects jogging at 9 km/h. The vertical acceleration signals were automatically segmented into strides and cross-correlation was used to align the strides. The peaks related to the heel strikes (square) and toe strikes (triangle) significantly differ in shape between the subjects.

2.1 Gait Analysis for Identification

Previous work has shown that gait differs between humans and that the gait style is fairly stable for a subject [3,18]. Bianchi et al. stated that the variability across humans depends on different kinematic strategies rather than on biomechanical characteristics [3]. Their study showed that subjects are different in the ability of minimising energy oscillations of their body segments for transferring mechanical energy.

In order to measure these inter-subject differences, severe intra-subject variations in the gait patterns between walking and jogging have to be taken into account. The intra-subject variations are a result of an adaptation of the gait to achieve different velocities. The velocity of a person is described by stride length and stride frequency. According to Weyand et al., longer strides are achieved by applying greater support forces to the ground which significantly increases the amplitude of the vertical acceleration signal, whereas the step frequency changes frequency components of the signal [24].

Samples of vertical acceleration data of one subject walking and jogging at different velocities on a treadmill are shown in Fig. 2. Strides are presented by two consecutive peaks corresponding to the heel and toe strikes. Significant changes in amplitude and an almost doubling of the step frequency can be observed between walking at 3 km/h and jogging at 9 km/h.

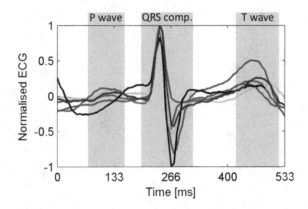

Fig. 4. Comparison of heartbeat segments of six subjects (different colours). The DC-offset was removed and the heartbeat segments were aligned using cross-correlation. We use inter-subject variations in the ECG waveform to identify subjects (color figure online).

Despite this intra-subject variability in the gait patterns, we observed inter-subject variations in acceleration signals recorded during walking and jogging [10]. In particular, heel and toe strikes differ in the vertical acceleration signal's shape between subjects (see Fig. 3). The peak acceleration of the heel strikes varies between the four subjects about 2 m/s².

2.2   ECG Analysis for Identification

Inter-subject variability is also found in the ECG's waveform. The variations depend on position, size and anatomy of the heart, age, sex, relative body weight, chest configuration and various other factors [13,22]. Figure 4 shows sample heartbeat segments from six subjects recorded with our WBS. The ECG reflects the electrical activity of the heart and consists of the P wave followed by the QRS complex and the T wave [11, chap. 2]. Discerning information on the subjects is found in the QRS complex, the P and the T wave.

Chan et al. observed a high degree of reproducibility of information extracted from the QRS complex of a person through several sessions of recording [7]. Furthermore, a higher identification accuracy was determined for the P wave than the T wave.

During physical exercise these characteristics can be superposed by motion related artefacts. These artefacts are caused by a slipping of the ECG electrodes and variations in the contact resistance during body movements [9]. Figure 7 shows disturbances in the ECGs of two subjects recorded during jogging on a treadmill.

2.3   Preprocessing of Acceleration and ECG Signals

ECGs recorded with our WBSs showed hardware-related differences in the DC-offset making an ECG associable to a WBS. Furthermore, using textile ECG

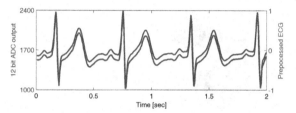

(a) ADC output and preprocessed ECG of subject 1.

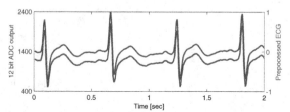

(b) ADC output and preprocessed ECG of subject 2.

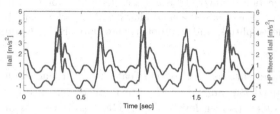

(c) Magnitude of the acceleration vector $\|\mathbf{a}\|$ and the result of the offset reduction by the high-pass filter.

Fig. 5. The 12 bit analog-to-digital converter (ADC) output and the preprocessed ECG in comparison. An offset of 300 between the ADC output of the two different subjects was removed by the preprocessing. In Fig. 5c the offset due to the static acceleration of gravity and a sensor-related zero-g-level offset are reduced after preprocessing.

electrodes, the skin contact resistance decreases over time because of an increased transpiration which results in changes in the DC-offset. In order to avoid classification errors, we removed the DC-offset using a 4th-order high-pass butterworth filter with a cutoff frequency of $f_c = 0.67$ Hz. Additionally, we applied a low-pass filter with a cutoff frequency of $f_c = 40$ Hz to remove noise in the ECG signal.

With a decrease in skin contact resistance after a few minutes of exercise, we observed an increase in the ECG signal's amplitude which improved the signal-to-noise ratio. We normalised the signal's amplitude to assure that ECG segments are comparable. The results of the ECG preprocessing are shown in Figs. 5a and b.

For the frequency analysis of the acceleration measurements, we approximated the dynamic accelerations by applying a 4th-order butterworth high-pass filter with a cutoff frequency of $f_c = 0.1$ Hz to the magnitude of the acceleration vector $\mathbf{a} = (a_{AP}, a_{ML}, a_V)$; a_{AP} denotes anteroposterior accelerations, a_{ML}

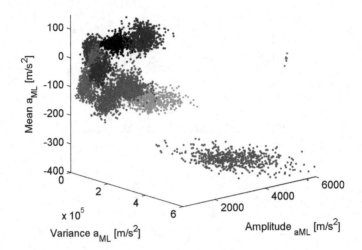

Fig. 6. Visualisation of time domain features extracted from mediolateral accelerations a_{ML} of ten subjects at 9 km/h. Clusters are observable for the different subjects. In our feature selection we obtained a good identification performance based on the mean, the variance, the amplitude and the root-mean-square (RMS) features (see Table 3).

mediolateral accelerations and a_V vertical (up-down) accelerations. The high-pass filter reduced the impact of the static acceleration due to gravity and a sensor-related offset (zero-g level offset). The results of this preprocessing step are shown in Fig. 5c.

2.4 Feature Extraction for Identification

In order to access characteristics of a subject in the acceleration and ECG measurements, we extracted features in the time and the frequency domain.

The features were calculated within a sliding window with no overlap and length N. Each window at time t consists of N measurements $x(t:t+N-1) = x(t), x(t+1), ..., x(t+N-1)$. We empirically determined an appropriate window length of two seconds ($N = 300$).

2.5 Time Domain Features

In the time-domain we calculated the variance, amplitude, mean and root mean square (RMS) along the three orthogonal axes a_{AP}, a_{ML} and a_V of the windowed acceleration signals. The variance, mean and amplitude of a_{ML} are visualised in Fig. 6. Discriminative clusters can be observed for the different subjects.

From the ECG signal we calculated a feature measuring the closeness of an unknown heartbeat segment to five reference patterns stored for each subject. This step requires a segmentation of the ECG signal into heartbeats. We used a QRS detection based on the algorithm of [1] in its implementation of Schloegl in the BioSig toolbox [23]. The five reference heartbeat segments were chosen

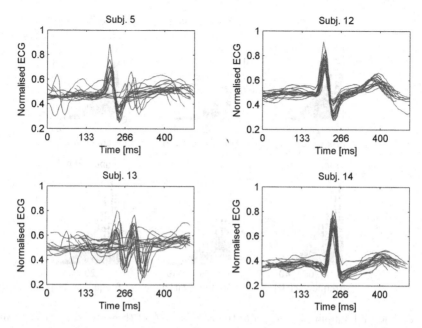

Fig. 7. Alignment of 20 heartbeat segments of four subjects recorded during jogging on a treadmill. A correct placement of the chest strap is important for an identification based on a similarity measure between heartbeat segments. Motion related artefacts and poor skin contact can disturb the ECG-signal (see subjects 5 and 13).

randomly from the ECG data of each subject. However, we assured that only heartbeat segments without severe disturbances were chosen. For identification, an unknown segment x was aligned to each reference segment y using cross-correlation:

$$R_{xy}(m) = \frac{1}{N} \sum_{j=0}^{N-m-1} y(j+m)x(j) \tag{1}$$

where N is the length of a segment and m the offset with $m = 0, 1, ..., 2N - 1$. We calculated the Pearson's correlation coefficient as a measure of similarity between the two segments. The Pearson's correlation coefficient is defined as the covariance (cov) of the two segments divided by the product of their standard deviation σ:

$$r(x,y) = \frac{\text{cov}(x,y)}{\sigma_x \sigma_y} \tag{2}$$

Figure 7 shows the alignment of 20 heartbeat segments of four subjects. The QRS-detection and the alignment are sensitive to motion-related artefacts (see subjects 5 and 13).

For heartbeat segments without major disturbances the alignment centred the segments around the QRS complex. The discerning information in this region of the ECG is fairly stable in relation to morphology changes in the ECG waveform during effort.

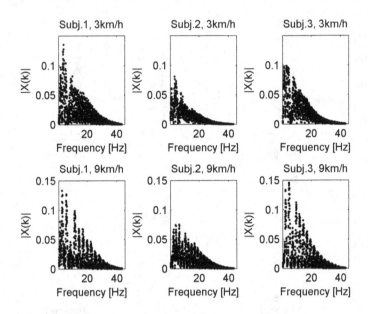

Fig. 8. The FFT amplitude spectra of the ECG signals of three subjects during walking (3 km/h) and jogging (9 km/h). The amplitude spectra show differences between the subjects but also vary with the velocity.

2.6 Frequency Domain Features

In the frequency domain we use the discrete Fourier transform (DFT) to extract frequency components of each window. The DFT is defined as:

$$X(k) = \sum_{j=t}^{t+K-1} x(j)e^{-i2\pi k \frac{j}{K}}, \quad k = 0, ..., K-1 \tag{3}$$

where K is the number of outputs $X(k)$. We used a 512-point fast Fourier transform (FFT) algorithm to compute the DFT efficiently for our windows of the length $N = 300$. Therefore, each window $\mathbf{x}(t:t+N-1)$ was padded with trailing zeros to the length of $K = 512$. Before calculating the FFT, a Hamming window function was applied to each window to reduce the spectral leakage.

Figure 8 shows the FFT amplitude spectra of ECGs of three subjects recorded during walking (3 km/h) and jogging (9 km/h). Despite velocity related variations in the amplitude spectra, differences can be observed between the three subjects.

We calculated additional frequency domain features from the amplitude spectrum (FFT features): the variance, the mean, the Fourier coefficient with the highest amplitude and the Shannon entropy SE:

$$SE = -\sum_{k=0}^{K-1} |X(k)| \log_2(|X(k)|) \tag{4}$$

where $X(k)$ is the output of the DFT of length K.

2.7   Methods for Classification

We used a standard classifier to identify the subject based on the extracted features. The identification performance was determined by evaluating three different classifiers: artificial neural network (ANN), support vector machine (SVM), and random forest (RF).

2.8   Artificial Neural Network (ANN)

We used a feed-forward ANN with 25 neurons with tangent sigmoid activation functions in one hidden layer to associate the extracted features with the subjects' identities. The ANN was trained using back-propagation which is a supervised learning method [14]. During training the prediction of the network is compared to the known target value (subject's identity) and the weights are modified to minimize the mean square error. These errors propagate backwards from the output layer to the hidden layer [14]. The network was trained using the scaled conjugate gradient algorithm described in [17]. The weights and bias values of the neurons were updated using a gradient descent with momentum.

2.9   Support Vector Machine (SVM)

We used a ν-SVM [20] with a sigmoid kernel in its implementation in the LIB-SVM[1] [8]. SVMs are fundamentally a two-class classifier. Various methods have been proposed how to use SVMs for multi-class problems [5, chap. 7]. We used a one-against-one method which constructs $n(n-1)/2$ classifiers where n is the number of classes to distinguish. Each classifier is trained on tuples from two classes. A voting strategy is then applied to determine the winning class [15].

2.10   Random Forest (RF)

A random forest is a classifier consisting of a combination of tree predictors. The growth of each tree is governed by independently and identically distributed random vectors [6]. Each tree votes for one class and the class which occurs most frequently is the output of the classifier. RF classifiers are fast in the training phase and the training time is linear to the number of trees used. The testing of an unknown tuple is performed on each tree independently and is therefore parallelisable. We used a RF consisting of 100 trees, with each tree being constructed of ten randomly chosen features.

3   Subjects and Data Collection

For the evaluation of our identification method, we recorded data within two experiments of subjects who volunteered to participate in the study. The subjects

---

[1] LIBSVM: library for support vector machines.

Table 1. Characteristics of the subjects who participated in the two experiments.

(a) First experiment with 22 subjects (15 men, 7 women).			(b) Second experiment with 9 male subjects.		
Characteristic	Mean ± SD	Range	Characteristic	Mean ± SD	Range
Age (yr)	26.6 ± 4.0	18-33	Age (yr)	27.0 ± 3.7	21-33
Height (cm)	179.8 ± 9.6	160-198	Height (cm)	180.2 ± 9.6	160-198
Weight (kg)	76.7 ± 11.1	58-108	Weight (kg)	77.3 ± 10.8	65-108

were informed verbally and in writing in advance and signed an informed consent document.

In the first experiment 22 healthy subjects participated (see Table 1(a)). The data was collected using the treadmills in the gymnasium of our university. Velocities between 3 to 9 km/h were chosen to cover slow, normal, and fast walking as well as jogging. The treadmill was set to no incline and the velocity was manually increased by 2 km/h every two minutes. This procedure was repeated twice for each subject (total: 16 min per subject).

In order to estimate the impact of variations in the gait and ECG patterns resulting from reapplying the body sensor (electrode placement and conductance, acceleration sensor orientation), we repeated the experiment with a smaller group of nine male subjects from which data was collected within two independent sessions of recording (see Table 1(b)). Both sessions, which were one week apart, covered data at velocities from 3 to 11 km/h (total: 20 min per subject).

The accelerations of the upper body and the ECG were recorded with a self-made WBS (see Fig. 1). The WBS measures accelerations within a range of $\pm 6 \text{m/s}^2$ along three orthogonally oriented axes using a commercial off-the-shelf accelerometer (ST LIS3LV02DL). The ECG is digitized using the analog-to-digital converter of a TI MSP430 microcontroller. Body accelerations and ECG were measured with a 150 Hz sampling rate and a 12 bit resolution (range 0 to 4095). The measurements were sent wirelessly to a nearby receiver for recording.

The subjects were given an explanation on how to place the chest strap with the WBS tightly around the chest. However, we didn't verify the correct placement of the WBS to assure real world conditions. Furthermore, no instructions were given on how to perform the exercise.

4   Results

This section describes the evaluation of the athlete identification on the data collected during walking and jogging on the treadmills.

4.1   Evaluation Methods

All features were calculated on windows of acceleration and ECG measurements of two seconds. No overlap of the windows was chosen to ensure fully discriminative training and testing data. We concatenated features of two consecutive

Table 2. Accuracy (ACC) and overall specificity ($\overline{S}$) of the identification of the 22 subjects. The results were determined with three different classifiers on a feature space combining acceleration and ECG features (combination C8, see Table 3).

Classifier	ACC	$\overline{S}$
ANN	94.2 %	99.8 %
SVM	90.4 %	99.5 %
RF	98.1 %	99.9 %

windows to have samples of four seconds of data to identify the subject. For the data of the first experiment with 22 subjects, we determined the identification performance using a ten-fold cross-validation. The two sessions of recording from the second experiment with 9 subjects were used for training the classifier (first session) and for evaluation of the resulting model (second session).

For the evaluation, we used three statistical measures: sensitivity, specificity and accuracy. In order to calculate the statistics we obtained the number of true positive samples TP_i, true negative samples TN_i, false positive samples FP_i, and false negative samples FN_i from the classifier's output. For a class i the sensitivity R_i is defined as:

$$R_i = \frac{\mathrm{TP}_i}{\mathrm{TP}_i + \mathrm{FN}_i} * 100 \tag{5}$$

The sensitivity (also referred to as recall) measures the percentage of correctly classified positive samples in relation to all positive samples. For negative samples the specificity S_i is defined as:

$$S_i = \frac{\mathrm{TN}_i}{\mathrm{TN}_i + \mathrm{FP}_i} * 100 \tag{6}$$

We calculated the overall sensitivity $\overline{R}$ and the overall specificity $\overline{S}$ as a class-based weighted average. For our multi-class problem we refer to the overall sensitivity as the accuracy of the classifier:

$$ACC = \overline{R} = \sum_{i=1}^{n} p_i R_i \tag{7}$$

where n denotes the number of classes and p_i the probability of the occurrence of the class in the test data. In our data from the two experiments the samples are equally distributed for the $n = 22$ and $n = 9$ subjects ($p_i = 1/n, \forall i$). The overall specificity $\overline{S}$ is calculated accordingly. The optimum of the statistical measures is 100 %.

4.2   Results of the Athlete Identification

The following results were obtained for the data from the first experiment with the 22 subjects. We determined the identification performance for three standard

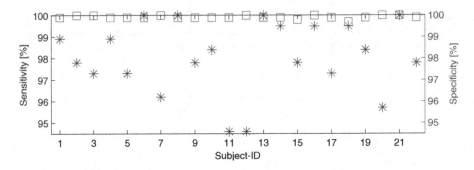

Fig. 9. Class-specific sensitivity (stars) and specificity (squares) results of the identification of the 22 subjects (RF classifier, feature combination $C8$). The sensitivity varied between 94.6 to 99.5 %. The specificity was over 99.7 % for all subjects.

classifiers: ANN, SVM and RF. The classifiers and their parametrization are described in Sect. 2.7. We achieved up to 98.1 % accuracy (see Table 2) with the RF classifier using a feature space combining acceleration and ECG based features. The lowest accuracy of 90.4 % was obtained with the SVM. For all three classifiers, we obtained an overall specificity $\overline{S}$ of more than 99 %.

The class-specific sensitivity (see Eq. 5) of the identification varied between 94.6 to 99.5 % for the different subjects (RF classifier, see Fig. 9). We observed only low deviations in the identification's specificity between the 22 subjects. A class-specific specificity (see Eq. 6) of more than 99.7 % was achieved for all subjects.

We performed a feature selection using the ANN classifier to determine the impact of the different features and to identify combinations C with a high classification performance (see Table 3). We obtained a similar identification accuracy based on acceleration (86.6 %, $C6$) and ECG (84.8 %, $C4$) features. In combination, the accuracy improved to 94.2 % ($C8$).

The ECG contained more information on the subject in the frequency domain than the acceleration measurements (12.3 % higher accuracy). Frequencies of up to 10 Hz contained the most discriminant information of the acceleration measurements. A reduction of the frequency band from 40 to 10 Hz reduced the identification accuracy by only 3.8 %. For the ECG measurements, a reduction from 40 Hz to 15 Hz resulted in a 8.8 % lower accuracy. Overall, we obtained an accuracy of 72.4 % ($C1$) for features extracted from the ECG in the frequency domain.

We found that correlation coefficients describing the similarity between heartbeat segments provide useful insights to identify subjects (80.3 % accuracy, $C2$). To reduce the dimensionality of the feature space, we averaged the correlation coefficients corresponding to the five reference segments per subject. This averaging resulted in a 7.7 % lower accuracy. However, in combination with other features this difference was negligible (0.4 % for $C8$).

Table 3. Identification accuracy for the 22 subjects using different feature combinations C. We obtained a similar accuracy with acceleration and ECG based features (see $C4$, $C6$). Combining both improved the accuracy (see $C8$). The feature selection was performed using the ANN classifier. * denotes the use of the average over the five correlation coefficients per subject.

C	Acceleration feat.			ECG feat.			ACC
	Time dom.	FFT coef.	FFT feat.	FFT coef.	FFT feat.	Corr. coef.	
$C1$	-	-	-	x	x	-	72.4%
$C2$	-	-	-	-	-	x	80.3%
$C3$	x	-	-	-	-	-	83.3%
$C4$	-	-	-	x	x	x	84.8%
$C5$	x	-	-	-	x	-	86.5%
$C6$	x	x	x	-	-	-	86.6%
$C7$	x	-	-	-	-	x	93.6%
$C8$	x	x	x	x	x	x*	94.2%

The time domain features calculated from the acceleration signals showed a good accuracy (83.3%, $C3$). Additional information on the gait in the frequency domain improved the identification accuracy to 86.6% ($C6$).

By combining the time domain features of the acceleration data with the correlation coefficients derived from the ECG, we achieved a high accuracy of 93.6% ($C7$), which is only 0.6% less than using the full feature set ($C8$).

For the time domain features extracted from the acceleration signals, we analysed the impact of the different acceleration axes on the subject's identification accuracy. The highest accuracy was obtained for the anteroposterior accelerations (a_{AP}). The mediolateral accelerations (a_{ML}) showed a 4.4% and the vertical accelerations (a_V) a 16.2% lower accuracy.

We additionally evaluated our approach using a hold-out validation for which the data set was split into 66% training data and 34% testing data. A hold-out validation avoids temporal proximity between training and testing data and allows therefore a more accurate estimation of the generalization performance. We noted only a slight decrease in accuracy by 0.9% for the RF classifier.

To estimate the impact of the number of subjects in the data set on the identification performance, we randomly selected eleven out of the twenty-two subjects and repeated the evaluation. With the smaller group of athletes to distinguish, the overall accuracy improved by 1.2% (RF classifier).

The above results were obtained for the data from the first experiment with one session of recording per subject. However, reapplying the body sensor can change the waveform of the signals due to a different position of the ECG electrodes in relation to the heart, differences in the conductance of the electrodes or alternations in the acceleration sensor's orientation. In order to estimate the

Table 4. Equal error ratio (ERR) and accuracy (ACC) of other gait and ECG based identification methods. For comparison, our results on the data sets from the first and second experiment are listed below. N denotes the number of subjects who participated in the experiments.

	Type	Velocities	N	ERR	ACC
Mäntyjärvi et al. (2005)	Gait	slow, normal and fast walking	36	7.0 %	-
Ailisto et al. (2005)	Gait	normal walking	36	6.4 %	-
Gafurov et al. (2006)	Gait	normal walking	21	5.0 %	-
Rong et al. (2007)	Gait	normal walking	21	5.6 %	-
Chan et al. (2008)	ECG	-	50	-	89.0 %
First experiment (1 session)	Gait & ECG	3, 5, 7 and 9 km/h	22	1.1 %	98.1 %
Second experiment (2 sessions)	Gait & ECG	3, 5, 7, 9 and 11 km/h	9	2.5 %	93.8 %

impact of this variability in the gait and ECG patterns, we evaluated the identification method on the two independent sessions of recording from the second experiment. The first session was used to train the RF classifier and the second session functioned as an independent test set. We obtained an accuracy of 93.8 % with a class-specific sensitivity between 83.6 to 89.7 % for three subjects and above 94.5 % for the other six subjects; the highest sensitivity was 99.4 %. The overall specificity was 99.2 % showing only minor variations between the nine subjects (range: 97.4 to 99.9 %). In comparison to the results from the first experiment, the overall accuracy decreased by 4.3 % (22 subjects) and 5.5 % (reduced subset of 11 subjects) mainly because of the outlying results of the three subjects. However, the results showed that identification is possible with a good accuracy despite reapplying the body sensors and an increased velocity of up to 11 km/h. Furthermore, the second evaluation points out that the identification method is stable for short-term physiological variations in the ECG and alternations in the gait patterns.

In order to compare our results with existing work, we additionally calculated the equal error rate (ERR) of the RF classifier on feature combination $C8$. The ERR is the rate at which both accept and reject errors are equal. For our data set from the first experiment (22 subjects), we obtained an ERR of 1.1 % and for the evaluation on the data from the second session of the second experiment (9 subjects) the ERR was 2.5 %. Compared to other approaches, which are based on only gait characteristics, our achieved ERR is lower (see Table 4). For a comparison of our approach with an ECG based identification we have chosen the method of Chan et al. because the results are also based on data from non-clinical ECGs [7]. With 98.1 % our accuracy is higher than Chan et al. results (89 %). However, with an identification on ECG characteristics only, we obtained a lower accuracy (84.8 %, $C4$). Overall, our high performance is achieved by combining ECG and gait characteristics. We believe that motion related artefacts in the ECG, and a high variability in the gait patterns between changing from slow walking to jogging, reduce the identification performance when we use only one source of information.

5   Discussion and Conclusions

This paper is concerned with the identification of humans during walking and jogging using a single wireless body sensor module attached to a chest strap. Our approach focuses on recognising a human using a biometric measure based on the characteristics in the gait style and the ECG of the human and is hence independent of the used hardware. Thus, our system overcomes the drawbacks of an identification based on the WBS's serial number or an radio-frequency based identification (RFID) which recognises the hardware but not the subject itself.

We have collected data from 22 subjects on a treadmill at velocities from 3 to 9 km/h using a WBS attached to a chest strap. To assure real world conditions, no advice was given on how to perform the exercise and the correct placement of the chest strap was not verified. Despite severe variations in the gait patterns and motion-related artefacts in the ECG, which occur due to real world conditions and physical exercise, our method achieves up to 98 % accuracy.

In order to estimate the impact of variations in the ECG and gait patterns resulting from reapplying the body sensor and short-term physiological alternations, we repeated the experiment with nine subjects with two sessions of recoding per subject which were one week apart. The first session was used for training the classifier and the second session served as test data. With 93.8 % the accuracy of the identification is still high considering also the extended range of velocity classes (3 to 11 km/h).

Our feature selection showed a good identification accuracy for time domain features extracted from the acceleration signals. By using simple and low-dimensional features on the acceleration signal our method can potentially be implemented on computationally constrained platforms, such as a microcontroller on a WBS.

Our identification method can presumably not be extended to an unlimited number of subjects. The individual characteristics in the subject's ECG and gait patterns are extremely difficult to capture and may change over time because of an adaptation to physical exercise. However, we believe our method is well suited to provide an automatic annotation of sensor measurements from several WBSs with the subject's identity for use in sports medicine and athletic training research. Moreover, our method helps to customize a training session by loading personal settings of the recognized athlete on the WBS or other sport equipment.

Our future work includes the evaluation of the identification method within team sports. In particular, we want to recognize handball players in order to support a real-time vision-based tracking of these players.

Acknowledgements. This research was supported by the DFG CoE 277: Cognitive Interaction Technology (CITEC).

References

1. Afonso, V.X., Tompkins, W.J., Nguyen, T.Q., Luo, S.: ECG beat detection using filter banks. Trans. Biomed. Eng. **46**(2), 192–202 (1999)
2. Ailisto, H.J., Lindholm, M., Mantyjarvi, J., Vildjiounaite, E., Makela, S.M.: Identifying people from gait pattern with accelerometers. In: Society of Photo-Optical Instrumentation Engineers, pp. 7–14 (2005)
3. Bianchi, L., Angelini, D., Lacquaniti, F.: Individual characteristics of human walking mechanics. Pflügers Arch. Eur. J. Physiol. **436**, 343–356 (1998)
4. Biel, L., Pettersson, O., Philipson, L., Wide, P.: ECG analysis: a new approach in human identification. IEEE Trans. Instrum. Meas. **50**(3), 808–812 (2001)
5. Bishop, C.M.: Pattern Recognition and Machine Learning (Information Science and Statistics). Springer-Verlag New York Inc., Secaucus (2006)
6. Breiman, L.: Random forests. Mach. Learn. **45**(1), 5–32 (2001)
7. Chan, A.D.C., Hamdy, M.M., Badre, A., Badee, V.: Wavelet distance measure for person identification using electrocardiograms. IEEE Trans. Instrum. Meas. **57**(2), 248–253 (2008)
8. Chang, C.-C., Lin, C.-J.: LIBSVM: a library for support vector machines. ACM Trans. Intell. Syst. Technol. **2**, 27:1–27:27 (2011)
9. Christ, P., Mielebacher, J., Haag, M., Rückert, U.: Detection of body movement and measurement of physiological stress with a mobile chest module in obesity prevention. In: Proceedings of the 10th Australasian Conference on Mathematics and Computers in Sport, pp. 67–74, July 2010
10. Christ, P., Werner, F., Rückert, U., Mielebacher, J.: An approach for determining linear velocities of athletes from acceleration measurements using a neural network. In: Proceedings of the 6th IASTED International Conference on Biomechanics, pp. 105–112. ACTA Press, November 2011
11. Conover, M.B.: Understanding electrocardiography. Mosby (2002)
12. Gafurov, D., Helkala, K., Søndrol, T.: Biometric gait authentication using accelerometer sensor. J. Comput. **1**(7), 51–59 (2006)
13. Green, L.S., Lux, R.L., Haws, C.W., Williams, R.R., Hunt, S.C., Burgess, M.J.: Effects of age, sex, and body habitus on QRS and ST-T potential maps of 1100 normal subjects. Circulation **71**(2), 244–253 (1985)
14. Han, J., Kamber, M.: Data mining: concepts and techniques. The Morgan Kaufmann series in data management systems. Elsevier (2006)
15. Hsu, C.W., Lin, C.J.: A comparison of methods for multiclass support vector machines. IEEE Trans. Neural Netw. **13**(2), 415–425 (2002)
16. Mäntyjärvi, J., Lindholm, M., Vildjiounaite, E., Mäkelä, S.-M., Ailisto, H.A.: Identifying users of portable devices from gait pattern with accelerometers. In: IEEE International Conference on Acoustics, Speech, and Signal Processing, vol. 2, pp. ii/973–ii/976, March 2005
17. Møller, M.F.: A scaled conjugate gradient algorithm for fast supervised learning. Neural Netw. **6**(4), 525–533 (1993)
18. Nixon, M.S., Tan, T., Chellappa, R.: Human identification based on gait, vol. 4. Springer-Verlag New York Inc., Secaucus (2006)
19. Rong, L., Jianzhong, Z., Ming, L., Xiangfeng, H.: A wearable acceleration sensor system for gait recognition. In: 2nd IEEE Conference on Industrial Electronics and Applications, pp. 2654–2659 (2007)
20. Schölkopf, B., Smola, A.J., Williamson, R.C., Bartlett, P.L.: New support vector algorithms. Neural Comput. **12**(5), 1207–1245 (2000)

21. Shen, T. W., Tompkins, W.J., Hu, Y.H.: One-lead ECG for identity verification. In: Proceedings of the 2nd IEEE International Joint Conference on Engineering in Medicine and Biology Society, vol. 1, pp. 62–63. IEEE (2002)
22. Simon, B.P., Eswaran, C.: An ECG classifier designed using modified decision based neural networks. Comput. Biomed. Res. 30(4), 257–272 (1997)
23. Vidaurre, C., Sander, T.H., Schlögl, A.: BioSig: the free and open source software library for biomedical signal processing. Comput. Intell. Neurosci. 2011, 12 (2011)
24. Weyand, P.G., Sternlight, D.B., Bellizzi, M.J., Wright, S.: Faster top running speeds are achieved with greater ground forces not more rapid leg movements. J. Appl. Physiol. 89(5), 1991–1999 (2000)

Statistical Modeling of Atrioventricular Nodal Function During Atrial Fibrillation Focusing on the Refractory Period Estimation

Valentina D.A. Corino[1]([✉]), Frida Sandberg[2], Federico Lombardi[3],
Luca T. Mainardi[1], and Leif Sörnmo[2]

[1] Dipartimento di Elettronica, Informazione e Bioingegneria,
Politecnico di Milano, Milan, Italy
{valentina.corino,luca.mainardi}@polimi.it
[2] Department of Biomedical Engineering and Center for Integrative
Electrocardiology (CIEL), Lund University, Lund, Sweden
{frida.sandberg,leif.sornmo}@bme.lth.se
[3] UOC Malattie Cardiovascolari, Fondazione IRCCS Ospedale Maggiore Policlinico,
Dipartimento di Scienze Cliniche e di Comunità, University of Milan, Milan, Italy
federico.lombardi@unimi.it

Abstract. We have recently proposed a statistical AV node model defined by a set of parameters characterizing the arrival rate of atrial impulses, the probability of an impulse passing through the fast or the slow pathway, the refractory periods of the pathways, and the prolongation of refractory periods. All parameters are estimated from the RR interval series using maximum likelihood (ML) estimation, except for the mean arrival rate of atrial impulses which is estimated by the AF frequency derived from the f-waves. In this chapter, we compare four different methods, based either on the Poincaré plot or ML estimation, for determining the refractory period of the slow pathway. Simulation results show better performance of the ML estimator, especially in the presence of artifacts due to premature ventricular beats or misdetected beats. The performance was also evaluated on ECG data acquired from 26 AF patients during rest and head-up tilt test. During tilt, the AF frequency increased (6.08 ± 1.03 Hz vs. 6.20 ± 0.99 Hz, $p < 0.05$, rest vs. tilt) and the refractory periods of both pathways decreased (slow pathway: 0.43 ± 0.12 s vs. 0.38 ± 0.12 s, $p = 0.001$, rest vs. tilt; fast pathway: 0.55 ± 0.14 s vs. 0.47 ± 0.11 s, $p < 0.05$, rest vs. tilt). These results show that AV node characteristics can be assessed non-invasively to quantify changes induced by autonomic stimulation.

Keywords: Atrial fibrillation · Atrioventricular node · Statistical modeling · Maximum likelihood estimation

1   Introduction

During atrial fibrillation (AF), atrial impulses cause summation and/or cancellation of wavefronts in the AV node, which in turn causes disorganization of the

© Springer-Verlag Berlin Heidelberg 2014
M. Fernández-Chimeno et al. (Eds.): BIOSTEC 2013, CCIS 452, pp. 258–268, 2014.
DOI: 10.1007/978-3-662-44485-6_18

penetrating impulses so that the ventricular rhythm is more irregular than during sinus rhythm. Although AV nodal properties such as refractoriness and concealed conduction determine the characteristics of the ventricular response [1], no evaluation is performed on a routine basis in clinical practice due to the lack of suitable noninvasive methodology.

Various nonparametric approaches to the analysis of AV coupling during AF have recently been proposed, e.g., [2–4]. The Poincaré surface profile is a histographic variant of the well-known Poincaré plot introduced to filter part of the AV node memory effects, with the overall aim to detect preferential AV nodal conductions [2]. The AV synchrogram was introduced for beat-to-beat assessment of AV coupling during AF, as well as for other atrial tachyarrhythmias. This technique involves a stroboscopic observation of the ventricular phase at times triggered by atrial activations [4]. The synchrogram was found useful for tracking the time course of AV coupling and for partially reconstructing the dynamics of AV response during AF.

A number of AV node models have been proposed during the last decade where it is assumed that the atrial electrogram is available, e.g., recorded during electrophysiological studies [5,6]. Thus, these models are less suitable for use in clinical routine where it is preferable to estimate all model parameters from the surface ECG. Simulation models represent another type of model which are useful for investigating certain AV nodal characteristics [7] or the effect of pacing [8,9]. These models offer detailed characterization of the underlying electrophysiological dynamics, but do not lend themselves to analysis of real data since the number of parameters is much too large to produce estimates with sufficient accuracy.

In a recent paper [10], we have shown that statistical model-based analysis, relying entirely on information derived from the surface ECG, can be employed for evaluating essential AV nodal characteristics during AF. The model is defined by a parsimonious set of parameters which characterizes the arrival rate of atrial impulses, the probability of an impulse passing through the fast or the slow pathway, the refractory periods of the pathways, and the prolongation of refractory periods. Maximum likelihood (ML) estimation was considered for estimating the parameters from the observed RR interval series, except for the shorter refractory period, estimated from the Poincaré plot of successive RR intervals, and the mean arrival rate of atrial impulses, estimated by the AF frequency derived from the f-waves of the ECG [11]. The results, determined from a total of 2004 30-min ECG segments, selected from 36 AF patients, showed that 88 % of the segments could be accurately modeled when the estimated probability density function (PDF) and an empirical PDF were at least 80 % in agreement. The study suggested that atrial activity is an important determinant of ventricular rhythm during AF.

In a subsequent paper, we have improved the AV node model to offer a more detailed characterization of the dual pathways [12]. The estimation procedure was also improved to become more robust with respect to artifacts in the RR interval series. The results for the improved model showed a significantly better fit between the estimated and the empirical PDF than previously reported for the original model in [10].

The goal of the present study is to compare different techniques for estimating the refractory period of the slow pathway. In particular, we compare three methods based on the Poincaré plot and one where the refractory period is estimated jointly with the ML estimation. The best-performing method (according to simulation results) is then studied on an ECG dataset recorded during rest and tilt testing.

2 Methods

2.1 AV Node Model

The AV node is treated as a lumped structure which accounts for concealed conduction, relative refractoriness, and dual AV nodal pathways [12]. Atrial impulses are assumed to arrive to the AV node according to a Poisson process with mean arrival rate λ. We assume that each arriving impulse is suprathreshold, i.e., the impulse results in ventricular activation unless blocked by a refractory AV node. The probability of an atrial impulse passing through the AV node depends on the time elapsed since the previous ventricular activation. The length of the refractory period is defined by a deterministic part τ and a stochastic part τ_p. The latter part models prolongation due to concealed conduction and/or relative refractoriness, and is assumed to be uniformly distributed in the interval $[0, \tau_p]$. Hence, all atrial impulses arriving to the AV node before the end of the refractory period τ are blocked. Then follows an interval $[\tau, \tau + \tau_p]$ with linearly increasing likelihood of penetration into the AV node. Finally, no impulses can be blocked if they arrive after the end of the maximally prolonged refractory period $\tau + \tau_p$. The mathematical characterization of refractoriness of the i:th pathway $(i = 1, 2)$ is thus defined by the positive-valued function $\beta_i(t)$,

$$\beta_i(t) = \begin{cases} 0, & 0 < t < \tau_i \\ \dfrac{t - \tau_i}{\tau_{p,i}}, & \tau_i \le t < \tau_i + \tau_{p,i} \\ 1, & t \ge \tau_i + \tau_{p,i}, \end{cases} \tag{1}$$

where t denotes the time elapsed since the preceding ventricular activation.

The probability of an atrial impulse to pass through the pathway with the shorter refractory period τ_1 is equal to α, and accordingly the other pathway is taken with probability $(1 - \alpha)$. For this model, the time intervals x_i between consecutive ventricular activations, i.e., corresponding to the RR intervals, are independent. It can be shown that the joint PDF is given by [10]

$$p_x(x_1, x_2, \ldots, x_M) = \prod_{m=1}^{M} (\alpha p_{x,1}(x_m) + (1 - \alpha) p_{x,2}(x_m)), \tag{2}$$

where M is the total number of intervals, and $p_{x,i}(x_m), i = 1, 2$, is given by

$$p_{x,i}(x) = \begin{cases} 0, & x < \tau_i \\ \frac{\lambda y_i}{\tau_{p,i}} \exp\left\{\frac{-\lambda y_i^2}{2\tau_{p,i}}\right\}, & \tau_i \le x < \tau_i + \tau_{p,i} \\ \lambda \exp\left\{\frac{-\lambda \tau_{p,i}}{2} - \lambda(y_i - \tau_{p,i})\right\}, & x \ge \tau_i + \tau_{p,i}. \end{cases} \tag{3}$$

where $y_i = x - \tau_i$.

2.2   Model Parameter Estimation

Interdependence of Consecutive RR Intervals. Since the property of statistical independence is not fully valid for RR intervals, a simple functional dependence of the refractory periods related to the previous RR interval is explored. The interdependence of consecutive RR intervals can be reduced by preprocessing the original RR interval series, denoted x'_m, with the linear transformation,

$$x_m = x'_m - \hat{s}_\tau x'_{m-1}, \tag{4}$$

where $\hat{s}_\tau$ is determined from the line that defines the lower envelope of the Poincaré plot.

Alternatively, the autocorrelation function of the RR intervals can be used for determining $\hat{s}_\tau$ [13]. During AF, the first lag of the autocorrelation is significant, whereas it is negligible for larger lags. Hence, decorrelation of the RR interval series is accomplished by (4), where $\hat{s}_\tau$ is taken as the smallest value in the interval $[0, 0.5]$ that makes the first lag negative.

Estimation of λ. The atrial impulses were assumed to arrive to the AV node according to a Poisson process at a rate λ. An estimate of λ is obtained by

$$\lambda = \frac{\lambda_{AF}}{1 - \delta \lambda_{AF}}, \tag{5}$$

where λ_{AF} is the dominant AF frequency estimated from the ECG (independently of the AV node parameters), and δ is minimum time interval between successive impulses arriving to the AV node. Equation (5) derives from the assumption that atrial impulses do not arrive to the AV node closer to each other than at a minimum interval δ.

Estimation of Dual Pathway Parameters. The model parameters related to the dual AV nodal pathways and the refractory period prolongation, except τ_1^{min}, are estimated by maximizing the log-likelihood function $\Lambda(\boldsymbol{\theta})$ with respect to the vector $\boldsymbol{\theta}$ that contains the unknown parameters [12],

$$\hat{\boldsymbol{\theta}} = \arg \max_{\boldsymbol{\theta}} \Lambda(\boldsymbol{\theta}), \tag{6}$$

where
$$\boldsymbol{\theta} = \left[\alpha \; \tau_2^{\min} \; \tau_{p,1} \; \tau_{p,2}\right]^T. \tag{7}$$

The parameter(s) defining both a single pathway model, i.e., $\theta = \tau_{p,1}$, and a dual pathway model, i.e., the vector $\boldsymbol{\theta}$ in (7), are estimated. The Bayes information criterion is used to determine which of these two models is the most appropriate one.

Since no closed-form solution can be found for the ML estimator, combined with the fact that the gradient is discontinuous, multi-swarm particle swarm optimization (MPSO) is employed for the maximization in (6). Briefly, a multi-initialization with N concurrent swarms is employed in MPSO [14,15]. Each swarm is moved within a search area to find the optimal solution. After a certain number of optimization epochs, particles are exchanged between swarms to avoid local maxima.

Estimation of $\tau_1^{\min}$. Four techniques for estimating the refractory period $\tau_1^{\min}$ of the slow AV pathway are compared, of which the first three methods explore the Poincaré plot in which each RR interval is plotted versus the preceding interval [16]. The resulting pattern may be used to distinguish AF from other supraventricular tachycardias such as atrial flutter with its much more regular ventricular response [17]. During AF, the irregularity of RR intervals results in a widely scattered distribution which is representative of disorganized atrial activity combined with atrioventricular conduction properties. The four techniques are now briefly described.

Linear fitting (LF) has been explored by plotting 512 points and dividing the horizontal axis into adjacent bins of 64 points [18]. The lower envelope results from a linear fit to the shortest RR intervals of all bins.

Modified linear fitting extends linear fitting by shifting the intercept of the fitted line until no points are below. This technique is motivated by the observation that the lower envelope represents the minimal refractory period.

The Hough transform is a technique for detecting straight lines in an image. Its application to the Poincaré plot in AF analysis was first pointed out in [19], see also [20]. Briefly, this plot is discretized (bin size of 20 ms) and edges are extracted using the Sobel approximation of the derivative. In the Hough space, a straight line is represented as a point, and the maximum value in this space corresponds to the most represented line in the input image. To find the lower envelope, the slope is constrained to 0–0.5 and the intercept to be positive. Among the lines satisfying these criteria, the one that is closest, in the mean square error sense, to the minimum points of the edge image is chosen.

Joint ML estimation of $\tau_1^{\min}$ and $\boldsymbol{\theta}$ was recently proposed [13]. Since the estimate of $\tau_1^{\min}$ is closely related to the shortest interval of the RR series, cf. the definition of $p_{x,i}(x)$ in (3), the handling of artifacts is important. The following iterative procedure is adopted to reduce the influence of artifactual intervals. Initially, 1 % of the shortest RR intervals are removed from the decorrelated RR interval series $\mathbf{x}$, after which ML estimation is performed on the truncated series, denoted $\tilde{\mathbf{x}}_0$. Since $\tilde{\mathbf{x}}_0$ is assumed to be free of incorrect RR intervals, the initial estimate

$$\tilde{\boldsymbol{\theta}}_0 = \left[\alpha(0)\ \tau_1^{\min}(0)\ \tau_2^{\min}(0)\ \tau_{p,1}(0)\ \tau_{p,2}(0)\right]^T \tag{8}$$

can serve as a reference. The removed RR intervals are then brought back to the truncated series one by one in order of size so that $\tilde{\mathbf{x}}_i = \left[\tilde{\mathbf{x}}_{i-1}\ x(i)\right]$, where $x(i)$ is the longest interval removed from $\tilde{\mathbf{x}}_{i-1}$; ML estimation is performed for each $\tilde{\mathbf{x}}_i$. The estimates corresponding to the maximum value of the log-likelihood function are chosen as the final ones.

3 Data

3.1 Simulated Data

Simulated 10-min RR interval series were generated, using the AV node model introduced in [12], to test the different methods for estimating $\tau_1^{\min}$. We used 5 different parameter settings (100 runs per setting), see Table 1. To test whether the estimation of $\tau_1^{\min}$ is robust to the presence of artifacts, we introduced a fixed percentage of artifacts (0, 0.3, 0.6, and 0.9 % of the RR series length). The occurrence time was evenly distributed in the range $0.2\,\mathrm{s}$–$\tau_1^{\min}$. The AF frequency λ was assumed to be known.

Table 1. Simulations parameter setting.

	Sim1	Sim2	Sim3	Sim4	Sim5
λ	6 Hz	6.5 Hz	5 Hz	6 Hz	8 Hz
α	0.8	0.5	0.4	0.3	0.7
$\tau_1^{\min}$	0.28 s	0.30 s	0.28 s	0.32 s	0.34 s
$\tau_2^{\min}$	0.35 s	0.37 s	0.40 s	0.40 s	0.43 s
$\tau_{p,1}$	0.10 s	0.20 s	0.05 s	0.10 s	0.10 s
$\tau_{p,2}$	0.15 s	0.10 s	0.05 s	0.15 s	0.10 s

3.2 Real Data

We analyzed 25 consecutive patients with persistent AF (67 ± 7 years, 16 females) who underwent electrical cardioversion, according to the international guidelines, at the Cardiology department of San Paolo Hospital, Milan, Italy. Recordings were acquired at rest and during a passive orthostatic stimulus ($75°$ tilting). One patient was excluded from analysis due to poor ECG quality which prevented the estimation of AF frequency. Hence, the results presented below are based on 24 patients.

The ECG was recorded at rest for 10 min and, when applicable, followed by tilting, using three orthogonal leads and a sampling rate of 1 kHz. All recordings were performed in the morning in a quiet environment following 15 min of adaptation. The study was approved by the Ethics Committee, and all patients gave their written informed consent to participate.

4   Results

4.1   Simulated Data

Figure 1 shows the mean and standard deviation of $\hat{\tau}_1^{\min}$ obtained with the four methods. It can be noted that the larger the percentage of inserted artifacts, the worse perform the methods based on the Poincaré plot. On the other hand, the estimates obtained by ML estimation remain quite stable and close to the true value.

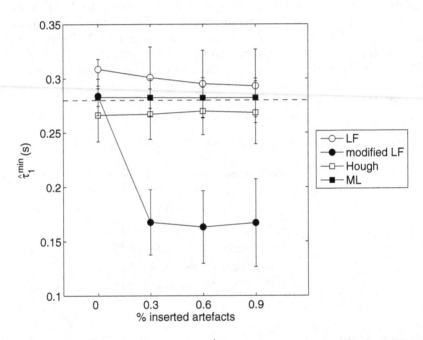

Fig. 1. Mean and standard deviation of $\hat{\tau}_1^{\min}$, computed for 100 RR series, for different percentages of inserted artifacts; the true value is indicated by the dashed line. The following model parameter values were used: $\lambda = 6\,\mathrm{Hz}$, $\tau_1^{\min} = 0.28\,\mathrm{s}$, $\tau_2^{\min} = 0.35\,\mathrm{s}$, $\alpha = 0.8$, $\tau_{p,1} = 0.1\,\mathrm{s}$, and $\tau_{p,2} = 0.15\,\mathrm{s}$.

Figure 2 shows the mean normalized absolute error between $\hat{\tau}_1^{\min}$ and the true value $\tau_1^{\min}$ averaged on the five simulation settings using the four analyzed methods. When estimating $\tau_1^{\min}$ using the ML estimation, it is observed that the error is well below 5 % even in the presence of a high percentage of artifacts.

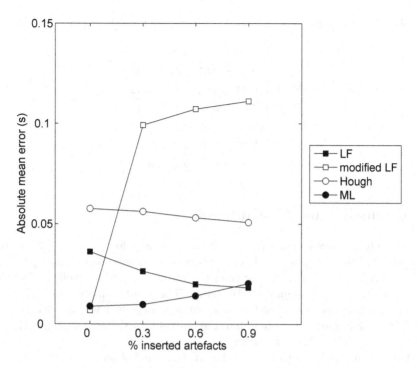

Fig. 2. The mean normalized absolute error between $\hat{\tau}_1^{\min}$ and the true value using the four methods.

4.2 Real Data

To assess whether the model parameters can capture changes due to increased sympathetic tone, e.g., observed during a tilt test, the parameter estimates obtained during rest were compared to those during tilt. Table 2 compares the model parameter estimates obtained at rest and during tilt, with significant changes due to sympathetic activation in both $\hat{\tau}_1^{\min}$ and $\hat{\tau}_2^{\min}$. The AF frequency was found to increase significantly during tilt. The probability of an atrial impulse to chose either pathway is almost equal during rest and tilt ($\alpha = 0.5$), although α spans the range from 0.05 to 1 in individual patients, thus making the involvement of the pathway with slower refractory period ($\alpha < 0.5$) in about half of all recordings. The refractory periods of both pathways are significantly shortened during tilt, whereas their prolongation remains almost unchanged.

Both the mean and standard deviation of RR intervals are significantly shortened during tilt due to sympathetic activation. The mean RR interval length was 763 ± 149 ms vs. 697 ± 135 ms (rest vs. tilt, $p < 0.0001$), and the related standard deviation was 161 ± 48 ms vs. 141 ± 32 ms (rest vs. tilt, $p < 0.0001$).

Table 2. Comparison of rest and tilt parameters (*p < 0.05, **p = 0.001).

	Rest	Tilt
$\hat{\alpha}$	0.53 ± 0.31	0.47 ± 0.33
$\hat{\tau}_1^{min}$ (s)	0.43 ± 0.12	0.38 ± 0.12 **
$\hat{\tau}_2^{min}$ (s)	0.55 ± 0.14	0.47 ± 0.11 *
$\hat{\tau}_{p,1}$ (s)	0.38 ± 0.32	0.31 ± 0.25
$\hat{\tau}_{p,2}$ (s)	0.22 ± 0.31	0.30 ± 0.20
$\hat{\lambda}$ (Hz)	6.08 ± 1.03	6.20 ± 0.99 *

5   Discussion and Conclusions

In this study we have compared four different methods for estimating the refractory period of the slow pathway in the presence of artifacts. As the most problematic artifacts are the ones shorter than the refractory period itself, we inserted only this type in the simulated RR series. The results showed that the estimation of refractory period of the slow pathway obtained jointly with the ML estimation offers better accuracy than the ones obtained from the Poincaré plot, independently of the ML estimation.

It is clearly desirable to include the AF frequency λ as well in the ML estimation procedure. However, the point process model is not easily extended from being entirely RR interval related to also account for information on f-waves because the f-waves need to be extracted from the ECG.

We described an AV node model defined by parameters characterizing the arrival rate of atrial impulses, the probability of an impulse choosing either one of the dual AV nodal pathways, the refractory periods of the pathways, and the prolongation of refractory periods. After the comparison made in this study, all model parameters are estimated from the RR interval series using ML estimation, except for the mean arrival rate of atrial impulses which is estimated by the AF frequency derived from the f-waves.

Considering the physiological aspects, our results indicate that tilting is associated with significant changes in AV conduction that are well-described by the model and reflected by shortening of both τ_1^{min} and τ_2^{min} during adrenergic activation. Thus, the present AV node model is adequate for studying and describing the functional characteristics of AV conduction in AF patients, e.g., to assess drug effect.

References

1. Fuster, V., Rydén, L.E., Cannom, D.S., Crijns, H.J., Curtis, A.B., et al.: ACC/AHA/ESC 2006 guidelines for the management of patients with atrial fibrillation: a report of the American College of Cardiology/American Heart Association task force on practice guidelines and the European Society of Cardiology committee for practice guidelines. Circ. **114**(2006), e257–e354 (2006)

2. Climent, A., de la Salud Guillem, M., Husser, D., Castells, F., Millet, J., Bollmann, A.: Poincaré surface profiles of RR intervals: a novel noninvasive method for the evaluation of preferential AV nodal conduction during atrial fibrillation. IEEE Trans. Biomed. Eng. **56**, 433–442 (2009)
3. Climent, A., Guillem, M., Husser, D., Castells, F., Millet, J., Bollmann, A.: Role of atrial rate as a factor modulating ventricular response during atrial fibrillation. PACE **15**, 1–8 (2010)
4. Masè, M., Glass, L., Disertoric, M., Ravelli, F.: The AV synchrogram: a novel approach to quantify atrioventricular coupling during atrial arrhythmias. Biomed. Signal Proc. Control **8**, 1008–1016 (2013)
5. Jørgensen, P., Schäfer, C., Guerra, P.G., Talajic, M., Nattel, S., Glass, L.: A mathematical model of human atrioventricular nodal function incorporating concealed conduction. Bull. Math. Biol. **64**, 1083–1099 (2002)
6. Mangin, L., Vinet, A., Page, P., Glass, L.: Effects of antiarrhythmic drug therapy on atrioventricular nodal function during atrial fibrillation in humans. Europace **7**, S71–S82 (2005)
7. Rashidi, A., Khodarahmi, I.: Nonlinear modeling of the atrioventricular node physiology in atrial fibrillation. J. Theor. Biol. **232**, 545–549 (2005)
8. Lian, J., Müssig, D., Lang, V.: Computer modeling of ventricular rhythm during atrial fibrillation and ventricular pacing. IEEE Trans. Biomed. Eng. **53**, 1512–1520 (2006)
9. Lian, J., Müssig, D.: Heart rhythm and cardiac pacing: an integrated dual-chamber heart and pacer model. Ann. Biomed. Eng. **37**, 64–81 (2009)
10. Corino, V.D.A., Sandberg, F., Mainardi, L.T., Sörnmo, L.: An atrioventricular node model for analysis of the ventricular response during atrial fibrillation. IEEE Trans. Biomed. Eng. **58**, 3386–3395 (2011)
11. Sandberg, F., Stridh, M., Sörnmo, L.: Frequency tracking of atrial fibrillation using hidden Markov models. IEEE Trans. Biomed. Eng. **55**, 502–511 (2008)
12. Corino, V.D.A., Sandberg, F., Mainardi, L.T., Sörnmo, L.: Atrioventricular nodal function during atrial fibrillation: model building and robust estimation. Biomed. Signal Proc. Control **8**, 1017–1025 (2013)
13. Corino, V.D.A., Sandberg, F., Mainardi, L.T., Sörnmo, L.: Statistical modeling of the atrioventricular node during atrial fibrillation: data length and estimator performance. In: Proceedings of 2013 35th Annual International Conference of the IEEE Engineering in Medicine and Biology Society (EMBC). vol. 35, pp. 2567–2570 (2013)
14. Van den Bergh, F., Engelbrecht, A.P.: A cooperative approach to particle swarm optimization. IEEE Trans. Evol. Comput. **8**, 225–239 (2004)
15. Niu, B., Zhu, Y., He, X., Wu, H.: MCPSO: a multi-swarm cooperative particle swarm optimizer. Appl. Math. Comput. **2**, 1050–1062 (2007)
16. Brennan, M., Palaniswami, M., Kamen, P.: Do existing measures of Poincaré plot geometry reflect nonlinear features of heart rate variability? IEEE Trans. Biomed. Eng. **48**, 1342–1347 (2001)
17. Anan, T., Araki, K.S.H., Nakamura, M.: Arrhythmia analysis by successive RR plotting. J. Electrocardiol. **23**, 243–248 (1990)
18. Hayano, J., Sakata, S., Okada, A., Mukai, S., Fujinami, T.: Circadian rhythms of atrioventricular conduction properties in chronic atrial fibrillation with and without heart failure-relation between mean heart rate and measures of heart rate variability. J. Am. Coll. Cardiol. **31**, 158–166 (1998)

19. Corino, V.D.A., Climent, A., Mainardi, L.T., Bollmann, A.: Analysis of ventricular response during atrial fibrillation. In: Mainardi, L.T., Sörnmo, L., Cerutti, S. (eds.) Understanding Atrial Fibrillation: The Signal Processing Contribution. Morgan and Claypool (2008)
20. Corino, V.D.A., Sandberg, F., Mainardi, L.T., Sörnmo, L.: Non-invasive, robust estimation of refractory period of atrioventricular node during atrial fibrillation. Int. J. Bioelectromagnetism **15**, 41–46 (2013)

Clinical Test for Validation of a New Optical Probe for Hemodynamic Parameters Assessment

T. Pereira[1(✉)], I. Santos[1], T. Oliveira[1], P. Vaz[1], T. Santos Pereira[2],
H. Santos[2], V. Almeida[1], H.C. Pereira[1,3], C. Correia[1], and J. Cardoso[1]

[1] Instrumentation Center, Physics Department, University of Coimbra,
Coimbra, Portugal
taniapereira@lei.fis.uc.pt
[2] Coimbra College of Health Technology, Coimbra, Portugal
[3] ISA- Intelligent Sensing Anywhere, Coimbra, Portugal

Abstract. The assessment of the cardiovascular system condition based on multiple parameters allows a more precise and accurate diagnosis of the heart and arterial tree condition. For this reason, the interest in non-invasive devices has presently increased in importance. In this work, an optical probe was tested in order to validate this technology for measuring multiple parameters such as Pulse Wave Velocity (PWV) or Augmentation Index (AIx), amongst others. The PWV measured by the optical probe was previously compared with the values obtained with the gold-standard system. Another analysis was performed in 131 young subjects to establish carotid PWV reference values as well as other hemodynamic parameters and to find correlations between these and the population characteristics. The results allowed us to conclude that this new technique is a reliable method to determine these parameters. The range of the obtained values for local PWV are in agreement with the values obtained by other studies, and significant correlations with age and smoking status were found. The AIx varied between −6.15 % and 11.46 % and exhibit a negative correlation with heart rate, and dP/dt$_{max}$ shows a significant decrease with age.

Keywords: Optical probe · Waveform distension · Hemodynamic parameters · Pulse wave velocity · Pulse waveform analysis

1 Introduction

The interest in non-invasive assessment of cardiovascular function has increased over the recent times, particularly around solutions able to perform multi parameter assessment for monitoring and early diagnosis of cardiovascular pathologies [1–4].

The pulse wave velocity and the parameters extracted from pulse waveform analysis (PWA) are widely used tools in the evaluation of the function of large arteries and cardiac activity, which have been shown to predict cardiovascular diseases [5].

The assessment of the cardiovascular system condition based on multi parameters allows more precise and accurate diagnosis of the heart and arterial tree condition. Risk

© Springer-Verlag Berlin Heidelberg 2014
M. Fernández-Chimeno et al. (Eds.): BIOSTEC 2013, CCIS 452, pp. 269–283, 2014.
DOI: 10.1007/978-3-662-44485-6_19

indicators that can be assessed from the distension waveforms acquired through the hereby presented system, can be determined from the main parameters extracted from measured and analysed waveform, as well as its time characteristics and the pulse wave velocity.

The PWV is defined as the speed at which the pulse pressure propagates along the arterial tree. It is known that PWV increases with age, blood pressure (BP) and arterial stiffness [5, 6]. The reference value for regional PWV in healthy young population is 6.2 m s^{-1} (range of 4.7–7.6 m s^{-1}) [6] and is usually measured from signals acquired in the femoral and carotid arteries. The time delay, or pulse transit time (PTT) between the two signals is determined through different algorithms depending on the commercial system used, while the distance (D) between the two arteries is externally measured. However, the large heterogeneity of the structure of the arterial walls at different sites constitutes an important limitation of PWV regional measurement [7]. In fact, the Expert Consensus Document in Arterial Stiffness states that the PWV increases from 4–5 m s^{-1} in the ascending aorta to 5–6 m s^{-1} in the abdominal aorta and 8–9 m s^{-1} in the iliac and femoral arteries [8]. A local PWV measurement technique is hence preferred.

Some studies [9] explored an ultrasound method for local PWV assessment in the carotid artery and obtained estimated PWV in the range of 4–9 m s^{-1}. In 2008 [10] an experimental method for the local determination of PWV in the carotid artery obtained values for PWV of 3–4 m s^{-1}.

The pulse wave analysis allows the non-invasive determination of main indices of cardiovascular function: Augmentation Index (AIx), Subendocardial Viability Ratio (SEVR), Maximum Rate of Pressure Change (dP/dt$_{max}$) and Ejection Time Index (ETI). The most important points of the pulse pressure waveform are presented in Fig. 1.

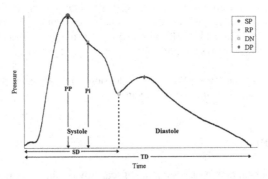

Fig. 1. Typical pressure waveform of a healthy subject and its main features used to compute the indices of cardiovascular function: pulse pressure (PP), systolic peak (SP), reflection point (RP), dicrotic notch (DN), dicrotic peak (DP), systolic duration (SD), total duration (TD) and pressure in the reflection point (Pi).

   The clinical definition for the referred parameters and the mathematical expressions for their determination are summarized in Table 1.

   If the backward wave arrives before the systolic peak, the AIx parameter, by definition, has a positive value due to the contribution of the increased systolic pressure. If the backward wave arrives later, it does not contribute to the increased pressure, turning the AIx value negative.

   Taking into account the currently available commercial devices, a new solution based on optical technology was developed and demonstrably benefits by a non-contact and more accurate measure and multi parameter assessment ability [14].

   The PWV assessment from the commercial systems consists always on a regional measurement, between two peripheral arteries as already described. With this work a local measure of PWV is proposed, where two measurements are taken at the carotid artery, simultaneously and separated by a 20 mm distance [15].

Table 1. Main parameters from pulse waveform analysis.

Parameter	Definition	Formula	Reference values
AIx	Describes the increase of systolic blood pressure due to an early backward wave, produced by the reflection of the forward systolic wave on the peripheral arterial tree structure.	$\pm \frac{Pi}{PP} \times 100$	- 22 to 40 (%) [11]
SEVR	Parameter that estimates the myocardial oxygen supply-demand relative to the cardiac workload. It is an indicator of subendocardial ischaemia.	$\frac{\int Diastole(t)dt}{\int Systole(t)dt} \times 100$	119 to 254 (%) [11]
dP/dt$_{max}$	The ventricular contractility can be evaluated by the maximum rate of pressure change, which gives information about the initial velocity of the myocardial contraction, which is also an index of myocardial performance.	dP/dt	772 ± 229 (mmHg/s) [12]
ETI	Ventricular systolic ejection time between the aortic valve opening and closing. It is an important component on the evaluation of the left ventricular performance.	SD/TD	30 to 42 (%) [13]
PWV	The velocity at which the pulse wave propagates along a length of artery.	PTT/D	6.2 (4.7 to 7.6) (m s^{-1}) [6]

The carotid artery is the natural probing site for pulse waveform measurement, due to the heart proximity and because it is easily accessible due to its proximity to the skin surface.

The purpose of this study was to validate the optical system for PWV and PWA measurements. Firstly, it was intended to assess the values for local PWV, to establish its reference values for the carotid artery in a young and healthy population and validate the technology for hemodynamic parameters assessment from the pulse pressure waveform. The study also aimed to find correlations between hemodynamic parameters with the population characteristics such as age, gender, smoking, body mass index, blood pressure or heart rate: main characteristics described in the literature and that have significant impact in the cardiovascular system evaluation.

2 Technology

The pressure wave, generated by the contraction of the left ventricle, originates a distension wave that propagates through the aorta and other proximal elastic arterial walls. The distensibility is determined as the ratio between the variation of volume from diastole to systole and the variation of pressure that origins that distension in the arterial wall [8].

Previous studies on comparison between pressure and distension waveforms have shown that these waves can be used interchangeably for many analysis due to their similar wave contour [8, 16–18].

The proposed probes were developed to measure the arterial pulse wave profile at the carotid site and are based on the reflectance fluctuations of the skin surface during the underlying pulse wave propagation. The propagation of the pulse pressure waveform causes distension in the artery wall. This distension, known as distension wave, changes the optical reflectance angle of the wall which produces a change in the reflection characteristics of skin, causing an amplitude modulation of the light. This effect can be used to generate an optical signal that correlates with the passing pressure wave.

The illumination source is provided by local, high brightness, 635 nm monochromatic light emitting diodes (LEDs) and the light detection is performed by two photodetectors, placed at a precise distance of 20 mm apart. This guarantees the local pulse wave profile assessment at two distinct spots, providing the precise local determination of pulse transit time (PTT) and thus of the local PWV. The probe structure is enclosed in a plastic box with an ergonomic configuration, comfortable to the patient and simple to use by the operator (Fig. 2).

The plastic case contacts with the patient skin although neither the LEDs nor the photodetectors do. The electronic components remain at a fixed distance, few millimetres from the skin ensuring a totally non-contact and non-invasive local PWV assessment.

Previous bench tests had shown that the optical probe is capable of accurately measure PTT as short as 1 ms with less than 1 % of error, one can guarantee the capability of the probes in truthfully determining local PWV [19].

Fig. 2. Structure of the optical probe with photodetector and visible light sources (LEDs), inside an ergonomic plastic box.

The comparison tests to evaluate the capability of the developed device in accurately detect the pulse waveform were carried out using an ultrasound imaging system, as source of reference data. When compared with ultrasound system, the optical sensors allow the reproduction of the arterial waveform with a much higher time resolution, adequate to feed feature extraction algorithms [14].

The signals from the photodetectors were digitized with a 16-bit resolution data acquisition system (National Instruments, USB6210®) with a sampling rate of 20 kHz and stored for offline analysis. All the algorithms were developed using Matlab® 7.8.0 (R2009a).

3   Preliminary Validation Study

In order to validate the data obtained by the developed optical system, a number of volunteers were previously submitted to a signal acquisition procedure, using simultaneously the proposed optical device and a gold-standard in the PWV assessment, a Complior Analyse® device. This preliminary study was undertaken in 14 healthy subjects (9 females, average age 23.2 ± 5.5 years).

The results showed a great consistency between the PWV obtained with the two devices. In spite of this comparison, it is worth to note that the nature of the PWV determination is different between the optical system, that is based on local assessment (carotid artery measure) and the Complior® system, which is based on a regional assessment (carotid-femoral measures).

Using a non-parametric correlation analysis between the values obtained from the two systems, the Pearson correlation value is 0.819, which is a strong correlation and significant at the 0.01 level (2-tailed).

The agreement between the PWV values obtained by the Complior® and the optical probe is shown in Fig. 3 (top). The values of PWV obtained by the two systems are correlated ($r^2 = 0.67$). The average difference between the two systems, Complior® and Optical probe, was -1.8557 m s^{-1} with a SD of 0.5744 m s^{-1} as shown in a Bland-Altman plot in Fig. 2 (bottom). As one can observe in Fig. 3 (top), there is a tendency to have systematic lower values from the optical probe device than Complior®. Again,

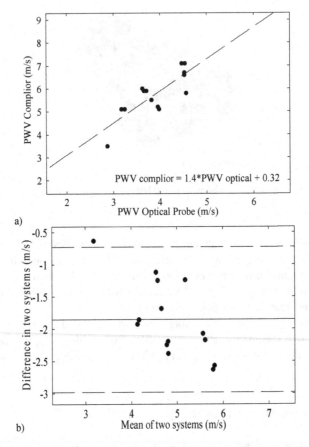

Fig. 3. Correlation between the two systems (Complior® and Optical probe) for PWV determination (a). Bland-Altman plot (b) displaying the difference between the two systems (Complior® and Optical probe) as a function of the average of the determined PWV.

it is important to bear in mind that the values obtained using both devices correspond to different PWV determination processes (local vs. regional) and lower values are expected for PWV in the carotid (local) than the PWV in a carotid-femoral measurement (regional). This issue could explain the associated deviate of final values.

Taken together, these results allow the use of this proposed optical system as a reliable method to determine local carotid PWV.

4   Study Protocol

The central purpose of these tests is to assess the main hemodynamic parameters extractable from the pressure waveform features and pulse wave velocity, in a young and healthy population using the optical system.

The complete study database contains 131 subjects constituting a representative cohort of 18–35 year old subjects randomly sampled. The study protocol was approved by the ethical committee of the Centro Hospitalar e Universitário de Coimbra, EPE Portugal. All the subjects were volunteers and gave a written informed consent.

Measurements were performed after a rest period in a temperature-controlled environment. Each exam procedure consisted in the acquisition of a set of cardiac cycles at the carotid artery during few minutes, with the patient lying in supine position.

The assessment of the arterial blood pressure (ABP) by conventional measurement using an automated digital oscillometric sphygmomanometer (Omron Matsusaka Co., Ltd., Japan) was performed prior and after the exam for reference purposes. The diastolic and systolic pressures of arm blood pressure were used to calibrate the system.

5   Signal Processing

The stored signal data were processed offline in order to parameterize the arterial pulse waveform and to calculate the corresponding cardiovascular performance indexes. A set of dedicated pre-processing algorithms were developed to segment the data stream in single cycles, generate an average pulse and identify the corresponding remarkable points in the waveform profile. Following this stage, the signal streams underwent the full processing sequence to determine all the significant APW features and indexes along with the local PWV.

The pulse wave velocity was determined by a cross-correlation method, based on the property of the peak of the cross correlogram, from which time delays can be calculated by subtracting the peak time position from the pulse length.

The pulse wave analysis is based on differential calculus, and uses zero-crossing of the three first derivatives. The remarkable points were found by means of an iterative third-order derivative method, with which the consecutive zero-crossing of the first, second and third derivatives are used to detect inflection points that correspond to the clinically interesting features of the waveform.

An assessment of ABP by conventional measurement using a sphygmomanometer was conducted prior and after the exam for calibration purposes. Both brachial diastolic (DBP) and mean arterial pressure (MAP) values were used to calibrate the system. Studies have reported that MAP is relatively constant along the arterial tree and that DBP do not vary considerably between the carotid and brachial arteries, whereas systolic blood pressure (SBP) increases along the arterial tree [7, 20]. Thus, we made the assumption that brachial DBP and MAP are approximately the same as carotid DBP and MAP. These values were used to calibrate the carotid pressure waveform as recommended and according to the calibration method proposed by Kelly and Fitchett [8, 21, 22].

6   Results

The characteristics of the volunteers are presented in Table 2. The group consisted of 131 subjects (62 men and 69 women), normotensive and with no documented history of cardiovascular disorders or diabetes, with mean (± SD) age of 22.6 ± 5.3 years old.

Table 2. Main characteristics of the volunteers.

Characteristics	
n, Males/Females	131 (62/69)
Age, year	22.6 ± 5.3
Height, cm	169.2 ± 0.1
Weight, kg	64.5 ± 13.2
BMI, kg/m^2	22.4 ± 3.2
Brachial SBP*, mmHg	113.3 ± 14.5
Brachial DBP*, mmHg	72.7 ± 9.9
Brachial MAP*, mmHg	86.2 ± 10.4
Estimated Carotid SBP**, mmHg	99.1 ± 12.2
Heart Rate*, bpm	69.0 ± 11.7

Values are numbers or means ±SD.
BMI indicates body mass index; SBP, systolic blood pressure;
DBP, diastolic blood pressure.
*Measure in brachial, with commercial sphygmomanometer
(blood pressure cuff).
**Determined using the calibration method.

The results for the parameters that were assessed by the optical probe are presented in the next sections. Data are reported as mean values (± SD) or 95 % confidence intervals, with $P < 0.05$ considered significant unless stated otherwise. The Shapiro-Wilk test of normality was used to assess the normality of the variables distribution. Mean differences between variables were assessed using ANOVA. The strength of the association between two variables was assessed using Pearson Correlation, for normal distributions, unless stated otherwise.

All statistical analyses were performed with Analytics Software Statistics 18.0.0 (SPSS, Inc, Chicago, IL).

6.1   Carotid PWV Results

In a total of 131 subjects the mean value for PWV is 3.33 ± 0.72 m s^{-1} (range of 2.00-5.13 m s^{-1}). The results obtained for PWV approximated a normal distribution. Statistically negligible differences between genders were found. For females the mean of PWV is 3.31 ± 0.64 m s^{-1} and for males is 3.35 ± 0.81 m s^{-1}.

It is well documented in other studies that PWV tends to increase with age [6]. Since the population under study is young we categorized it in three groups: under 20 years (22 subjects), 21–29 years (93 subjects) and over 30 years old (16 subjects).

The correlation between PWV and age is weak (0.244) but significant with a 2-tailed significance value of 0.008 ($p < 0.01$). The comparison of PWV mean values among the different age categories using ANOVA shown this statistically significant difference, with a significance value = 0.024 ($p < 0.05$). The PWV mean value (± SD) for each age category is 2.96 ± 0.52 m/s for the less than 20 years old group, 3.38 ± 0.7 m/s for the 20–29 years group and 3.58 ± 0.7 m/s for subjects above 30 years old (Fig. 4).

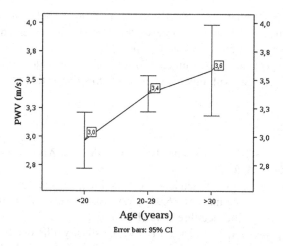

Fig. 4. PWV by age category. The continuous line connects mean values; error bars indicate the 95 % confidence interval of the sample mean.

It was hypothesized whether there is statistically significant difference for PWV, between smoker and non-smoker subjects. The data analysis shows that smoking influences significantly the PWV.

These results show a small but significant correlation between PWV and smoking with a 2-tailed significance value of 0.016 ($p < 0.05$). Non-smoker subjects presented a PWV mean value ($\pm$ SD) of 3.29 ± 0.72 m s^{-1} while smoker subjects showed a mean PWV value of 3.81 ± 0.6 m s^{-1}, represented in Fig. 5.

The comparison of PWV mean values among these two groups using ANOVA confirms the statistically significant difference between the obtained PWV for smokers and non-smokers, with a significance value = 0.021 ($p < 0.05$).

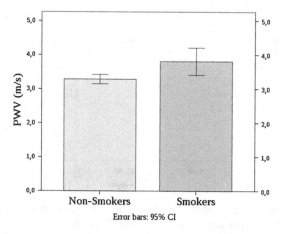

Fig. 5. Bar plot of PWV by smoking status. The error bars indicate the 95 % confidence interval of the sample mean.

The data suggests that, in spite of having a young population, smoking would significantly influence the arterial stiffness, thus leading to increased PWV. Similar findings have been reported by N. Jatoi et al., [23].

Contrary to expectations, no significant correlation between blood pressure and PWV was found. This contrasts with other studies where this correlation is verified [6, 24].

6.2   AIx Results

For the total of the subjects in study, the mean value for AIx is -6.151 ± 11.46 % (range -44.31 % to 24.26 %).

Small differences between genders were verified, as the female mean of AIx is -5.59 ± 1.36 % and the male is -6.93 ± 1.59 %. However, this difference did not reach the statistical significance threshold ($P > 0.05$).

Also for this parameter, as well as for PWV, statistically differences between smoker and non-smoker subjects were found. AIx is higher for smokers (-4.33 ± 4.42 %) than for non-smokers (-6.35 ± 1.04 %), this shows that for smokers the reflected wave arrives earlier in time, which is consistent with a slight higher the arterial stiffness.

The negative correlation between the AIx and the heart rate was described in other tests [4, 25] and was confirmed in this study (Fig. 6). The results of Person-Correlation test were compatible with a significant negative correlation, between heart rate (HR) and AIx at a 0.05 level (2-tailed). Nevertheless, the strength of the relationship found between these two variables is medium, since the Pearson Correlation is significant with a 2-tailed significance value of -0.226 ($p < 0.05$). The comparison of AIx mean values among these three groups using ANOVA confirms a statistically significant difference, with a significance value = 0.024 ($p < 0.05$).

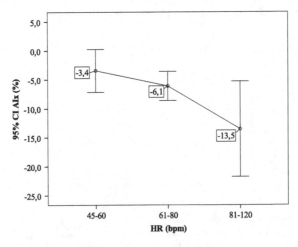

Fig. 6. The continuous line connects plots of AIx mean by heart rate category, with error bars representing the 95 % confidence interval of the sample mean.

The negative correlation between these two parameters is explained due to the early return of the reflected wave in systole when HR is lower, and the long period of heart contraction.

6.3   Other Hemodynamic Parameter Results

The Ejection Time Index, Subendocardial Viability Ratio and Maximum Rate of Pressure Change are other hemodynamic parameters assessed by the optical system, and the results for the population in study are shown in Table 3.

It is remarkable that the mean values of SEVR and ETI stay within the range presented by other studies (Table 1) even though it is clear that they show a wide variation. For the SEVR parameter it was also verified a decrease with the heart rate (Fig. 7) with a significant variance (ANOVA, $P = 0.01$) and a significant Pearson correlation at 0.01 level (2-tailed).

Table 3. Hemodynamic Parameters obtained with the Optical System.

	Min.	Max.	Mean	SD
SEVR (%)	86.41	412.25	176.86	53.44
$dP/dt_{max.}$ (mmHg/s)	212.59	953.33	443.92	151.68
ETI (%)	14.33	47.17	33.96	6.37

The derived values for dP/dt_{max} differ substantially from the ones presented by other studies, a wider range than the expected and a smaller mean value is evident probably originated by the differences in the calibration method used. However, a relation between $dP/dt_{max.}$ and gender, was found, since the Pearson Correlation is -0.408. The female subjects showed lower values for this parameter, which are represented in Fig. 8.

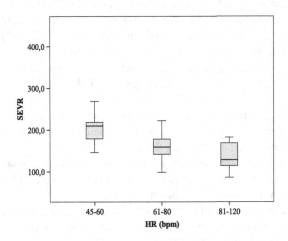

Fig. 7. Box plot of data from the determined SEVR versus heart rate categories.

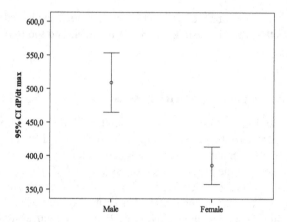

Fig. 8. Error plots of dP/dt$_{max}$ mean values according to gender. The error bars represent the 95 % confidence interval of the sample mean.

The dP/dt$_{max}$ values decrease significantly with subjects' age, this allows the inferring that with the increasing age the velocity of the myocardial contraction is lower, meaning that there is a decrease in the myocardial performance.

For the ETI parameter there were verified slight differences between gender, the mean vales for the females is 36.80 ± 0.72 % and for males 33.60 ± 0.85 %. This suggests that there is a shorter systole during cardiac cycle in males than in females. Also for this parameter, a significance correlation with heart rate was found since the Pearson Correlation is -0.408.

7 Conclusions

These trial tests were carried out in a large group of young and healthy subjects. This study has attempted to validate the proposed optical system as a reliable method to assess non-invasively local PWV in the carotid artery, to establish reference values of the local PWV as well as other mentioned parameters in this type of population.

Previously, this new system had also proved to be reliable in reproducing the arterial waveform with a higher resolution, adequate to feed feature extraction algorithms, when compared to an ultrasound image system that was used as a reference for comparison.

Besides of the ability of the optical system in acquiring non-invasively a carotid distension waveform with high definition, the presented optical system shows other advantages over the actual commercial devices.

One advantage of this optical approach relies on being a non-contact technique that avoids the drawback of pressure application as happens with piezoelectric transducers that could result in the distortion of the signal of interest.

Another benefit is the possibility of local PWV assessment, in a single arterial segment, avoiding coarse approximations of the distance between test points constituting an important advance in the PWV assessment.

The validation test, in which the obtained values using the optical probes were compared with those obtained with a commercial device are supported by all the previous bench test results and allowed to consider this technique as a valid method to assess PWV and analyse PWA parameters.

The range of the obtained values for local CA PWV are in agreement with the values obtained by other studies that also assessed the carotid local PWV. The expected and obtained lower values of the carotid PWV assessment when compared to those obtained with Complior® device are due to the different PWV determination processes (local vs. regional). The lack of compatibility between methods limits the inferences that can be made and thus, more studies of repeatability, comparing the local PWV obtained by the optical system with the values estimated by ultrasound data, are necessary to ensure accuracy of the proposed system.

The PWV measured by the optical probe confirmed a significant increase of PWV with age. Even though the sample consists of young subjects with a relatively narrow age distribution, this result generally agrees with those obtained in other studies.

The PWA parameters revealed the important relations between some characteristics of the population and the arterial system status. The negative correlation between the AIx and the heart rate were verified in this study, and differences between gender and smokers/non-smokers were found.

For SEVR parameter was also verified a decrease with heart rate. The $dP/dt_{max.}$ analysis showed a difference with gender, with lower values for female subjects, and this parameter decreased significantly with age.

All the correlations obtained in this study showed that age contributes to the arterial stiffness as reflected in the values found for various parameters. Smokers appear to have higher arterial age, and therefore increased cardiovascular risk.

The optical system under study proved to be a good choice for determination of hemodynamic parameters in a non-invasive and non-contact assessment, which allows a better knowledge about the cardiovascular condition and management of many disease states. Further studies must compare the non-invasive measurement of distension waveform in carotid artery with the invasive profile of the pulse pressure acquired by an intra-arterial catheter in ascending aorta, in patients who have undergone cardiac catheterization.

Another study should be designed to evaluate the behaviour of the carotid artery in atherosclerosis. In this case the comparison between invasive data (catheter) and Ecodoppler signals will be important to prove the agreement with non-invasive signals acquired by the optical probe in this particular disease.

Although this method provides a set of parameters strongly related to pulse waveform morphology it needs a proper and extensive gold standard comparison with an invasive system.

Finally, it is worth mentioning that, in spite of its lack of maturity (meaning that many engineering aspects of the probe can still be improved), the proposed method exhibits a very high patient hit success. In fact, for 119 out of 131 patients (90 %) it took around 1 to 2 mins to start acquiring reasonable quality signals, each take lasting for 30 seconds. Although some training is required, the method is much less demanding on the operator skills as compared to ultrasound, Complior® or tonometer.

Acknowledgements. The authors acknowledge the clinical collaboration from Dr. Rui Providência and Coimbra College of Health Technology, and acknowledge the support from Fundação para a Ciência e Tecnologia (FCT) for funding (SFRH/BD/79334/2011).

References

1. Willum-Hansen, T., Staessen, J.A., Torp-Pedersen, C., Rasmussen, S., Thijs, L., Ibsen, H., Jeppesen, J.: Prognostic value of aortic pulse wave velocity as index of arterial stiffness in the general population. Circulation **113**(5), 664–670 (2006)
2. Blacher, J., Asmar, R., Djane, S., London, G.M., Safar, M.E.: Aortic pulse wave velocity as a marker of cardiovascular risk in hypertensive patients. Hypertension **33**(5), 1111–1117 (1999)
3. Hayashi, T., Nakayama, Y., Tsumura, K.: Reflection in the arterial system. Am. J. Hypertens. **15**(5), 405–409 (2002)
4. Weber, T., Auer, J., O'Rourke, M.F., Kvas, E., Lassnig, E., Berent, R., Eber, B.: Arterial stiffness, wave reflections, and the risk of coronary artery disease. Circulation **109**(2), 184–189 (2004)
5. Huck, C.J., Bronas, U.G., Williamson, E.B., Draheim, C.C., Duprez, D.A., Dengel, D.R.: Noninvasive measurements of arterial stiffness: repeatability and interrelationships with endothelial function and arterial morphology measures. Vasc. Health Risk Manage. **3**(3), 343–349 (2007)
6. Vermeersch, S.J., Dynamics, B., Society, L.: Determinants of pulse wave velocity in healthy people and in the presence of cardiovascular risk factors: 'establishing normal and reference values'. Eur. Heart J. **31**(19), 2338–2350 (2010)
7. Safar, M.E.: Arterial stiffness: a simplified overview in vascular medicine. Atherosclerosis, Large Arteries Cardiovasc. Risk **44**, 1–18 (2007)
8. Laurent, S., Cockcroft, J., Van Bortel, L., Boutouyrie, P., Giannattasio, C., Hayoz, D., Pannier, B., Vlachopoulos, C., Wilkinson, I., Struijker-Boudier, H.: Expert consensus document on arterial stiffness: methodological issues and clinical applications. Eur. Heart J. **27**(21), 2588–2605 (2006)
9. Rabben, S.I., Stergiopulos, N., Hellevik, L.R., Smiseth, O.A., Slørdahl, S., Urheim, S., Angelsen, B.: An ultrasound-based method for determining pulse wave velocity in superficial arteries. J. Biomech. **37**(10), 1615–1622 (2004)
10. Sørensen, G.L., Jensen, J.B., Udesen, J., Holfort, I.K.: Pulse wave velocity in the carotid artery, vol. 1(1), pp. 1386–1389 (2008)
11. Siebenhofer, A., Kemp, C., Sutton, A., Williams, B.: The reproducibility of central aortic blood pressure measurements in healthy subjects using applanation tonometry and sphygmocardiography. J. Hum. Hypertens. **13**(9), 625–629 (1999)
12. Payne, R.A., Hilling-Smith, R.C., Webb, D.J., Maxwell, S.R., Denvir, M.A.: Augmentation index assessed by applanation tonometry is elevated in Marfan Syndrome. J. Cardiothorac. Surg. **2**, 43 (2007)
13. Istratoaie, O., Mustafa, R., Donoiu, I.: Central aortic pressure estimated by radial applanation tonometry in hypertensive pulmonary oedema. J. Hyperten. **28**(p e149) (2010)
14. Pereira, T., Oliveira, T., Cabeleira, M., Matos, P., Pereira, H.C., Almeida, V., Borges, E., Santos, H., Pereira, T., Cardoso, J., Correia, C.: Signal analysis in a new optical pulse waveform profiler for cardiovascular applications. In: Signal and Image Processing and Applications/716: Artificial Intelligence and Soft Computing, Sipa, pp. 19–25 (2011)

15. Pereira, T., Oliveira, T., Cabeleira, M., Almeida, V., Borges, E., Cardoso, J., Correia, C., Pereira, H.C.: Visible and infrared optical probes for hemodynamic parameters assessment. In: 2011 IEEE SENSORS Proceedings, pp. 1796–1799, Oct 2011

16. Avolio, P., Van Bortel, L.M., Boutouyrie, P., Cockcroft, J.R., McEniery, C.M., Protogerou, A.D., Roman, M.J., Safar, M.E., Segers, P., Smulyan, H.: Role of pulse pressure amplification in arterial hypertension: experts' opinion and review of the data. Hypertension 54(2), 375–383 (2009)

17. Kips, J., Vanmolkot, F., Mahieu, D., Vermeersch, S., Fabry, I., de Hoon, J., Van Bortel, L., Segers, P.: The use of diameter distension waveforms as an alternative for tonometric pressure to assess carotid blood pressure. Physiol. Meas. 31(4), 543–553 (2010)

18. Vermeersch, S.J., Rietzschel, E.R., De Buyzere, M.L., De Bacquer, D., De Backer, G., Van Bortel, L.M., Gillebert, T.C., Verdonck, P.R., Segers, P.: Determining carotid artery pressure from scaled diameter waveforms: comparison and validation of calibration techniques in 2026 subjects. Physiol. Meas. 29(11), 1267–1280 (2008)

19. Pereira, T., Cabeleira, M., Matos, P., Borges, E., Cardoso, J., Correia, C.: Optical methods for local pulse wave velocity assessment. In: 4th International Joint Conference on Biomedical Engineering Systems and Technologies, Rome, Italy, pp. 74–81 (2011)

20. Lamia, B., Chemla, D., Richard, C., Teboul, J.-L.: Clinical review: interpretation of arterial pressure wave in shock states. Crit. Care (London, England) 9(6), 601–606 (2005)

21. Kelly, R., Fitchett, D.: Noninvasive determination of aortic input impedance and external left ventricular power output: a validation and repeatability study of a new technique. J. Am. Coll. Cardiol. 20(4), 952–963 (1992)

22. Proudfoot, N.A.: The acute effects of moderate intensity exercise on vascular stiffness in children with repaired coarctation of the aorta (2009)

23. Jatoi, N.A., Jerrard-Dunne, P., Feely, J., Mahmud, A.: Impact of smoking and smoking cessation on arterial stiffness and aortic wave reflection in hypertension. Hypertension 49(5), 981–985 (2007)

24. Padilla, J.M., Berjano, E.J., Sáiz, J., Fácila, L., Díaz, P., Mercé, S., De Morelia, I.T., De Castellón, H.P., Electro, M.V.: Assessment of relationships between blood pressure pulse wave velocity and digital volume pulse. Comput. Cardiol. 33, 893–896 (2006)

25. Wilkinson, B., MacCallum, H., Flint, L., Cockcroft, J.R., Newby, D.E., Webb, D.J.: The influence of heart rate on augmentation index and central arterial pressure in humans. J. Physiol. 525(Pt. 1), 263–270 (2000)

Influence of Iris Template Aging
on Recognition Reliability

Adam Czajka[1,2]($\boxtimes$)

[1] Institute of Control and Computation Engineering,
Warsaw University of Technology, Warsaw, Poland
[2] Biometrics Laboratory, Research and Academic Computer Network (NASK),
Warsaw, Poland
aczajka@elka.pw.edu.pl
http://www.pw.edu.pl

Abstract. The paper presents an iris aging analysis based on comparison results obtained for four different iris matchers. We collected an iris aging database of samples captured even eight years apart. To our best knowledge, this is the only database worldwide of iris images collected with such a large time distance between capture sessions. We evaluated the influence of the intra- vs. inter-session accuracy of the iris recognition, as well as the accuracy between the short term (up to two years) vs. long term comparisons (from 5 to 9 years). The average genuine scores revealed statistically significant differences with respect to the time distance between examined samples (up to 14 % of degradation in the average genuine scores is observed). These results may suggest that the iris pattern ages to some extent, and thus appropriate countermeasures should be deployed in application assuming large time distances between iris template replacements (or adaptations).

Keywords: Biometric template aging · Iris recognition · Biometrics.

1   Introduction

The statement of a high temporal stability of iris features, in the context of personal identification, appeared as early as in the Flom's and Safir's US patent granted in 1987 [1]. A claim, drawn from a clinical evidence, said that 'significant features of the iris remain extremely stable and do not change over a period of many years'. Although these 'significant features' were not clearly specified in the patent, its context (i.e., recognition of one's identify) suggests that said features should relate to all the iris characteristics having a power to individualize a human within a population. The pioneering work by John Daugman [2] includes more precise suggestion and relates to the high stability of the iris trabecular pattern (as the iris texture is used directly to calculate an iris code). The stability of the iris meshwork is put in contrast to possible changes of other characteristics of the eye, not commonly used in iris recognition, e.g., a melanin concentration

© Springer-Verlag Berlin Heidelberg 2014
M. Fernández-Chimeno et al. (Eds.): BIOSTEC 2013, CCIS 452, pp. 284–299, 2014.
DOI: 10.1007/978-3-662-44485-6_20

responsible for an eye color. Flom's and Daugman's statements are thus very often cited in the iris recognition literature, fueling a common belief that iris templates, ones determined, are useful for unspecified, yet very long time periods.

Recently this highly desired attribute of biometrics seems to be undermined for iris modality by experimental results revealing an increasing deterioration of a recognition accuracy when the time distance between capturing the gallery and probe images extends significantly, e.g., to a few years. This suggests that the initial claim related to the stability of iris texture might be inaccurate.

However, one should aware that the stability of iris texture only partially contributes to the stability of the iris templates, and many factors may influence the template lifespan. Iris is not exposed to the external environment and it is covered by a transparent fluid (an aqueous humor) that fills the space (an anterior chamber) between the cornea and the iris. Hence, capturing the iris image relies on registering this complicated, three-dimensional structure constituting the frontal, visible part of the eye. Although an iris is the most apparent element of this structure, the aging related to the aqueous humor or the cornea may also influence the aging of iris templates, even if the iris tissue is immune to a flow of time. Moreover, the equipment flux should be considered as an influential element of the template aging phenomenon. This may relate to replacement of the camera between the gallery and probe image capture, or wearing the camera components out. Next, the consequences of the template aging should be distinguished from effects related to inter- and intra-session matching scores. Intra-session comparison scores will typically exhibit a better match among images when compared to the corresponding inter-session results. However, the inter-session changes in imaging conditions, e.g., environmental parameters or subject's interaction with the equipment, usually blur more subtle aging effects related to the eye biology. From the technological point of view, the iris biometric features, not images, are used to finally judge on the extent to which the aging occurs, as we are interested in how this phenomenon transforms from the image space (possible to inspect visually) to the feature space (natural for biometric matching). However, the transformation between these spaces is always proprietary to an iris coding method, and the strength of the template aging effect may depend on the feature extraction methodology. Last but not least, we have no guarantee that the aging effect will be evenly observed across the subjects of different populations. Experiment results obtained with a particular database of images may be a weak predictor of this phenomenon for people of different race, health or dietary culture.

The expected stability of the iris pattern may also be regarded (in a broader sense) as a demand for *stationarity*. Stationary time series is characterized by temporal stability of its statistical properties. However, stability of one property (e.g., the average value) does not guarantee the stationarity, as other properties (e.g., sample variance) may still vary over time. This makes the research of the aging phenomenon even more complicated, as the judgment should not be based solely on properties of a single statistics (e.g. monotonic behavior of the average matching score).

Above aspects related to discovering the truth about the iris aging urgently call for experiments carried out for different populations across the world, performed in different environments, for as many feature extraction methods as possible and for the longest possible time lapse between measurements. Answering this call, we present the iris aging analysis based on comparison results obtained for four iris matchers and the biometric samples captured even eight years apart. To our best knowledge, eight years is the longest time interval characterizing samples used in the iris aging analysis up to date worldwide.

2   Related Work

Iris recognition is relatively young discipline, and thus there is still a shortage of data-bases of iris images collected with adequate time intervals to observe the template aging phenomenon. This is why the literature devoted to the iris template aging is still limited.

Gonzalez et al. first addressed a possibility of influence of the time lapse onto iris recognition accuracy [3]. They estimated coding method parameters using a part of the multimodal BioSec database, containing samples of 200 subjects. Final results were generated with the use of the BioSecurID database containing iris images captured for more than 250 volunteers. Although the databases used were reasonably large, the time lapse between image captures in the test database was very short (one to four weeks). As the observation of the aging effects in such a short time period may be difficult, the authors focused on inter- vs. intra-session recognition accuracy analysis. According to the expectations, the genuine intra-session comparisons revealed a better match (e.g. FNMR $\in (0.085, 0.113)$ @FMR $= 0.01$)[1] when compared to the inter-session results (e.g. FNMR $\in (0.224, 0.258)$ @FMR $= 0.01$). No significant differences in comparison scores can be observed for inter-session results with respect to these very short time intervals. Thus any conclusions on the iris aging cannot be drawn based on this work, yet it supports the intuition related to the importance of the enrollment procedure that should produce the enrollment samples predicting, to the maximum possible extent, the inter-session variations.

The intra- vs. inter-session variations in iris matching scores were also studied by Rankin et al., who used a database of images captured for 238 subjects [4]. The sessions were separated by three and six months time periods. Results are presented separately for irises grouped in classes depending on the iris texture density, and support a claim on the increase of false rejections when time interval between samples increases. However, this study lies slightly next to the main course of biometrics technology, as the images were captured in visible, not near-infrared light and by a specialized biomicroscope, not typically used in iris recognition.

---

[1] FNMR (False Non-Match Rate) is an empirical estimate of the recognition method error relying on falsely rejecting a genuine sample; FMR (False Match Rate) is an empirical estimate of the recognition method error relying on falsely accepting an impostor sample.

Baker *et al.* presents the first known to us analysis of the iris aging under long, four-year time lapse [5]. A small database consisted of images captured for 13 volunteers was used in the analysis with the iris segmentation inspected manually (this excludes the segmentation errors from the source of matching score deterioration). As opposed to Gonzalez *et al.*, they eliminated intra-session scores, and the analysis was focused on comparison between short-time-lapse matches (i.e., for images taken a few months apart) and long-time-lapse-matches (i.e., for images taken four years apart). The authors found a statistically significant difference in the average comparison scores between short-time-lapse matches and long-time-lapse matches, namely the genuine comparison scores (based on the Hamming Distance) increased by 3–4 % for long-time-lapse-matches, and the simultaneous change in the impostor scores was not observed. Bowyer *et al.* continue Baker's work presenting the results for slightly enlarged database of iris images captured for 26 subjects [6]. Again, the comparisons between short-time-lapse matches (i.e. for images separated by less than 100 days) and long-time-lapse-matches (i.e. for images taken at least 1000 days apart) are analyzed, and statistically significant deterioration in the genuine comparison scores is reported (increase of the Hamming Distance by approximately 4 %). Later, Baker *et al.* expand their initial work by the use of additional matcher submitted by the University of Cambridge to the Iris Challenge Evaluation 2006 [7]. The authors report an increase in false rejection rate for longer time lapse between images, supporting an evidence of an iris template aging effect. Simultaneously, they concluded that pupil dilation, contact lenses and amount of iris occlusions were not significant factors influencing the aging-related results.

Fenker and Bowyer [8] presented the first study based on comparison results obtained by more than one coding method, with one being a well-recognized commercial product (VeriEye; used also in this paper). The database, consisted of images separated by two years interval, was built for 43 volunteers. The authors, similarly to the previous studies, generated short-time-lapse (from 5 to 51 days) comparisons and long-time-lapse (from 665 to 737 days) comparisons, and the aging effect is studied through observation of the increase of false rejections as a function of time interval. Although we expected an increase of FNMR when the time interval grows, the reported numbers are surprisingly large and alarming. Namely, FNMR increased by 157 % to 305 % for the authors' matcher, and by 195 % to 457 % for a commercial matcher, depending on the acceptance threshold set optimally for short-time-lapse comparison scores. The authors created data subsets with images presenting homogeneous pupil dilation and captured for eyes not wearing contact lenses, yet the results obtained for these data subsets did not show a clear evidence of the significant influence that these factors might have onto the original conclusions. The same authors have broadened their research and used images separated by one-, two- and three-year time intervals captured for 322 subjects [9]. They evaluated four different matchers to select the most accurate one, used finally in their evaluations. Similarly to the prior work, the reduction in the recognition accuracy was observed, as the average false non-match rate increased by 27 %, 82 % and 153 % for one-, two- and three-year intervals between compared samples, respectively.

Current literature delivers also a claim that the iris aging – if exists – is of a negligible significance [10]. However, one should be careful as the linear regression models used in this work explain only partially possible sources and nature of matching scores non-stationarity, as they try to find monotonic deterioration in the selected statistical property (the average matching score). The lack of linear trend in one statistics does not guarantee the statistical stationarity, as still the remaining (and important) statistical properties (e.g. score variance) may vary. Moreover, non-monotonic changes of statistical properties may also be a consequence of aging and might be interesting to the biometric community.

3 Aging Database

3.1 Database Summary

BioBase-Aging-Iris database prepared for this work is a part of a larger set – *BioBase*. The BioBase contains biometric samples of five characteristics collected for the same persons, namely: iris, fingerprints, hand geometry and face images, as well as on-line handwritten signatures registered on the graphical tablet. The BioBase was collected mostly in 2003 and 2004 for more than 200 volunteers. We had repeated in 2010 and 2011 the data collection process for all biometric characteristics a few times for 31 individuals, who agreed to participate in the re-enrollment, building five *BioBase-Aging* datasets, separately for each biometric characteristics. Certainly, we had frozen our database collection environment to use exactly the same equipment and the software, configured identically to minimize the influence of environmental factors onto the biology-related aging effects. To capture samples for all five characteristics, a single measurement session lasted approximately 30 min. In particular, during each session, we realized three iris capture attempts, and each attempt consisted of as many presentations as it was necessary to capture two iris images (per eye). We intentionally did not capture the iris images in immediate series, to introduce some between-attempt variability in intra-session samples. This scenario yields to six iris images for each eye obtained in each measurement session.

The iris images were captured for both subject's eyes. To minimize the influence of poor image quality on the aging-related conclusions, we decided to manually remove poor quality samples, e.g. those showing less than 50 % unoccluded iris texture. Hence, the *BioBase-Aging-Iris* contains 571 iris images for 58 different eyes. The shortest time interval between sessions is 30 days while the longest is 2960 days (i.e., more than 8 years). The resolution of the resulting iris images is 768×576 pixels, and the image quality highly exceeds minimum requirements suggested by the ISO/IEC 19794-6:2011 and ISO/IEC 29794-6 standards, Fig. 1.

3.2 Equipment Used

Limited availability of commercial cameras offering raw iris images in 2003, when the experiment was initiated, encouraged us to construct a complete hardware

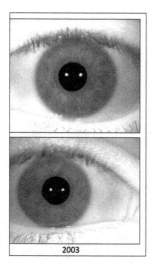 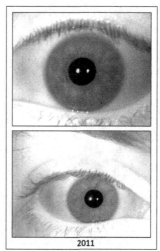

Fig. 1. Example iris images captured in 2003 (left) and the corresponding images of the same eyes captured in 2011 (right). Visual inspection of the upper samples reveal no serious differences within the iris pattern, yet the bottom samples show slight differences in the iris size and in the distribution of illumination. These possible differences may also influence the matching scores, incorrectly obscuring an aging phenomenon resulted from the biology. Images originate from *BioBase-Aging-Iris* database.

setup for the iris capture: the IrisCUBE camera. This prototype equipment captures the iris from a convenient distance of approximately 30 cm with the desired speed and quality. The camera was equipped with optics that actively compensates for small depth-of-field, typical in iris recognition systems, through an automatic focus adjustment. Two illuminants of near infrared light (with maximum power set at 850 nm) are placed horizontally and equidistantly to the lens, what guarantees consistent and sufficient scene illumination. IrisCUBE uses TheImagingSource DMK 4002-IR B/W camera that embeds SONY ICX249AL 1/2" CCD interline sensor with enlarged sensitivity to infrared light. Camera parameters such as shutter speed and gain may be adjusted manually or automatically.

During lifespan of the database collection project, new equipment with iris capture capability emerged on the market, and nowadays we may select among dozen of iris capture devices. However, due to high quality of the images captured by the constructed camera and due to the aim of guarantying homogeneous data collection environment, we used IrisCUBE to capture all the images in *BioBase-Aging-Iris*, thus also in 2010 and 2011 re-captures.

3.3   Database Variants

We observed that iris images captured after a few years might have different iris diameters when compared to these captured at the beginning of this project. Although each recognition methodology should be iris-diameter agnostic and normalize its size prior to feature extraction, we prepare a second variant of the

290   A. Czajka

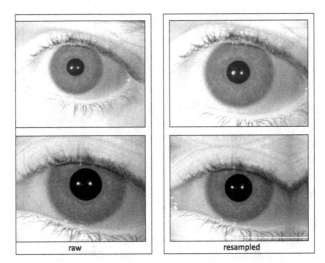

raw   resampled

Fig. 2. Examples of raw database images (left) and the corresponding size normalization results (right) after an increase (top) and a decrease (bottom) of the image resolution through bicubic interpolation. We may observe the effects of cropping and filling up with neighboring elements when changing the image resolution. Normalization is performed to center the iris within the image, what should help the matchers in correct data segmentation.

database with iris diameters normalized to the intra-class average using bicubic interpolation.

Images with iris diameter smaller that the intra-class average diameter are enlarged and cropped to the original resolution (768 × 576 pixels). If the iris diameter is larger that the average, the image resolution is decreased and the missing parts at the image borders are filled up with a mirror copy of the neighboring parts, again to keep the original resolution. We use the iris segmentation parameters which were calculated at this stage, and the cropping or filling up the image are realized to to center the iris within the image (Fig. 2). Further in the paper we refer to these two variants as the *raw* and *resampled* versions.

4   Iris Coding Methods

We use four different, commercially or publicly available iris matchers in this work, namely Neurotechnology VeriEye [11], Monro Iris Recognition Library – MIRLIN [12], Open Source for IRIS – OSIRIS [13], as well as the BiomIrisSDK, which is based on the methodology developed by this author [14]. In the following paragraphs we briefly characterize all four methods and provide rationale for their employment.

Neurotechnology VeriEye employs a proprietary and not published iris coding methodology. The manufacturer claims a correct off-axis iris segmentation with the use of active shape modeling, in contrast to typical circular approximation of the iris boundaries. VeriEye was tested for a few standard iris image databases, it was used in the NIST IREX project and presents pretty good accuracy. It is also the only – known to us – commercial matcher available for free (for a month period). The resulting score corresponds to the similarity of samples, i.e. a higher score denotes a better match. For the sake of simplicity, we use further the *NT* acronym for the VeriEye matcher.

MIRLIN derives the iris features from the zero-crossings of the differences between Discrete Cosine Transform (DCT) calculated in rectangular iris image subregions [15]. The coding method yields to binary iris codes, thus the comparison requires to calculate a Hamming Distance (a lower score denotes a smaller distance between samples, i.e. a better match). The *ML* acronym is used for the MIRLIN matcher further in the paper.

BiomIrisSDK employs the Zak-Gabor wavelet packets and the binary iris feature vectors are derived by one-bit coding of the Zak-Gabor transform coefficients' signs. The uniqueness of this method relies on the fact, that it does not employ image filtering, popular in iris recognition, and produces iris features that reveal global character with respect to the iris regions used in coding. As for the MIRLIN matcher, a Hamming Distance is used to calculate the matching score. We use the *ZG* acronym further in the paper when referring to the Zak-Gabor-based matcher.

OSIRIS implements Daugman's iris recognition idea of using 2D Gabor filtering to calculate a binary feature vector (iris code). Pairs of bits code one of the four phase quadrants, thus roughly approximating the signal phase information. A Hamming Distance, normalized by the number of non-occluded code bits, is used to calculate the dissimilarity between irises. A few programming bugs had to be fixed prior to the OSIRIS application, and after these improvements it finally offers a good reliability.

This is noteworthy that we had a full control over the ZG and OS methods. This allows to separate segmentation and coding processes, and in consequence to feed the matchers by images correctly segmented. This possibility is highly beneficial, as the results depend only on the properties of iris pattern fluctuations, and not on segmentation errors. We thus manually inspected the segmentation results achieved by ZG method for all samples in *BioBase-Aging-Iris*, and – if needed – corrected positions of the pupil, iris and sector positions used in ZG to generate the feature vector. For OS matcher, we transformed all iris images to the polar coordinates applying Daugman's rubber sheet model [2], and manually corrected occlusion masks. As the OS matcher employs normalized images expressed in polar coordinate, there is no use to examine its performance for resampled database variant, thus only raw subset was used for this method.

We had however no chance to separate the segmentation and the coding procedures in the commercial matchers (NT and ML), thus this part of the aging assessment encompasses the entire performance of the methods (i.e., eventual segmentation errors and changes in the iris tissue).

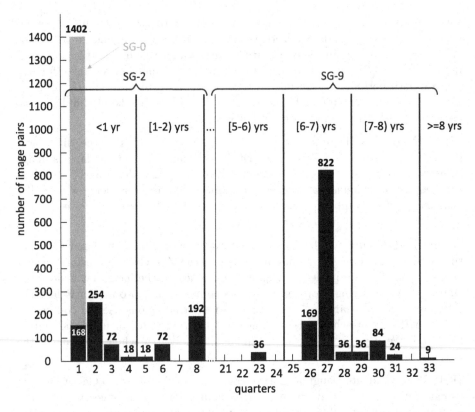

Fig. 3. Numbers of possible genuine pairs that can be constructed with the images of *BioBase-Aging-Iris* database plotted as a function of time lapse between the samples (gathered in quarters, i.e., three-month periods). Different-session pairs are marked with dark blue color, and the light blue color shows the number of pairs including same-session ones. Segmentation of pairs into three groups (SG-0, SG-2 and SG-9) is also shown, where SG-0 contains all the scores for intra-session comparisons, SG-2 groups all the scores generated for time lapse not greater than 2 year (excluding the intra-session scores) and SG-9 gathers all the scores calculated for samples distant by at least 5 years.

5 Results

5.1 Matching Score Generation

We inspected the *BioBase-Aging-Iris* database to construct a distribution of all possible pairs of the same-eye images with respect to the time lapse between image captures, Fig. 3. The number of possible genuine comparisons is equal to twice the number of possible iris image pairs, as the matchers may not return a symmetrical scores (i.e., the score between the iris image A and the iris image B may be unequal to the score between B and A). We may generate 3 244 image pairs in *BioBase-Aging-Iris*, thus the total number of all genuine comparison is

6 488. Among these comparisons, we have 2 468 results of comparing the iris images captured in the same session, and 4 020 scores of matching inter-session images. *BioBase-Aging-Iris* allows to construct 51 654 impostor comparisons for all the time intervals observed during genuine comparison generation.

NT, ZG and OS matchers allow to generate all the above mentioned genuine and impostor scores for both database variants (raw and resampled). The ML matcher generated a smaller number of scores (5 948 and 44 514 of genuine and impostor scores, respectively) due to the template generation errors, yet the numbers of scores for resampled database is slightly greater (6 328 and 49 162 of genuine and impostor scores, respectively), what may mean that normalization of the iris size increases the accuracy of the ML matcher for this database.

5.2 Matching Score Grouping

It was impossible to encourage all volunteers to participate in the experiment on each day we organized the re-capture, thus the number of sample pairs with respect to the time interval is uneven, Fig. 3, yielding highly uneven numbers of comparison scores possible to be generated in short periods. To obtain statistically significant results, we decided to gather comparison scores into three groups that can be identified when observing the distribution of sample pairs shown in Fig. 3. The first score group, denoted by *SG-0*, contains all the genuine and impostor scores for intra-session comparisons. The second – inter-session – subset, denoted by *SG-2*, groups all the scores generated for the samples with time lapse not greater than 2 year, certainly *excluding* the intra-session scores. The third subset, referred to as *SG-9*, gathers all the scores calculated for samples distant by at least 5 years and up to 2960 days, i.e. more than eight years. Table 1 details the numbers of genuine and impostor scores obtained for all the matchers used in this work with respect to all three score groups.

5.3 Evaluation Results

To answer the question related to the existence of aging effects in iris recognition, we present the average genuine and impostor scores calculated for each score group: SG-0, SG-2 and SG-9. Note that the conclusions related to aging should be based solely on inter-session comparisons (i.e. scores classified as SG-2 and SG-9), and the intra-session results are presented only for a completeness to show the potential influence of intra- vs. inter-session captures on the recognition accuracy.

The NT matcher was the most accurate for images in *BioBase-Aging-Iris* as it reached zero EER[2] for samples of raw database variant, and only for one time period a non-zero EER was observed after resampling the data. We may clearly observe a deterioration in the average genuine scores within the SG-2 and SG-9 subsets, Fig. 4 on the left. The genuine scores are approximately 14 % lower when the time lapse between samples starts from 5 years and reaches more than

---

[2] EER (Equal Error Rate) is the value of FNMR (or FMR) at the operating point of Receiver Operating Curve yielding equal values of FMR and FNMR.

Table 1. Number of genuine ξ_g and impostor ξ_i scores in score groups (SG-0, SG-2 and SG-9) for four iris recognition methods used in this work, namely VeriEye (NT), Zak-Gabor-based (ZG), MIRLIN (ML) and OSIRIS (OS). As the latter matcher (ML) behaves differently for raw and resampled data, numbers for ML are presented separately for these database variants.

	Score group → Coding method ↓	Same session (SG-0)	≤ 2 years (SG-2)	From 5 to 9 years (SG-9)
$\|\xi_g\|$	NT, ZG &OS for each DB	2468	1588	2432
	ML for raw DB	2292	1548	2108
	ML for re-sampled DB	2394	1588	2346
$\|\xi_i\|$	NT, ZG &OS for each DB	7988	10 186	33 480
	ML for raw DB	7188	9362	27 964
	ML for re-sampled DB	7690	9646	31 826

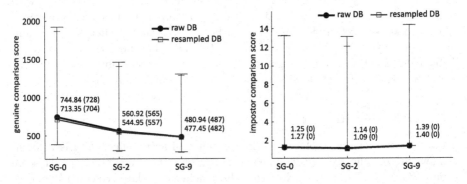

Fig. 4. Average and median scores (in brackets) for raw (circles, black color) and resampled (rectangles, blue color) database variants, shown with respect to the score groups for NT matcher. The whiskers show the 95 % boundaries of the sample distributions in each combination of the SG and database variant. Result for genuine and impostor scores are shown on the left and right, respectively. A higher score denotes a better match.

8 years, and this observation is supported by the outcome of one-way unbalanced analysis of variance (ANOVA). Namely, we cannot accept the null hypothesis that all samples in SG-2 and SG-9 subsets are drawn from populations with the same mean, as they obtained p-value is near to zero ($p < 10^{-47}$ for raw database

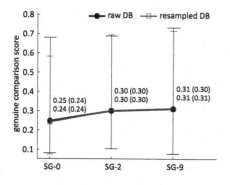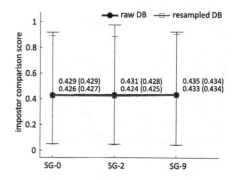

Fig. 5. Same as in Fig. 4 for the ZG matcher. A lower score denotes a better match.

variant and $p < 10^{-37}$ for resampled database variant). When comparing the intra- vs. inter-session scores, we encounter even higher accuracy deterioration, namely 25 % decrease of average score in SG-2 when compared to the SG-0 average, and a decrease by 35 % of SG-9 scores when compared to the SG-0 average. Certainly, also in these cases the analysis of variance casts doubt on the null hypothesis, as the obtained p-values are below machine accuracy when compared SG-0 vs. SG-2 and SG-0 vs. SG-9 scores for both variants of the database. Note that resampling of the iris images has a little influence on the average genuine scores, what may suggest that NT matcher is iris-size agnostic and the aging of the NT templates seems to occur independently of this factor.

Analogously to the presentation of the NT matcher results, we show the average genuine scores for ZG method, Fig. 5 on the left. The intra- vs. inter-session scores show 16 % and 19 % increase in the average Hamming Distance for raw datasets (20 % and 22 % increase for resampled datasets) when compared SG-0 average score with the SG-2 and SG-9 averages, respectively. These changes are statistically significant, as obtained p-values are below machine accuracy (for all combinations of a database variant and a time period). Comparing SG-2 and SG-9 scores show only 3 % of the average score increase, yet still $p < 10^{-8}$ that suggests rejecting of the null hypothesis on equal means. We may observe that the ZG matcher is robust to the absolute iris size, as the genuine scores for raw and resampled database variants do not differ significantly.

The ML matcher average genuine scores are presented in Fig. 6 on the left. As for NT and ZG methods, we may encounter statistically significant differences when comparing average intra- vs. inter-session comparison scores (p-value not exceeding 10^{-10} for all combinations of SG and a database variant), namely the decrease reaches 27 % and even 45 % when compared SG-0 vs. SG-2 and SG-0 vs. SG-9 average scores, respectively. However, when compared the average scores between SG-2 and SG-9 we obtain a low p-value ($p < 10^{-15}$) only for the raw database variant, and $p = 0.19$ for the resampled data, although the increase of the average score in the latter case reaches 4 %. This may suggest, that the aging effect related to the ML templates is somehow compensated by the unifying of the iris diameters inside the iris classes (but in a sense of statistical significance

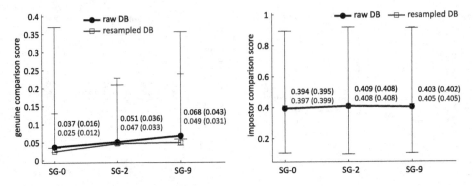

Fig. 6. Same as in Fig. 4 for the ML matcher. As for ZG matcher, a lower score denotes a better match.

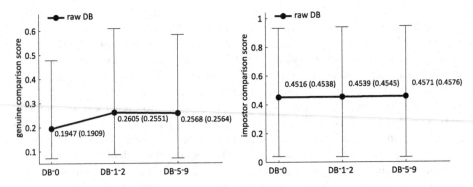

Fig. 7. Same as in Fig. 4 for the OS matcher. As for ZG and ML matchers, a lower score denotes a better match.

of the ANOVA test). One, however, should note that resampling allowed for a better accuracy of the ML matcher, manifesting by a greater number of correctly calculated templates and a lower average of the genuine scores in each score subgroup.

As for all the matchers presented above, also the OS matcher presents statistically significant differences in average genuine scores when comparing intra- and inter-session results (p-value near zero). The genuine average scores increases by about 25 % in both cases, i.e. SG-0 vs. SG-2 and SG-0 vs. SG-9). However, unexpectedly the OS method seems to be resistant to aging effects as the average of inter-session genuine scores (i.e. SG-2 vs. SG-9) slightly decreases by 1.5 % (opposite to the expectations), yet the median slightly increases by 0.5 % (suggesting aging effect), Fig. 7 on the left. These changes are statistically significant as $p < 0.0064$. To investigate this result in more details, we plot cumulative distributions of the genuine scores, Fig. 8. We can see, that the OS matcher behaves differently in regions of higher and lower similarity of iris images. Namely, the stronger is the comparison score (higher similarity between images), the more

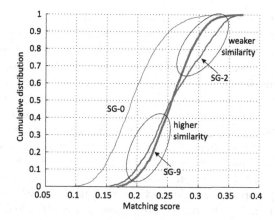

Fig. 8. Cumulative distribution of genuine scores for OS matcher. Regions of weaker and higher similarities between images are intentionally shown to highlight different behavior related to the sensitivity of the OS matcher to aging effects.

aging effect is visible. However, if the images are not matched well (image similarity is lower), then the order of average comparison score is opposite to the expectation: average SG-2 result is greater than average SG-9 score. This suggests that more accurate matching present the aging effect to the more extent, and supports the observation related to the most accurate (on this DB) NT matcher, presenting the highest sensitivity to the time lapse.

As of now we discussed the genuine scores only, and now we turn toward the matching between impostor samples. Figures 4, 5, 6 and 7 on the right present the average impostor scores for NT, ZG and ML matchers, respectively. As it was already suggested in the literature related to the iris aging effect, we may observe lower differences in average scores in groups when compared to the result of genuine comparisons. Namely, the changes range from 0.5 % for ZG and OS matchers (intra- vs. inter-session averages for raw samples) to 22 % for NT matcher (inter-session averages for resampled images). The ANOVA test casts doubt on accepting the hypothesis that investigated sample subsets are drawn from populations with the same mean ($p < 0.021$ for all combinations of SG and database variant). We may also see that iris absolute diameter size has no influence on the impostor scores.

6   Discussion

Average values of the genuine scores obtained for all the tested matchers may suggest that iris templates age, what partially supports earlier findings determined for different matchers and different databases, yet collecting samples with a shorter time lapse between captures than in this work. The extent to which the template aging phenomenon is observed is however uneven across different matchers, in particular we observe a higher influence of the time flow for more accurate methods. This observation may be explained in two ways. On the one hand, less accurate matcher may encounter a higher number of segmentation errors, or improper iris image mapping, when comparing irises with different pupil dilation, iris diameter or occlusion extent. These factors may be classified as 'aging factors' (as the resulting template 'ages', independently of the 'aging' source), yet we rather would like to answer the question if the aging relates also (or first of all) to the iris texture, i.e. a direct donor of the biometric features. If the answer is affirmative, then assessment to what extent it tackles the elements of the complicated iris tissue would be of a great value. So, on the other hand, we may assume that high accuracy of the matcher relates to a higher accuracy of the segmentation process. If so, more comparison scores result from an appropriate matching of the iris patterns (with occlusions appropriately removed and iris texture appropriately mapped), which – according to the experimental results – exhibits significantly different nature after a few year time lapse. Note that the aim of each coding method is to be sensitive for iris features which guarantee individualization of subjects (e.g., frequency bands in wavelet-based coding routines). Collecting these thoughts, we would hazard a guess that the iris aging relates also to the iris characteristics that are responsible for our individual biometric features, i.e. the iris pattern. Simultaneously, we stress again that difference in average comparison scores is only one indicator of the inter-session variability, suggesting the non-stationarity in iris recognition.

The fact of iris aging, if finally confirmed by a series of additional experiments exploiting a large number of matchers and big, heterogeneous datasets, should under no circumstances devalue the strength of the iris recognition. Next research step should be focused on the assessment of the extent to which the aging phenomenon deteriorates the accuracy of this modality, allowing for introducing precise rules of template usage, in particular adequate time periods which call for re-enrollment, what may only increase an accuracy of this prominent and very accurate authentication technology.

Acknowledgements. The author would like to thank Mr. Mateusz Trokielewicz for initial validation of the segmentation results calculated for the *BioBase-Aging-Iris* samples. The author is also cordially indebted Dr. Joanna Putz-Leszczyska, Dr. Lukasz Stasiak, Mr. Marcin Chochowski and Mr. Rafał Brize, with whom he had collectively built the *BioBase-Aging-Iris* database. The author also appreciates Mr. Marcin Chochowski for necessary bugs fixing in the original OSIRIS matcher, and making a new, corrected version available for this work. This work was partially funded by The National Centre for Research and Development (grant No. OR0B002701: "Biometrics and PKI techniques for modern identity documents and protection of information systems – BIOPKI").

References

1. Flom, L., Safir, A.: Iris recognition system. US Patent 4,641,349, February 1987
2. Daugman, J.: High confidence visual recognition of persons by a test of statistical independence. IEEE Trans. Pattern Anal. Mach. Intell. **15**(11), 1148–1161 (1993)

3. Tome-Gonzalez, P., Alonso-Fernandez, F., Ortega-Garcia, J.: On the effects of time variability in iris recognition. In: IEEE Conference on Biometrics: Theory, Applications and Systems, pp. 1–6. IEEE (2008)
4. Rankin, D., Scotney, B., Morrow, P., Pierscionek, B.: Iris recognition failure over time: the effects of texture. Pattern Recognit. **45**, 145–150 (2012)
5. Baker, S., Bowyer, K.W., Flynn, P.J.: Empirical evidence for correct iris match score degradation with increased time lapse between gallery and probe images. In: International Conference on Biometrics, pp. 1170–1179 (2009)
6. Bowyer, K.W., Baker, S.E., Hentz, A., Hollingsworth, K., Peters, T., Flynn, P.J.: Factors that degrade the match distribution in iris biometrics. Identity Inf. Soc. **2**(3), 327–343 (2009)
7. Baker, S., Bowyer, K.W., Flynn, P.J., Phillips, P.J.: Template aging in iris biometrics. In: Burge, M., Bowyer, K.W. (eds.) Handbook of Iris Recognition. Advances in Computer Vision and Pattern Recognition, pp. 205–218. Springer, London (2013)
8. Fenker, S.P., Bowyer, K.W.: Experimental evidence of a template aging effect in iris biometrics. In: IEEE Computer Society Workshop on Applications of Computer Vision, pp. 232–239 (2011)
9. Fenker, S.P., Bowyer, K.W.: Analysis of template aging in iris biometrics. In: CVPR Biometrics Workshop, pp. 1–7 (2012)
10. Shchegrova, S.: Analysis of iris stability over time using statistical regression modeling. In: Biometric Consortium Conference & Technology Expo, September 18–20, 2012, Tampa, Florida, USA (2012)
11. Neurotechnology: VeriEye SDK, version 4.3, revision 87298, July 2012
12. SmartSensors: MIRLIN SDK, version 2.23.2, August 2012
13. BioSecure: OSIRIS, version 2.01 (2009)
14. Czajka, A., Pacut, A.: Iris recognition system based on Zak-Gabor wavelet packets. J. Telecommun. Inf. Technol. **4**, 10–18 (2010)
15. Monro, D.M., Rakshit, S., Zhang, D.: DCT-based iris recognition. IEEE Trans. Pattern Anal. Mach. Intell. - Special Issue on Biometrics: Progress and Directions **29**(4), 586–595 (2007)

Application of Electrode Arrays for Artifact Removal in an Electromyographic Silent Speech Interface

Michael Wand[(✉)], Matthias Janke, Till Heistermann, Christopher Schulte,
Adam Himmelsbach, and Tanja Schultz

Cognitive Systems Lab, Karlsruhe Institute of Technology,
Adenauerring 4, 76137 Karlsruhe, Germany
{michael.wand,matthias.janke,tanja.schultz}@kit.edu,
{till.heistermann,christopher.schulte,adam.himmelsbach}@student.kit.edu

Abstract. An electromygraphic (EMG) *Silent Speech Interface* is a system which recognizes speech by capturing the electric potentials of the human articulatory muscles, thus enabling the user to communicate silently. This study deals with the introduction of multi-channel electrode arrays to the EMG recording system, which requires meticulous dealing with the resulting high-dimensional data. As a first application of the technology, Independent Component Analysis (ICA) is applied for automated artifact detection and removal. Without the artifact removal component, the system achieves optimal average Word Error Rates of 40.1 % for 40 training sentences and 10.9 % for 160 training sentences on EMG signals of audible speech. On a subset of the corpus, we evaluate the ICA artifact removal method, improving the Word Error Rate by 10.7 % relative.

Keywords: Electromyography · EMG · Silent speech interface · Electrode array · Emg-based speech recognition

1 Introduction

Speech is the most convenient and natural way of human communication and interaction. However all but 150 years ago, the lack of technical devices for amplifying, processing, transmitting, or storing acoustic signals limited spoken communication to face-to-face talk or speeches in front of at most medium-sized audiences.

Nowadays, mobile phone technology has made speech a wide-range, ubiquitous means of communication, and talking to any person, worldwide, has become a reality. Furthermore, many speech-controlled devices and services have been developed, including telephone-based information retrieval systems, voice-operated car navigation systems, and large-vocabulary dictation and translation assistants.

© Springer-Verlag Berlin Heidelberg 2014
M. Fernández-Chimeno et al. (Eds.): BIOSTEC 2013, CCIS 452, pp. 300–312, 2014.
DOI: 10.1007/978-3-662-44485-6_21

Unfortunately, voice-based communication and interaction suffers from several challenges which arise from the fact that the speech needs to be clearly audible and cannot be masked, including lack of robustness in noisy environments, disturbance for bystanders, privacy issues, and exclusion of speech-disabled people. These challenges may be alleviated by Silent Speech Interfaces, which are systems enabling speech communication to take place without the necessity of emitting an audible acoustic signal, or when an acoustic signal is unavailable [1].

Over the past few years, we have developed a Silent Speech Interface based on surface electromyography (EMG): When a muscle fiber contracts, small electrical currents in form of ion flows are generated. EMG electrodes attached to the subject's face capture the potential differences arising from these ion flows, which are then used to retrace the corresponding speech. This allows speech to be recognized even when it is produced silently, i.e. mouthed without any vocal effort.

Many EMG-based speech recognizers rely on small sets of less than 10 EMG electrodes attached to the speaker's face [2–6]. This setup is based on standard Ag-AgCl gelled electrodes as used in medical applications and imposes some limitations, for example, small shifts in the electrode positioning between recordings are difficult to compensate, and it is impossible to separate superimposed signal sources, thus single active muscles or motor units cannot be discriminated.

In this study we present experiments on using *electrode arrays* for the recording of EMG signals of speech. In a first step, we establish that our existing EMG-based continuous speech recognizer [2] is able to deal with the increased number of signal channels. This requires extending standard Linear Discriminant Analysis with a prior Principal Component Analysis step. Secondly, we present a first application of the EMG array methodology, namely, we show that Independent Component Analysis (ICA) can be used to remove artifacts from the multi-channel EMG signal.

This article is organized as follows: In the following Sect. 2, we describe our new recording system, and Sect. 3 contains a description of the underlying recognizer. Section 4 establishes the multi-channel recording system and presents our experiments on artifact detection and removal. The final Sect. 5 concludes the article.

2   Recording System Setup and Corpus

EMG signals were recorded with the multi-channel biosignal amplifier *EMG-USB2*, which is produced and distributed by *OT Bioelettronica*, Torino, Italy http://www.otbioelettronica.it. The EMG-USB2 device allows to record and process up to 256 EMG channels, supporting a selectable gain of 100 - 10.000 V/V and a recording bandwidth of 3 Hz - 4400 Hz. For line interference reduction, we used the integrated DRL circuit [7]. The electrode arrays were acquired from *OT Bioelettronica* as well. Electrolyte cream was applied to the EMG arrays in order to reduce the electrode/skin impedance.

We used two different EMG array configurations for our experiments, see Fig. 1. In *setup A*, we unipolarly recorded 16 EMG channels with two EMG

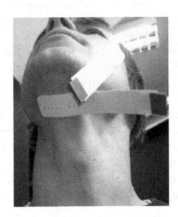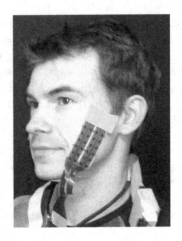

Fig. 1. EMG array positioning for setup A (left) and setup B (right).

arrays each featuring a single row of 8 electrodes, with 5 mm inter-electrode distance (IED). One of the arrays was attached to the subject's cheek, capturing several major articulatory muscles [3], the other one was attached to the subject's chin, in particular recording signals from the tongue. A reference electrode was placed on the subject's neck.

In *setup B*, we replaced the cheek array with a larger array containing four rows of 8 electrodes, with 10 mm IED. The chin array remained in its place. In this setup, we achieved a cleaner signal by using a *bipolar* configuration, where the potential difference between two adjacent channels in a row is measured. This means that out of $4 \cdot 8$ cheek electrodes and 8 chin electrodes, we obtained $(4+1) \cdot 7 = 35$ signal channels. For both setups, we chose an amplification factor of 1000, a high-pass filter with a cutoff frequency of 3 Hz and a low-pass filter with a cutoff frequency of 900 Hz, and a sampling frequency of 2048 Hz. The audio signal was parallely recorded with a standard close-talking microphone. We used an analog marker system to synchronize the EMG and audio recordings, and according to [8], we delayed the EMG signal by 50ms compared to the audio signal.

The text corpus which we recorded is based on [2]. We used two different text corpora for our recordings: Each session contains a set of ten "BASE" sentences which is used for testing and kept fixed across sessions. Furthermore, each session contains 40 training sentences, which vary across sessions. For reference, we call this basic text corpus "Set 1". A subset of our sessions has been extended to 160 different training sentences and 20 test sentences, where the 20 test sentences consist of the BASE set repeated twice. This enlarged text corpus is called "Set 2".

The recording proceeded as follows: In a quiet room, the speaker read English sentences in normal, audible speech. Note that we also recorded silently mouthed speech, which was not used in this study. The set of sentences was taken from the Broadcast News Domain. The recording was supervised by a member of the research team in order to detect errors (e.g. detached electrodes) and to assure a consistent pronunciation. The training and test sentences were always recorded in randomized order. Thus we finally have four setups to investigate, namely, setups A-1 and A-2 (with 16 EMG channels) and B-1 and B-2 (with 35 EMG channels). The suffixes "1" and "2" refer to the recorded text corpus, e.g. the A-1 sessions consist of 40 + 10 sentences, the A-2 sessions consist of 160 + 20 sentences, etc. At this point we remark that the results on the four setups are not directly comparable, since the number of training sentences, the set of speakers and the number of sessions per speaker differ. Also, our experience indicates that even for one single speaker, the recognition performance may vary drastically between sessions, possibly due to variations in electrode positioning, skin properties, etc. However, it is certainly plausible to compare the effects of different feature extraction methods on the recognition performance of *each* of the setups. Note that the test sets of the four setups consist of identical sentence lists, so their characteristics in terms of perplexity and vocabulary are the same.

The following table summarizes the properties of our corpus.

Setup	# of Speakers / Sessions	Average data length in sec.		
		Training	Test	Total
A-1	3 / 6	144	37	181
A-2	2 / 2	528	74	602
B-1	6 / 7	149	42	191
B-2	4 / 4	570	83	653

3   Feature Extraction, Training and Decoding

The feature extraction is based on *time-domain features* [8]. We first split the incoming EMG signal channels into a high-frequency and a low-frequency part, after this, we perform framing and compute the features. The whole feature extraction proceeds as follows.

For any given feature $\mathbf{f}$, $\bar{\mathbf{f}}$ is its frame-based time-domain mean, $\mathbf{P_f}$ is its frame-based power, and $\mathbf{z_f}$ is its frame-based zero-crossing rate. $S(\mathbf{f}, n)$ is the stacking of adjacent frames of feature $\mathbf{f}$ in the size of $2n + 1$ ($-n$ to n) frames.

For an EMG signal with normalized mean $x[n]$, the nine-point double-averaged signal $w[n]$ is defined as

$$w[n] = \frac{1}{9} \sum_{k=-4}^{4} v[n+k], \quad \text{where} \quad v[n] = \frac{1}{9} \sum_{k=-4}^{4} x[n+k].$$

The high-frequency signal is $p[n] = x[n] - w[n]$, and the rectified high-frequency signal is $r[n] = |p[n]|$. Mathematically, the computation of $w[n]$ from $x[n]$ is an application of a weighted moving average filter.

The final feature **TD**n is defined as follows:

$$\mathbf{TD}n = S(\mathbf{TD}0, n), \text{where } \mathbf{TD}0 = [\bar{\mathbf{w}}, \mathbf{P_w}, \mathbf{P_r}, \mathbf{z_p}, \bar{\mathbf{r}}]. \quad\quad (*)$$

Frame size and frame shift are set to 27 ms respective 10 ms.

The **TD**0 feature is the most basic feature used in this study, consisting of a channel-wise combination of five standard time-domain features, which are computed frame by frame. In the final step defined by Eq. (*) a stacking of adjacent feature vectors with context width $2 \cdot n + 1$ is performed, with varying n. This process is performed for each channel, and the combination of all channel-wise feature vectors yields the final **TD**n feature vector. For the stacking context widths, different n have been used in prior work, e.g. [8] uses a context width of 5, however on a different corpus, [9] shows that increasing the context width to 15 frames, i.e. 150 ms per side, yields improved results.

In all cases, we apply Linear Discriminant Analysis (LDA) on the **TD**n feature. The LDA matrix is computed by dividing the training data into 136 classes corresponding to the begin, middle, and end parts of 45 English phonemes, plus one silence phoneme. From the 135 discriminant dimensions which are yielded by the LDA algorithm, we always retain 32 dimensions. As shown in Sect. 4.2, it may be necessary to perform Principal Component Analysis (PCA) before computing the LDA matrix, see Sect. 4.2 for further details. In the experiments described in Sect. 4.3, Independent Component Analysis (ICA) and possible artifact suppression is applied *before* the feature extraction step, on the raw EMG data.

The recognizer is based on three-state left-to-right fully continuous Hidden-Markov-Models. All experiments use bundled phonetic features (BDPFs) [2] for training and decoding. In order to obtain phonetic time-alignments as a reference for training, the parallely recorded acoustic signal is forced-aligned with an English Broadcast News (BN) speech recognizer. Based on these time-alignments, the HMM states are initialized by a merge-and-split training step [10], followed by four iterations of Viterbi training.

For decoding, we use the trained acoustic model together with a trigram Broadcast News language model. The test set perplexity is 24.24. The decoding vocabulary is restricted to the words appearing in the test set, which results in a test vocabulary of 108 words. Note that we do *not* use lattice rescoring for our experiments.

Further information can be found in [2], the recognizer presented therein serves as the baseline for this study.

4   Experiments and Results

In this section we first present results on applying our baseline system towards the new, high-dimensional signals. Subsequently, we present our ICA-based automated artifact removal method, and evaluate its performance.

4.1   Baseline Recognition System

In the first experiment, we use our baseline recognizer, as described in Sect. 3, and feed it with the EMG features from the array recording system. Figure 2 shows the Word Error Rates for different stacking widths, averaged over all sessions of each setup.

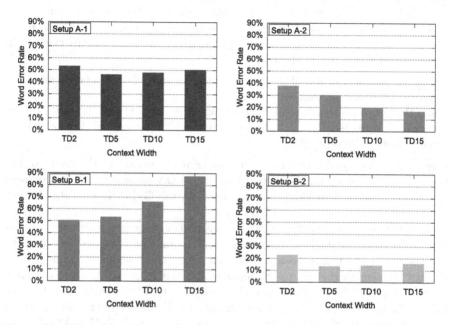

Fig. 2. Word Error Rates for the baseline system with different stacking context widths. PCA or ICA were not applied.

We consider the optimal context widths for the four setups. Our observations for the four distinct setups are very different: For setup A-1, with 16 channels and 40 training sentences, the Word Error Rate (WER) varies between 46.3 % and 53.2 %, with the optimum reached at a context width of 5 (i.e. **TD**5). For the B-1 setup, with 35 channels but the same amount of training data, the optimal context width appears to be **TD**2 with a WER of 50.5 %, widening the context causes deteriorating results, the worst WER is 87.6 % for the **TD**15 stacking.

For the setups with 160 training sentences, the recognition performance is consistently better due to the increased training data amount. With respect to context widths, we observe a behavior which vastly differs from the results above: For 16 EMG channels (setup A-2), the optimal context width is 15, with a WER of only 13.2 %. For setup B-2, **TD**5 stacking is optimal, with a WER of 10.6 %.

The behavior described in this section is quite consistent across recording sessions. The variation in optimal stacking width indicates a deep inconsistency between our setups, which leads us to the series of experiments described in the following section.

4.2   PCA Preprocessing to Avoid LDA Sparsity

Machine learning tasks frequently exhibit a challenge known as the "Curse of Dimensionality", which means that high-dimensional input data, relative to the amount of training data, causes undertraining, diminishes the effectiveness of machine learning algorithms, and reduces in particular the generalization capability of the generated models. The maximal feature space dimension which allows robust training depends on the amount of available training data.

The dimensionality of the feature space in our experiments depends on the number of EMG channels and the stacking width in the feature extraction. From the results of Sect. 4.1, we observe

- that for both setups A and B, increasing the amount of training data increases the optimal context width
- and that for both the 40-sentence training corpus (set 1) and the 160-sentence training corpus (set 2), the optimal context width with setup B is lower than the optimum for setup A.

This strongly suggests that the "Curse of Dimensionality" is the cause of the discrepancy we observed. However, since the LDA algorithm *always* reduces the feature space dimensionality to 32 channels, the GMM training itself is not affected by varying feature dimensionalities.

We propose that the deterioration of recognition accuracy for small amounts of training data and high feature space dimensionalities is caused by the LDA computation step. It has been shown that when the amount of training data is small relative to the sample dimensionality, the LDA within-scatter matrix becomes sparse, which reduces the effectivity of the LDA algorithm [11]. This may be the case in our setup, since with only a few minutes of training data, we may have a sample dimensionality before LDA of up to $35 \cdot 5 \cdot 31 = 5425$ for the 35-channel system with a **TD**15 stacking.

The following set of experiments deals with coping with the LDA sparsity problem. We expect an improved recognition accuracy and, in particular, a more consistent result regarding the optimal feature stacking width. Our method is to apply an additional PCA dimension reduction step before the LDA computation, as advocated for visual face recognition [12]. This step should allow an improved LDA estimation, however, if the PCA cutoff dimension is chosen too low, one will lose information which is important for discrimination.

The computation works as follows: On the training data set, we first compute a PCA transformation matrix. We apply PCA and keep a certain number of components from the resulting transformed signal, where the components are, as usual, sorted by decreasing variance. Then we compute an LDA matrix of the PCA-transformed training data set, finally keeping 32 dimensions. The resulting PCA + LDA preprocessing is now applied to the entire corpus, normal HMM training and testing is performed, and we use the Word Error Rate as a measure for the quality of our preprocessing.

Figure 3 plots the Word Error Rates of the recognizer for setups A and B and different stacking widths versus the number of retained dimensions after the PCA step. In all cases, we jointly plot the WERs for training data sets 1 and 2.

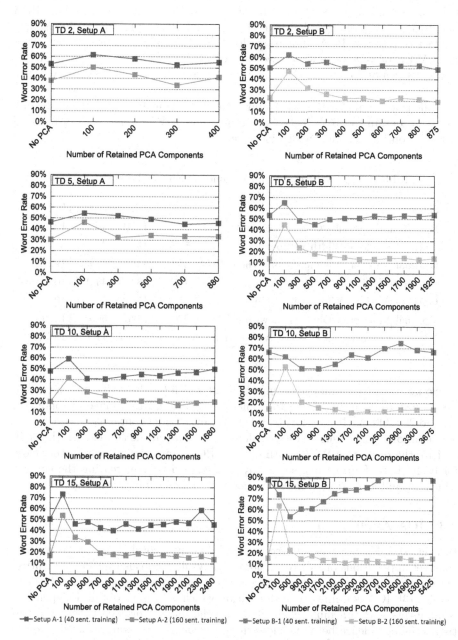

Fig. 3. Word Error Rates for different PCA dimension reductions. Observe that the feature space dimension *before* the PCA step increases from left to right and from top to bottom.

The figures show that the PCA step indeed helps to overcome LDA sparsity. For example, in the A-1 setup, the optimal context width without PCA application is 5, yielding a WER of 46.3 %. With PCA application, the optimal number of retained PCA dimensions for the **TD5** context width is 700, yielding a WER of 44.8 %. However, we can still do better: With a vastly increased context width of 15, we get the best WER of 40.1 %, at a dimensionality of 900 after PCA application.

This is also true for the other four setups, see table 1 for an overview. In all cases, we obtain WER reductions of more than 10 % relative, and also, in all cases the optimal context width increases.

Table 1. Optimal Results and Parameters with and without PCA.

Setup	A-1	A-2	B-1	B-2
Best Result without PCA ("Baseline")	46.3 %	17.0 %	50.5 %	13.4 %
Optimal Stacking Width without PCA	5	15	2	5
Optimal Number of Dimensions without PCA	880	2480	875	1925
Best Result with PCA	40.1 %	13.9 %	44.9 %	10.9 %
Optimal Stacking Width with PCA	15	15	5	10
Optimal Number of Dimensions with PCA	900	2480	500	1500
Relative Improvement by PCA Application	13.4 %	18.2 %	11.1 %	18.7 %

So far, we have found the optimal context width for the EMG speech classification task to lie around 10 to 15 frames on each side, which makes a context of around 200-300 ms. It may be possible to try even wider contexts, however, close examination of the results in Fig. 3 show that between the context widths of 10 and 15, the respective WERs with optimal PCA dimensionality are rather close for each of the four setups, so it may be assumed that increasing the context width beyond around 15 frames will not yield further improvements.

4.3 ICA Application

Having established a baseline recognizer for array recordings of EMG, we now consider applications of this technology. One well-established means of identifying signal sources in multi-channel signals is *Independent Component Analysis (ICA)* [13]. ICA is a linear transformation which is used to obtain independent components within a multi-channel signal; the underlying idea is that the statistical independence between the estimated components is maximized. We use the Infomax ICA algorithm according to [14], as implemented in the Matlab EEGLAB toolbox [15].

In this study, we apply ICA for *artifact removal*. For this purpose, we interpret ICA as a method of (blind) source separation: We run ICA on the raw EMG signal, obtaining a set of 16 resp. 35 ICA components. The ICA decomposition matrix is computed separately for the two arrays. We then develop a heuristical measure to determine whether any detected component should be considered an artifact or a "target" EMG signal component. This method has been presented in detail in [16]. For our experiments, we choose the B-1 setup with 40 training sentences.

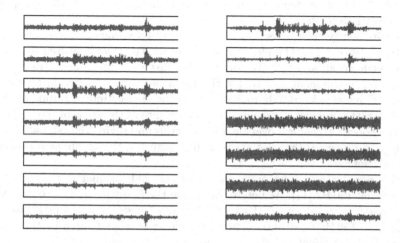

Fig. 4. EMG signals of the chin array before ICA processing (left) and after ICA processing (right). The ICA decomposition shows visibly distinct EMG signal components and artifact noise.

As an example, Fig. 4 shows a short extract of EMG signals recorded with the chin array, together with their ICA decomposition. The example shows that artifact components may be vastly different from target components: the 7 original EMG channels of the chin array of one utterance (left) are decomposed into three "target" components which look like EMG signals, and four "noise" components. Therefore we expect that the removal of the noise channels improves the recognition results. In [16] we present two strategies:

- **Direct Method:** We take the ICA components, identify and remove artifact components, and then compute the EMG features on the *remaining components*.
- **Backprojection:** We take the ICA components, identify and remove artifact components as before, and then back-project these components to the original signal. Mathematically, this can be described as applying the ICA decomposition, setting the artifact ICA components to zero, and then multiplying the altered set of ICA components with the *inverse* of the ICA matrix.

In [16], it is shown that the direct method yields better results than backprojection. Therefore in this article we present results on artifact removal with the direct method, i.e. we compute EMG features on the remaining ICA components.

Artifact components are identified by the following three measures, which are computed on the ICA decomposition:

- Autocorrelation Measure: This method typically identifies very regular (periodic) artifacts, like power line noise. We compute the autocorrelation sequence of the input component and then take the value of the *first maximum* after the first zero of the sequence. If this value is greater than 0.5, this component is deemed an artifact.
- High-frequency noise detection: The surface EMG signal has a frequency range of 0 Hz - 500 Hz [17]. Therefore a component with distinct high-frequency parts is considered an artifact. We compute the discrete-time Fourier transform of the input component and divide the frequency axis into two intervals: The "signal" interval from 0 Hz to 500 Hz, and the "noise" interval from 500 Hz to 1024 Hz (the Nyquist frequency). We then compute the areas of the amplitude of the Fourier transform over the two intervals and divide the "signal" area by the "noise" area. If the quotient is smaller than 1.3, this component is deemed an artifact.
- EMG signal range: The main energy of the EMG signal is found between 50 Hz and 150 Hz [17]. As before, we divide the frequency axis into two parts: A "signal" interval from 50 Hz to 150 Hz, and "noise" part from 0 Hz to 50 Hz and from 150 Hz to 1024 Hz. Then we divide the "signal" area by the "noise" area. If the quotient is below 0.25, we deem this component an artifact. For this measure, we found that the power spectral density yielded slightly more robust estimates than a standard Fourier transformation.

Our measures are first computed on each ICA component of *each utterance* of the training data set. In a second step, we combine the results: For a component to be considered an artifact, we require that *at least one* of the three methods considers this component an artifact on *a minimum percentage* of (training) utterances. This "threshold percentage" is varied between 10 % and 50 %, where a lower value causes more components to be removed. We observed that the threshold makes a difference when components vary across utterances, e.g. when the contact between electrode and skin deteriorates over time.

For our experiments, we used the *optimal setup* of the B-1 corpus, namely TD5 stacking with PCA dimensionality reduction to 500. Figure 5 shows Word Error Rates for this experiment with different strategies: The baseline is attained without ICA application, here we obtain 44.9 % WER. ICA increases the WER to 50.6 %—this result is in line with [18], where it is shown that direct ICA application may cause both improvement or deterioration of the WER. With a slightly different preprocessing, in [16] we obtained an insignificant improvement with direct ICA application.

Using our noise removal strategies clearly improves the WER: With a threshold percentage of 50 %, we obtain 44.9 % WER, about as much as without ICA. But we can still improve this result: The best WER is achieved with a 10 % noise threshold, where the WER is reduced to 40.1 %. Compared with the baseline, this is a relative improvement of 10.7 %.

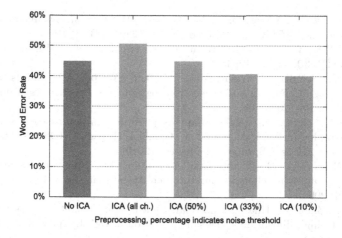

Fig. 5. Results for ICA with noise removal with different noise threshold percentages on the B-1 setup. See text for details.

5   Conclusions

In this study we introduced a new EMG-based speech recognition technology, based on electrode arrays instead of single electrodes. We presented two basic recording setups and evaluated their potential on data sets with different amount of training data. An unexpected inconsistency with respect to the optimal feature stacking width led us to the introduction of a PCA preprocessing step before the LDA matrix is computed, which gives us consistent relative Word Error Rate improvements between 11% to 19%, even for small training data sets of only 40 sentences. As a first application of the new array technology, we showed that Independent Component Analysis (ICA) can be applied for artifact detection and removal, improving the Word Error Rate by 10.7% relative. However, we observed that application of ICA without the noise detection heuristics did not improve our result.

References

1. Denby, B., Schultz, T., Honda, K., Hueber, T., Gilbert, J.: Silent speech interfaces. Speech Commun. **52**(4), 270–287 (2010)
2. Schultz, T., Wand, M.: Modeling coarticulation in large vocabulary EMG-based speech recognition. Speech Commun. **52**(4), 341–353 (2010)
3. Maier-Hein, L., Metze, F., Schultz, T., Waibel, A.: Session independent non-audible speech recognition using surface electromyography. In: Proceedings of IEEE ASRU, San Juan, Puerto Rico (2005)
4. Freitas, J., Teixeira, A., Dias, M.S.: Towards a silent speech interface for Portuguese. In: Proceedings of Biosignals, pp. 91–100 (2012)
5. Jorgensen, C., Dusan, S.: Speech interfaces based upon surface electromyography. Speech Commun. **52**(4), 354–366 (2010)

6. Lopez-Larraz, E., Mozos, O.M., Antelis, J.M., Minguez, J.: Syllable-based speech recognition using EMG. In: Proceedings of IEEE EMBS (2010)
7. Winter, B.B., Webster, J.G.: Driven-right-leg circuit design. IEEE Trans. Biomed. Eng. **BME–30**, 62–66 (1983)
8. Jou, S.-C., Schultz, T., Walliczek, M., Kraft, F., Waibel, A.: Towards continuous speech recognition using surface electromyography. In: Proceedings of Interspeech, Pittsburgh, PA (2006)
9. Wand, M., Schultz, T.: Speaker-adaptive speech recognition based on surface electromyography. In: Fred, A., Filipe, J., Gamboa, H. (eds.) BIOSTEC 2009. CCIS, vol. 52, pp. 271–285. Springer, Heidelberg (2010)
10. Ueda, N., Nakano, R., Ghahramani, Z., Hinton, G.E.: Split and merge EM algorithm for improving Gaussian mixture density estimates. J. VLSI Sig. Proc. **26**, 133–140 (2000)
11. Qiao, Z., Zhou, L., Huang, J.Z.: Sparse linear discriminant analysis with applications to high dimensional low sample size data. Int. J. Appl. Math. **39**, 48–60 (2009)
12. Belhumeur, P.N., Hespanha, J.P., Kriegman, D.J.: Eigenfaces vs fisherface: recognition using class-specific linear projection. IEEE Trans. Pattern Anal. Mach. Intell. **19**, 711–720 (1997)
13. Hyvrinen, A., Oja, E.: Independent component analysis: algorithms and applications. neural networks **13**, 411–430 (2000)
14. Bell, A.J., Sejnowski, T.I.: An Information-Maximization Approach to Blind Separation and Blind Deconvolution. Neural Comput. **7**, 1129–1159 (1995)
15. Makeig, S., et al.: EEGLAB: ICA Toolbox for Psychophysiological Research. WWW Site, Swartz Center for Computational Neuroscience, Institute of Neural Computation, University of San Diego California (2000). www.sccn.ucsd.edu/eeglab/ (World Wide Web Publication)
16. Wand, M., Himmelsbach, A., Heistermann, T., Janke, M., Schultz, T.: Artifact removal algorithm for an EMG-based silent speech interface. In: Proceedings of IEEE EMBC (2013)
17. Zhao, H., Xu, G.: The research on surface electromyography signal effective feature extraction. In: Proceedings of 6th International Forum on Strategic Technology (2011)
18. Wand, M., Schulte, C., Janke, M., Schultz, T.: Array-based electromyographic silent speech interface. In: Proceedings of Biosignals (2013)

Stable EEG Features for Biometric Recognition in Resting State Conditions

Daria La Rocca[1]([✉]), Patrizio Campisi[1], and Gaetano Scarano[2]

[1] Section of Applied Electronics, Department of Engineering,
University of Roma Tre, Via Vito Volterra 62, 00146 Roma, Italy
{daria.larocca,patrizio.campisi}@uniroma3.it
[2] DIET, Sapienza Università di Roma, Via Eudossiana 18, 00184 Roma, Italy
gaetano.scarano@uniroma1.it

Abstract. In this paper electroencephalogram (EEG) signals are studied to extract biometric traits for identification of users. Different recording sessions separated in time are considered in order to infer about usability of EEG biometrics in real life applications. The aim of this work is to provide a representation of the data and a classification approach which would show repeatability of the EEG features employed in the proposed framework. The brain electrical activity has already shown some potentials to allow automatic user recognition, but an extensive analysis of EEG data aiming at retain stable and distinctive features is still missing. In this contribution we test the invariance over time of the discriminant power of the employed EEG features, which is a relevant property for a biometric identifier to be employed in real life applications. The enrolled healthy subjects performed resting state recordings on two different days. Combinations of different electrodes and spectral subbands have been analyzed to infer about the distinctiveness of different topographic traits and oscillatory activities. Autoregressive statistical modeling using reflection coefficients has been adopted and a linear classifier has been tested. The observed results show that a high degree of accuracy can be achieved considering different acquisition sessions for the enrollment and the testing stage. Moreover, a proper information fusion at the match score level showed to improve performance while reducing the sample size used for the testing stage.

Keywords: EEG · Biometrics · Repeatability · Resting · Fusion

1 Introduction

EEG signals have been widely studied since the beginning of the last century, mainly for clinical applications, to investigate brain diseases like Alzheimer, epilepsy, Parkinson and many others. Specifically, EEG signals, acquired by means of scalp electrodes, sense the electrical activity related to the firing of specific collections of neurons during a variety of cognitive tasks such as response to audio or visual stimuli, real or imagined body movements, imagined speech, etc.

© Springer-Verlag Berlin Heidelberg 2014
M. Fernández-Chimeno et al. (Eds.): BIOSTEC 2013, CCIS 452, pp. 313–330, 2014.
DOI: 10.1007/978-3-662-44485-6_22

Table 1. State of the art on EEG biometrics using a resting state protocol.

Paper	Protocol	# Subj	# Ch.	Features
[4]	CE	4	1	AR (6th)
[5]	CE	48	3	AR (6th)
[6]	CE, OE	40	1	AR (21th)
[7]	CE	5	6	AR (6th)
[8]	CE	10	1	AR (6th) + PSD
[9]	CE, OE	10	4	AR (21th)
[10]	OE	10	2-3	AR (11th)
[11]	CE	23	1	FFT

The most relevant cerebral activity falls in the range of $[0.5, 40]$ Hz. In details five wave categories have been identified, each associated to a specific bandwidth and to specific cognitive tasks: *delta waves* (δ) $[0.5, 4]$ Hz which are primarily associated with deep sleep, loss of body awareness, and may be present in the waking state; *theta waves* (θ) $[4, 8]$ Hz which are associated with deep meditation and creative inspiration; *alpha waves* (α) $[8, 13]$ Hz which indicate either a relaxed awareness without any attention or concentration; *beta waves* (β) $[13, 30]$ Hz usually associated to active thinking; *gamma waves* (γ) $[30, 40]$ Hz usually used to locate right and left side movements.

In the last decades, the brain activity, registered by means of EEG, has been heavily employed in brain computer interfaces (BCI) [1] and more recently in brain machine interface (BMI) [2] for prosthetic devices. In the last few years EEG signals have also been proposed to be used in biometric based recognition systems [3].

EEG signals present some peculiarities, which are not shared by the most commonly used biometrics, like face, iris, and fingerprints. Specifically, brain signals generated on the cortex are not exposed like face, iris, and fingerprints, therefore they are more privacy compliant than other biometrics since they are "secret" by their nature, being impossible to capture them at a distance. This property makes EEG biometrics also robust against the spoofing attack at the sensor since an attacker would not be able to collect and feed the EEG signals. Moreover, being brain signals the result of cognitive processes, they cannot be synthetically generated and fed to a sensor, which also addresses the problem of liveness detection. Also, the level of universality of brain signals is very high. In fact people with some physical disabilities, preventing the use of biometrics like fingerprint or iris, would be able to get access to the required service using EEG biometrics.

However, the level of understanding of the physiological mechanisms behind the generation of electric currents in the brain, not yet fully got, makes EEG a biometrics at its embryonic stage. Nevertheless, some preliminary, but promising, results have already been obtained in the recent literature, see for example [4, 5, 12–14] where a review on the state of the art of EEG biometrics is also given, and [15]. Due to the early stage of research dealing with EEG as biometrics, currently, the deployment of convenient and accurate EEG based applications in real world are limited with respect to well established biometrics like fingerprints, iris, and face. However the brain electrical activity has already shown some potentials to allow automatic user recognition. Answers to practical and theoretical questions addressed for the development of a usable system can be found in [16] where promising results are obtained from the implementation of a portable EEG biometric framework for applications in real world scenarios. Improvements in EEG signal acquisition and technological advances in the use of wireless and dry sensors, easy to wear and robust with respect to noise [17] could represent the cue for outlining guidelines for practical systems implementation.

In Table 1, an extensive although not exhaustive list of research studies which have already been published using a resting state acquisition protocol, either closed eyes (CE) or open eyes (OE), is provided. It is evident that the database dimension is quite limited in almost all of these contributions. This is also due to the lack of a public EEG database suitably collected for the biometric recognition purpose, where acquisitions and protocols would be designed according to the specific requirements. In fact most of the works in this field test the implemented techniques on datasets recorded in BCI contexts. Moreover the issue of the repeatability of EEG biometrics in different acquisition sessions has never been systematically addressed in any of the aforementioned contributions and it has never received the required attention from the scientific community. Nevertheless, its understanding is propaedeutic towards the deployment of EEG biometrics in real life. Although in some referred works different acquisition sessions have been provided, they were considered to assort a single dataset where randomly selected EEG segments were used for training or testing a classification algorithm for the recognition purpose. On the other hand, in [12] the session-to-session variability was tested on a dataset of 6 subjects performing imagined speech. The entire set of 128 channels was used to extract features, and results show a decreasing performance when considering sessions temporally apart, which led to assess that the imagined speech EEG does not show to have a reliable degree of repeatability. Therefore, in this paper we further speculate on the use of EEG as a biometric characteristic by focusing on the analysis of repeatability of its features, thus starting filling a gap in the existing literature. More in details, we rely on two simple acquisition protocols, namely "resting states with eyes open" and "resting states with eyes closed" to acquire data from nine healthy subjects in two acquisition sessions separated in time. Different configurations for the number of electrodes employed and for their spatial placement have been taken into account. Specifically, sets of three and five electrodes have been considered to acquire the signals, and several frequency bands have been analyzes. The so acquired signals, after proper preprocessing, are then modeled using autoregressive stochastic modeling in the feature extraction stage. Linear classification is then performed. The paper is organized as follows. The acquisition protocol is detailed in Sect. 2.1, and in Sect. 2.2 the template extraction procedure is described. In Sect. 3 classification is performed, while experimental results are given in Sect. 4. Finally conclusions are drawn in Sect. 5.

2 Experimental Setup

2.1 EEG Data

Nine healthy volunteers have been recruited for this experiment. Informed consent was obtained from each subject after the explanation of the study, which was approved by the local institutional ethical committee. During the experiment, the participants were comfortably seated in a reclining chair with both arms resting on a pillow in a dimly lit room properly designed minimizing external sounds and noise in order not to interfere with the attention and the relaxed state of subjects. The subjects were asked to perform one minute of "resting state with eyes open" and one minute of "resting state with eyes closed" [18] in two temporally separated sessions, from 1 to 3 weeks distant from each other, depending on the subject.

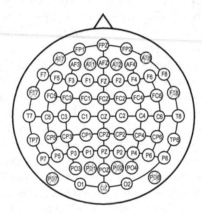

Fig. 1. Scalp electrodes positioning in the employed protocol according to an extension of the standard 10-20 montage.

Brain activity has been recorded using a BrainAmp EEG recording system operating at a sampling rate of 200 Hz. The EEG was continuously recorder from 54 sites positioned according to the International 10-20 system as shown in Fig. 1. Such configuration is not meant to be a user convenient solution, but allows to investigate about a proper electrode placement, able to catch distinctive features according to the employed protocol. Before starting the recording session, the electrical impedance of each electrode was kept lower than 10 kOhm through a dedicated gel maximizing the skin contact and allowing for a low-resistance recording through the skin. After the EEG recording sessions, the EEG signals have been band pass filtered in the band [0.5, 30] Hz, before further analysis.

2.2   Methods

The template is generated by considering the signals acquired by a properly chosen set of electrodes. We have tested different acquisition configurations. Specifically, sets of three and five electrodes have been employed in our tests to understand, at a first stage, the proper number of electrodes to employ, limiting it, and at a later stage, the proper electrodes positioning to use in order to capture repeatable and stable features, if present. The signals so acquired are preprocessed as described in Sect. 2.2 in order to perform denoising and to select the proper subbands. Then, the EEG signals in the selected subbands are AR modeled as described in Sect. 2.2. The template is obtained by concatenating the reflection coefficients vectors related to the different channels in the sets under analysis. Specific brain rhythms are mainly predominant in certain scalp regions during different mental states. Therefore, we expected a certain variability of recognition performance spanning the entire scalp through specific configurations of electrodes, and considering the closed or open eyes condition, being different the capability to catch distinctive features.

Preprocessing. Before performing feature extraction, each acquired raw EEG signal has been processed as described in the following. Neural activity reflected in resting state EEG signals shows to contain frequency elements mainly below 30 Hz. Hence, a decimation factor has been applied to the collected raw signals, after filtering them through an anti-aliasing FIR filter. A sampling rate of $S_r = 60$ Hz was selected to retain spectral information present in the four major EEG subbands referring to the resting state (δ [0.5, 4] Hz, θ [4, 8] Hz, α [8, 13] Hz and β [13, 30] Hz). The γ subband [30, 40] Hz is not considered, given that it is known not to be relevant in a resting condition. A further stage of zero-phase frequency filtering was applied to discriminate the different EEG rhythms. The single δ, θ, α and β subbands and their combinations (frequency components from 0.5 Hz up to 30 Hz, and from 0.5 Hz up to 14 Hz) have been considered in our experiments.

A spatial filter has been then applied to the acquired signals. When sufficiently large numbers of electrodes are employed, potential at each location may be measured with respect to the average of all potentials, approximating an inactive reference. Specifically, a common average referencing (CAR) filter has been employed in the herein proposed analysis by subtracting the mean of the entire $C_T = 54$ electrodes montage (*i.e.* the common average) from the channel c of interest, with $c = 1, 2, \cdots, C_T$, at any one instant, according to the formula:

$$^{CAR}V_u^c[n] = V_u^c[n] - \frac{1}{C_T}\sum_{j=1}^{C_T} V_u^j[n], \tag{1}$$

where $V_u^j[n]$ is the potential between the j-th electrode and the reference electrode, for the user u, with $u = 1, 2, \cdots, U$. CAR filtering has been employed to reduce artifacts related to inappropriate reference choices in monopolar recordings [19] or not expected reference variations, as well as to provide measures as

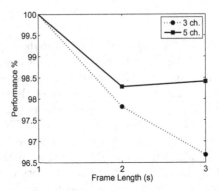

Fig. 2. Classification results in % obtained for the best performing set of three (P7-Pz-P8) and five channels (Cz-CP5-CPz-CP6-Pz), considering AR order $Q = 10$ and different values of frame length (T_f).

independent as possible from the recording session. This results in an increased signal-to-noise ratio, since artifacts related to a single reference electrode are better controlled, as showed in [20], where authors compared spatial filter methods with a conventional ear reference in an EEG-based system.

A set of instances to be used for the training and the testing stages has been obtained from the signal segmentation. A range from 1 up to 3 seconds of EEG frame length has been spanned stepwise, in order to best characterize each user brain signal for the identification purpose. The one second frame length has been experimentally selected as it has shown to best catch distinctive features of users' EEG segments for the recognition purpose. This can be observed in Fig. 2, where best performance is achieved both for sets of three and five electrodes, considering one second EEG segments, in the band $\delta \cup \theta \cup \alpha \cup \beta = [0.5, 30]$ Hz shown to be the best performing. These results refer to 10 order AR modeling and best sets of three and five channels, and show averaged performance obtained training the classifier on each acquisition session and testing it on the other one. Such framework has been employed to increase the number of trials used to study the repeatability of EEG biometrics in terms of recognition performance over the investigated period.

In this stage an overlap interval between adjacent frames was set to increase the sample size. Overlapping percentages of 25 %, 50 % and 75 % have been tested. Subsequently the DC component jointly to the linear trend has been removed from each EEG segment. The so obtained data-set have been further processed to extract the distinctive features from each user brain signal, as described below.

Modeling and Feature Extraction. After the preprocessing stage, detailed in Sect. 2.2, each acquired signal is modeled as a realization of an AR stochastic process. A realization $x[n]$ of an AR process, of order Q, can be expressed as:

$$x[n] = -\sum_{q=1}^{Q} a_{Q,q} \cdot x[n-q] + w[n] \tag{2}$$

where $w[n]$ is a realization of a white noise process of variance σ_Q^2, and $a_{Q,q}$ are the autoregressive coefficients. The well known Yule-Walker equations [21], which allow calculating the Q coefficients, can be solved recursively, employing the Levinson algorithm and introducing the concept of reflection coefficients. Specifically:

$$\begin{cases} a_{Q,q} = a_{Q-1,q} + K_Q \cdot a_{Q-1,Q-q}, & q = 1, \cdots, Q-1 \\ \sigma_Q^2 = \sigma_{Q-1}^2 \left(1 - K_Q^2\right), \end{cases} \tag{3}$$

where the factor K_Q is the so-called *reflection coefficient* of order Q which is calculated as follows [21]:

$$K_Q = -\left(R_x[Q] + \sum_{q=1}^{Q-1} R_x[q] \cdot a_{Q-1,Q-q}\right) / \sigma_{Q-1}^2 \tag{4}$$

where the generic $R_x[m]$ is the signal autocorrelation function, defined as $R_x[m] = \mathrm{E}\{x[n]\,x[n-m]\}$, for all $m \geq 0$.

Among the possible estimation approaches, the Burg method [21] estimates the reflection coefficients K_q, for $q = 1, \ldots, Q$, operating directly on the observed data $x[n]$ rather than estimating the autocorrelation samples $R_x[m]$. Therefore, the Burg's reflection coefficients, which have been shown in [5] to achieve better performance than the most commonly employed AR coefficients, are here employed.

Given the generic user u, and the generic channel c, let us indicate with $\zeta^{(u,c)}$ the vector, of length Q, composed by the AR model reflection coefficients K_q, for $q = 1, \ldots, Q$, using the Burg method:

$$\zeta^{(u,c)} = [K_1^{(u,c)}, K_2^{(u,c)}, \cdots, K_Q^{(u,c)}]^T. \tag{5}$$

The model order Q has been selected according to the Akaike Information Criterion (AIC) to minimize the information loss in fitting the data. It can be observed in Fig. 3(a), that the AIC(Q) function, averaged among subjects and channels, reaches minimum plateau zone for values of Q from 6 to 12. The feature vector **x** for the user u is obtained by concatenating the AR coefficients vectors related to the signals obtained from the channels in the set under analysis. The 10 AR order has been experimentally selected since it has shown to best fit the EEG data for the recognition purpose, as it can be observed in Fig. 3(b), where correct classification percentage is reported considering one second EEG segmentation. Averaged results are shown, obtained training on each session and testing on the remaining one, and considering the best performing sets of three and five channels.

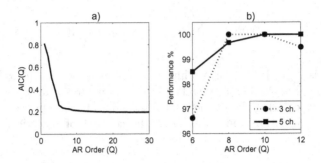

Fig. 3. (a) AIC function, averaged on all subjects and channels, for the frequency band $[0.5, 30]$ Hz. (b) Classification results in % obtained for the best performing set of three (P7-Pz-P8) and five channels (Cz-CP5-CPz-CP6-Pz), considering $T_f = 1s$ and different AR orders Q.

3   Classification

The classifier we propose estimates the class (user identity) to which the observed feature vector $\mathbf{x}$ belongs to by means of a linear transformation $\hat{\mathbf{y}}^T(\mathbf{G}) = \mathbf{x}^T\mathbf{G}$, where the transformation matrix $\mathbf{G}$ is obtained by minimizing the mean square error (MMSE) thus obtaining:

$$\mathbf{G} = \arg\min_{\boldsymbol{\Gamma}} \sum_{i=1}^{N} \mathcal{P}_i \cdot \mathrm{E}_{\mathbf{x}|\mathcal{H}_i}\{[\mathbf{y}_i - \hat{\mathbf{y}}(\boldsymbol{\Gamma})]^T[\mathbf{y}_i - \hat{\mathbf{y}}(\boldsymbol{\Gamma})]\} \qquad (6)$$

where $\mathcal{H}_i$ indicates the hypothesis $\mathbf{x}$ belongs to the i-th class, with $i = 1, 2, \cdots, N$. Here, assuming the hypothesis $\mathcal{H}_i$ holds, the vector $\mathbf{y}_i = [0, \ldots, 0, 1, 0, \ldots, 0]$ with the unique one in the i-th position, indicates the class i $\mathbf{x}$ belongs to, while $\hat{\mathbf{y}}^T(\mathbf{G})$ represents its estimation. $\mathcal{P}_i$ denotes the prior probability that $\mathbf{x}$ belongs to the i-th class. It can be easily shown that the employed optimization criterion given in (6) brings to the normal equations:

$$\mathbf{R}_{\mathbf{x}} \cdot \mathbf{G} = \mathbf{R}_{\mathbf{xy}} \qquad (7)$$

where $\mathbf{R}_{\mathbf{x}} = \mathrm{E}\{\mathbf{x}\mathbf{x}^T\}$ is the auto-correlation matrix for the elements of the feature vector $\mathbf{x}$, while $\mathbf{R}_{\mathbf{xy}} = \sum_{i=1}^{N} \mathcal{P}_i \cdot \mathrm{E}_{\mathbf{x}|\mathcal{H}_i}\{\mathbf{x}\} \cdot \mathbf{y}_i^T$ turns out to be the matrix whose columns are the probabilistically averaged conditional mean values of the observations $\mathbf{x}$.

Dataset. As pointed out in Sect. 2.2, different sets of $N_c = 3, 5$ channels, to acquire the signals from which the feature vectors $\mathbf{x}$ is extracted, have been considered for both the employed protocols. Given a chosen set of channels, each of the signals so acquired has been pre-processed, as described in Sect. 2.2, segmented into N_f frames, and modeled by resorting to the reflection coefficients of an AR model of order Q. Therefore, considering, for each user, EEG signals of duration of 60 s, segmented into frames of 1, 2 and 3 s, with an overlap factor of 75 %, a number of $N_f = 237, 117$ and 77 frames has been obtained respectively, each of which is represented by the feature vector $\mathbf{x}$ of $Q \times N_c$ elements.

Such a set of feature vectors has been collected for each of the two temporally separated recording sessions, 1-3 weeks distant from each other, and each protocol, *i.e.* closed and open eyes resting conditions. It is worth pointing out that the vectors used in the training stage and in the recognition stage have been obtained from the two different acquisition sessions in order to infer about the repeatability over the considered interval of the EEG features for the acquired dataset and the employed acquisition protocols. Hence, we applied the classification algorithm selecting the train and the test datasets without shuffling the EEG frames belonging to different sessions, as performed in other works with user recognition aims. In order to achieve our goal, each one of the two sessions has been sequentially considered for the training dataset while the remaining session has been used to obtain the test dataset, thus obtaining two couples of temporally separated datasets, (training set, recognition set) to train and test the classifier. This kind of validation framework has been provided just to encrease the statistical significance of the results. They show that we can't assess a perfect symmetry of changes over time, but that the features keep stable over the considered interval (1-3 weeks). Some of the results of each test are shown in subsequent columns of Table 2 where each set has been acquired at a different time.

3.1 Training

The training stage consists in the estimation of the matrix $\mathbf{G}$ in (7) computed as $\mathbf{G} = \widehat{\mathbf{R}}_x^{-1} \cdot \widehat{\mathbf{R}}_{xy}$, where the matrices $\mathbf{R}_x$ and $\mathbf{R}_{xy}$ are estimated through MonteCarlo runs, considering equal prior probabilities $\mathcal{P}_i$ for all the classes (users identities) to distinguish between. The estimation was obtained performing the following two sample averages:

$$\widehat{\mathbf{R}}_x = \frac{1}{NM} \sum_{i=1}^{N} \sum_{m=1}^{M} \mathbf{x}_{m,i} \mathbf{x}_{m,i}^T$$
$$\widehat{\mathbf{R}}_{xy} = \frac{1}{NM} \sum_{i=1}^{N} \sum_{m=1}^{M} \mathbf{x}_{m,i} \mathbf{y}_i^T, \tag{8}$$

where $\mathbf{x}_{m,i}$ is the m-th observed feature vector belonging to the i-th class, with M being the number of instances of $\mathbf{x}$ for each class, and $\mathbf{y}_i = [0, \ldots, 0, 1, 0, \ldots, 0]^T$, with the unique 1 in the i-th position. The considered matrices can be simply upgraded in case of enrollment of N' new users, summing the related matrices $\mathbf{x}_{m,i} \mathbf{x}_{m,i}^T$ to $\mathbf{R}_x$, and adding new columns i to $\mathbf{R}_{xy}$ given by $\frac{N}{M(N+N')} \sum_{m=1}^{M} \mathbf{x}_{m,i}$, where $i = N + 1, \ldots, N + N'$. To avoid failures and to control accuracy in the estimation of $\mathbf{R}_x^{-1}$, the singular value decomposition based pseudoinversion has been used for the matrix inversion.

Table 2. Classification results in % for CE protocol, obtained using the acquisition session t for training and the acquisition session r for recognition, with $t, r = S_1, S_2$ and $t \neq r$, for the subband $\delta \cup \theta \cup \alpha \cup \beta = [0.5, 30]$ Hz, for sets of three electrodes. For each test $t \rightarrow r$ 2 results are provided, considering 75 % of the training dataset while 25 % (first column) and 75 % (second column) of the test dataset.

	Closed eyes							
	Spatial filtering (CAR)				No spatial filtering			
Electrodes	$S_1 \rightarrow S_2$		$S_2 \rightarrow S_1$		$S_1 \rightarrow S_2$		$S_2 \rightarrow S_1$	
Fp1 Fpz Fp2	51.66	57.57	65.03	69.43	59.87	70.18	65.31	73.42
AF3 AFz AF4	63.43	65.40	66.15	70.79	76.89	91.00	79.23	87.15
F7 Fz F8	50.54	57.76	73.89	82.09	70.32	71.26	78.86	88.65
F3 Fz F4	53.26	59.54	66.34	67.84	59.45	65.54	66.71	69.39
F1 Fz F2	60.81	66.43	74.03	82.56	73.14	85.98	72.95	76.14
FC3 FCz FC4	73.61	81.15	80.45	91.09	77.12	91.14	84.29	94.05
T7 Cz T8	68.26	74.59	70.56	77.12	65.64	59.77	63.90	69.29
C3 Cz C4	78.15	82.51	74.82	85.56	78.57	93.48	84.29	87.39
C1 Cz C2	78.43	88.98	81.81	94.19	80.78	87.01	92.50	99.91
TP7 CPz TP8	65.78	75.34	62.82	57.01	75.06	79.75	80.97	85.61
CP3 CPz CP4	63.62	66.85	59.12	72.95	69.25	71.26	80.03	87.25
P7 Pz P8	93.44	100	94.56	99.62	95.50	100	97.47	100
P5 Pz P6	89.69	99.62	93.25	99.06	80.54	87.01	92.31	100
P3 Pz P4	79.00	79.89	86.08	89.22	67.84	67.93	70.84	78.48
P1 Pz P2	65.07	70.60	62.82	70.79	63.90	63.85	63.06	69.48
PO3 POz PO4	70.98	69.06	77.64	80.87	68.07	81.20	74.07	85.51
O1 POz O2	69.57	74.82	70.70	67.79	69.67	78.43	66.24	70.75

3.2 Recognition

In the recognition stage, a linear transformation is applied to each of the $M \times N$ observations from the test dataset. For the i-th user a score vector $\hat{\mathbf{y}}_i$ was obtained for each instance of $\mathbf{x}_i$ in the dataset applying the discrimination matrix $\mathbf{G}$ to $\mathbf{x}_{m,i}$:

$$\hat{\mathbf{y}}_{m,i} = \mathbf{G} \cdot \mathbf{x}_{m,i} \qquad (9)$$

with $m = 1, \ldots, M$. Subsequently, the M score vectors related to each tested user were summed together to reduce the misclassification error, obtaining

$$\hat{\mathbf{y}}_i = \sum_{m=1}^{M} \hat{\mathbf{y}}_{m,i}. \qquad (10)$$

Finally the estimation of the index representing the user identity is obtained locating the maximum of the score vector $\hat{\mathbf{y}}_i = [\hat{y}_i(1), \ldots, \hat{y}_i(N)]^T$ according to the criterion

$$\hat{i} = \arg_l \max y_i(l). \qquad (11)$$

As previously pointed out, to solve the classification problem we separately considered different symmetrical sets of sensors and different brain rhythms for the template extraction. In order to improve accuracy, an information fusion integrating multiple sensors distributions and brain rhythms was then performed at the match score level, which is the most common approach in multibiometric systems [22]. The aim was to determine the best sets of channels configurations and frequency bands that could optimally combine the decisions rendered individually by each of them. The basic hypothesis was that different representations of brain rhythms show different distinctive traits, which may increase variability between-subjects if efficiently combined together. In this regard the proper level for information fusion in the multibiometric approach is an important issue to dramatically improve the classification performance. We observed that different subjects showed advantagious scores for different sets of channels and different rhythms, that is each of them presented particular spactral distribution and topography of distinctive traits (Table 4). Hence, for each tested user the proposed score fusion was obtained through the sum

$$\frac{1}{N_B N_{Ch}} \sum_{b}^{B} \sum_{Ch}^{S} {}_{Ch}^{b} \hat{\mathbf{y}}_i \qquad (12)$$

of scores vectors $\hat{\mathbf{y}}_i$ related to specific N_B bands $b \in B$ and N_S selected sets $Ch \in S$ composed of three electrodes. All tests performed and obtained results are reported in the next Section.

4 Experimental Results

The results of the performed analysis are reported for all the experiments carried out. The aim of the study was to test the repeatability of the considered EEG features, needed to recognize users previously enrolled in a biometric system. Repeatability and stability represent properties of paramount importance for the use of EEG biometrics in real life systems. For this purpose we have selected two simple tasks to be performed by a set of users, and a classification problem has been set up, where the training set and the one to be used in the recognition stage have been chosen belonging to temporally separated sessions 1-3 weeks distant from each other.

More in detail, given the "resting state" acquisition protocols here considered and the 54 employed channels shown in Fig. 1, we have selected different subsets of them in order to find the best performing spatial arrangements of the electrodes while minimizing their number. Although the considered acquisition technique doesn't result user convenient, not being this the focus of the paper, in a preliminary study, as this is, it allows to detect on the scalp the brain rhythms which provide the best distinctive features, according to the employed protocol. In order to achieve this goal we have considered sets of three and five electrodes, the former listed in Table 2. An information fusion approach combining match scores obtained for the selected distributions of sensors and the selected brain rhythms is also proposed to improve performance.

Table 3. Classification results in % for OE protocol. See caption of Table 2 for description.

| Electrodes | Open eyes | | | | | | | |
| | Spatial filtering (CAR) | | | | No spatial filtering | | | |
	$S_1 \to S_2$		$S_2 \to S_1$		$S_1 \to S_2$		$S_2 \to S_1$	
Fp1 Fpz Fp2	57.48	68.03	67.65	64.65	43.46	49.79	55.41	56.59
AF3 AFz AF4	56.12	56.96	47.12	56.17	53.87	59.21	69.39	73.98
F7 Fz F8	63.57	63.48	67.28	70.18	62.17	67.14	69.67	74.54
F3 Fz F4	66.10	68.17	72.25	70.28	65.78	73.00	60.76	69.01
F1 Fz F2	66.01	66.67	67.23	70.98	54.62	58.37	67.28	69.10
FC3 FCz FC4	83.97	87.15	78.57	87.11	66.99	68.45	67.56	65.96
T7 Cz T8	87.06	83.68	84.11	89.59	76.79	77.03	75.25	80.97
C3 Cz C4	80.08	90.53	78.20	80.87	76.84	81.58	69.48	77.92
C1 Cz C2	72.39	73.65	73.65	82.37	65.45	65.17	69.85	81.20
TP7 CPz TP8	53.26	53.91	58.74	66.29	52.23	55.09	58.09	62.59
CP3 CPz CP4	63.24	72.53	62.96	65.17	75.43	82.00	60.29	63.90
P7 Pz P8	55.32	56.54	59.82	64.70	59.17	60.85	51.24	47.30
P5 Pz P6	53.21	54.99	62.59	64.32	65.07	66.67	53.59	54.71
P3 Pz P4	53.73	57.99	68.59	76.32	79.93	85.33	67.74	76.65
P1 Pz P2	55.79	51.34	59.45	66.76	68.40	74.87	56.82	60.29
PO3 POz PO4	51.76	49.41	55.37	52.23	56.02	63.01	56.26	60.76
O1 POz O2	52.13	48.76	54.85	54.15	54.81	56.02	56.49	61.46

Template extraction has been performed as described in Sect. 2.2, by first preprocessing the EEG signals, which includes decimation with sampling rates $S_r = 60\,\mathrm{Hz}$, CAR spatial filtering, segmentation into frames of T_f s with an overlapping factor O_f between consecutive frames, and eventually band pass filtering in order to analyze the subbands δ, θ, α and β, which are the ones interested by the "resting state" protocols, and some of their combinations. A value of $O_f = 75\,\%$ has been here employed since we have experimentally proven it is able to guarantee good performance as it provides an adequate sample size to assort the training and recognition datasets. Then the so obtained frames are modeled using an AR model, whose tested orders $Q = \{6, 8, 10, 12\}$ have been estimated by means of the AIC function (see Fig. 3(a)). Value of $T_f = 1\,\mathrm{s}$ and $Q = 10$ have been selected as they showed to best characterize the users' EEG for the recognition task (see Fig. 2).

The template is then obtained by concatenating the reflection coefficients of the signals acquired by means of the electrode set under analysis, thus generating feature vectors of length $3Q, 5Q$ for the sets of three, and five electrodes respectively.

Table 4. Classification results in % of correct identification reported for each subject I_i in the cross validation framework (averages over 237 runs). Results refer to CE condition, when training on session S_1 (75 % of frames) and performing identification tests on S_2 (25 % of frames).

Electrodes	Subjects								
	I_1	I_2	I_3	I_4	I_5	I_6	I_7	I_8	I_9
Fp1 Fpz Fp2	78.06	75.53	27.00	85.23	100	91.56	33.76	47.68	0
AF3 AFz AF4	100	70.46	48.95	84.39	86.50	69.62	76.37	55.70	100
F7 Fz F8	100	9.70	13.50	80.17	34.60	100	100	94.94	100
F3 Fz F4	100	3.38	0	74.26	20.25	100	79.32	57.81	100
F1 Fz F2	73.84	72.57	0	79.75	100	100	74.26	58.65	99.16
FC3 FCz FC4	98.31	48.52	32.07	60.76	100	100	75.53	79.75	99.16
T7 Cz T8	7.59	100	100	100	0.00	100	34.60	96.62	51.90
C3 Cz C4	48.10	40.08	100	94.94	90.72	100	58.65	80.59	94.09
C1 Cz C2	37.97	81.43	94.94	100	100	100	27.00	85.65	100
TP7 CPz TP8	26.58	74.26	100	91.56	83.12	100	100	100	0
CP3 CPz CP4	23.63	36.29	100	100	49.37	100	35.86	87.34	90.72
P7 Pz P8	100	100	97.47	100	100	100	100	100	62.03
P5 Pz P6	100	48.95	100	97.47	100	100	37.13	100	41.35
P3 Pz P4	35.44	37.13	100	100	98.31	100	38.82	68.35	32.49
P1 Pz P2	42.19	0	100	100	64.56	100	62.45	85.65	20.25
PO3 POz PO4	100	39.66	65.40	85.65	82.28	100	4.22	64.98	70.46
O1 POz O2	100	12.66	64.56	78.06	100	100	37.97	81.86	51.90

In Tables 2 and 3 the results obtained for sets of three electrodes when using the MMSE classifier, described in Sect. 3, are given for both employed protocols CE and OE. It is worth pointing out that the signals employed to obtain the templates to be used in both the training and the recognition stage are disjoint in time. Therefore two different combinations of training (t) and recognition (r) sessions, (t, r) with $t, r \in \{S_1, S_2\}$ and $t \neq r$, have been tested. Such kind of tests varying the sequence of sessions in the recognition framework are provided to validate the results about repeatability of the considered EEG features over the interval under analysis, for a real usability of an EEG-based biometric system. The results for the different tests performed are reported separately, not expecting to make assumptions on symmetry of changes over time. Moreover, from the analysis of different spatial configurations of electrodes we could observe that triplets of channels allow achieving about same performance then configurations employing sets of five channels. This is due to a good spatial localization achieved by configurations of only three sensors, which allow to well capture the underlying phenomena, reducing the problem dimensionality. Moreover, in Tables 2 and 3 results are shown considering the band $F = \delta \cup \theta \cup \alpha \cup \beta = [0.5, 30]$ Hz which is

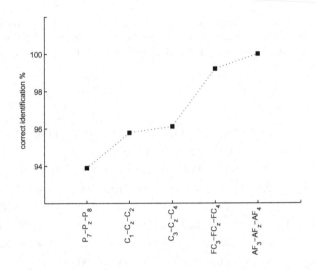

Fig. 4. Improvement of the correct recognition rate obtained performing subsequent score fusions (see Sect. 4), within the CE condition. Curves refer to the combination of different electrodes sets (x-axis). Labels in the x-axes refer to the score added at each subsequent step. Results refer to the training on S_1 and test on S_2.

Table 5. Classification results in % for both CE and OE protocols, obtained using the same acquisition session S for training and for recognition, with $S = S_1, S_2$, for the subband $\delta \cup \theta \cup \alpha \cup \beta = [0.5, 30]$ Hz, for the best performing sets of three and five electrodes. Results are provided considering 75 % of the dataset for training while 25 % of the dataset for recognition.

Electrodes	Closed eyes				Open eyes			
	Spatial filt. (CAR)		No spatial filt.		Spatial filt. (CAR)		No spatial filt.	
	$S_1 \rightarrow S_1$	$S_2 \rightarrow S_2$	$S_1 \rightarrow S_1$	$S_2 \rightarrow S_2$	$S_1 \rightarrow S_1$	$S_2 \rightarrow S_2$	$S_1 \rightarrow S_1$	$S_2 \rightarrow S_2$
P7 Pz P8	96.81	100	98.03	100	95.22	96.11	94.56	93.25
Cz CP5 CPz CP6 Pz	98.87	100	100	100	100	100	97.00	91.80

the one that allows obtaining the best results, and considering a preprocessing including CAR filtering or not.

Provided performance refers to a cross-validation framework, obtained selecting for each user 75 % of feature vectors **x** related to cyclically subsequent frames from the training dataset, while 25 % and 75 % from the test dataset, as reported in subsequent columns of Tables 2 and 3. Numerical results are obtained averaging over 237 independent cross-validation runs to improve the statistical analysis. As previously pointed out, recognition tests have been carried out keeping independency between the training and the test datasets, acquired in different sessions, for the classification purpose. This aspect is highlighted in Tables 2 and 3 denoting with $S_i \rightarrow S_j$ the result achieved training on S_i and testing on S_j.

It should be noticed that applying the CAR filter in the preprocessing stage doesn't yield a general improvement in the performance for all employed sets of channels and protocols, while it appears to provide best results for some selected channels (FC3-FCz-FC4, C3-Cz-C4, T7-Cz-T8) for sets of 3 channels in the OE protocol. This is likely to be due to artifacts which more affected the open-eyes condition, removed by the spatial filtering. As regards differences between the two employed protocols it is evident, by observing the reported results, that in this experiment the CE protocol provides best performance considering the adopted EEG feature extraction for the recognition task. In fact, within the CE condition 100 % of correct classification is achieved for instance employing channels P7-Pz-P8 and 75 % of the test feature vectors for each user in the cross-validation framework. This has been observed to be due both to being the open-eyes signal more affected by the eyes movement artifacts, and to distinctive traits contained in the α rhythm which is mainly detected on the posterior head when resting with eyes closed. In this regard is was noticed that the combination of channels affected in a different way the recognition results for CE and OE protocols. In fact, referring to sets of three channels the parietal region has proven to best perform in CE condition, while the central region achieved best results in OE condition. Moreover it has been observed, individually analyzing the extracted brain rhythms, that in CE the α band most contributed to the best performance obtained combining all bands ([0.5, 30] Hz). The results just pointed out are in agreement with the fact that in resting state with eyes closed the dominant brain rhythm α can be detected mainly in the posterior area of the scalp, while it is attenuated when opening eyes.

Repeatability over the considered interval of the analyzed EEG features can be inferred by observing that users enrolled in a session have been recognized in a different one, disjoint in time from 1 to 3 weeks. Besides, it is also evident that by swapping the training and recognition roles of the session datasets, that is by considering (t, r) or (r, t), quite coherent performance are obtained.

Table 5 shows results obtained training and testing the classifier on the same session. It should be noticed that very high correct recognition rate is achieved considering just 25 % of the test dataset (100 % for CE and S_2), while a greater number of feature vectors for each user are needed in the inter-sessions framework. This evidence proves the importance of speculating about the stability and repeatability over time of EEG features for biometric systems. The performance significantly decreases for the case of disjoint training and test datasets when considering just few frames for the identification test (25 % of the test dataset). On the other hand, the match score fusion obtained as discussed in Sect. 3.2 has led to a dramatic increase of the recognition accuracy, especially for the otherwise poorly performing case just mentioned, as observed in Fig. 4 where the CE condition is analyzed. Same improvements are observed within the OE condition, not reported in here due to space limitations, still remaining the CE condition the best performant one. For the selection of the electrodes configurations to combine, the best combination of rhythms was considered, and subsequent score fusions were performed. To this aim the electrode sets were sorted in descending order of performance achieved individually (see Table 2), and sequentially combined within a forward-backward stepwise approach, retaining in the information fusion only those sets which improved the correct classification. Results reported in Fig. 4 showed that a significant improvement was obtained combining rhythms and sets of channels. Within the multi-session framework a perfect recognition percentage of 100 % could be achieved when the sets of three inter-hemispheric channels $P7 - Pz - P8, C1 - Cz - C2, C3 - Cz - C4, FC3 - FCz - FC4, AF3 - AFz - AF4$, and the rhythms $[0.5, 30], \delta, \theta, \alpha$ containing most information, were combined into the match score fusion. It should be noticed that the selected channels result located all over the head, showing that antero-posterior differences could be distinctive as well as iter-hemispheric asymmetry. Figure 4 reports the improvements obtained across the subsequent steps of the information fusion, when considering 75 % of training frames from S_1 and just 25 % of test frames from S_2, for the CE condition. Accordingly, this approach allows to obtain high accuracy while significantly reducing the recording time needed for the recognition tests.

5 Conclusions

In this paper the problem of repeatability over time of EEG biometrics, for the same user, within the framework of EEG based recognition, has been addressed. Simple "resting state" protocols have been employed to acquire a database of nine people in two different sessions separated in time from 1 to 3 weeks, depending on the user. Although the dimension of the database employed is contained, we would like to stress out that this contribution represents the first systematic analysis on the repeatability issue in EEG biometrics. As such, this contribution paves the road to more refined analysis which would include more sessions separated in time as well as different acquisition protocols. Extensive simulations have been performed by considering different sets of electrodes both with respect to their positioning and number. A combination of match scores obtained from different analysis have shown to significantly reduce the frames needed for test, still maintaining high recognition accuracy. In summary in our analysis a very high degree of repeatability over the considered interval has been achieved with a proper number of electrodes, their adequate positioning and by considering appropriate subband related to the employed acquisition protocol.

Acknowledgements. We would like to thank Prof. F. Babiloni, University of Rome "Sapienza", and the Neuroelectrical Imaging and BCI Lab RCCS Fondazione Santa Lucia, Rome, for providing the dataset used in the presented analysis.

References

1. Dornhege, G., del R. Millán, J., Hinterberger, T., McFarland, D., Móller, K.R. (eds.): Towards Brain-Computing Interfacing. MIT Press, Cambridge, MA (2007)
2. Carmena, J.M.: Becoming bionic. IEEE Spectr. **49**(3), 24–29 (2012)

3. Campisi, P., La Rocca, D., Scarano, G.: EEG for automatic person recognition. Comput. **45**(7), 87–89 (2012)

4. Poulos, M., Rangoussi, M., Chrissikopoulos, V., Evangelou, A.: Parametric person identification from the EEG using computational geometry. In: The 6th IEEE International Conference on Electronics, Circuits and Systems, 1999, ICECS '99, vol. 2, pp. 1005–1008, Sep 1999

5. Campisi, P., Scarano, G., Babiloni, F., De Vico Fallani, F., Colonnese, S., Maiorana, E., L., F.: Brain waves based user recognition using the eyes closed resting conditions protocol. In: IEEE Int. Workshop on Information Forensics and Security (WIFS'11), November 2011

6. Paranjape, R., Mahovsky, J., Benedicenti, L., Koles, Z.: The electroencephalogram as a biometric. In: Canadian Conference on Electrical and Computer Engineering, pp. 1363–1366 (2001)

7. Palaniappan, R., Patnaik, L.M.: Identity verification using resting state brain signals. In: Quigley, M. (ed.) Enc. of Information Ethics and Security, IGI Global, pp. 335–341 (2007)

8. Zhao, Qinglin, Peng, Hong, Hu, Bin, Liu, Quanying, Liu, Li, Qi, YanBing, Li, Lanlan: Improving individual identification in security check with an EEG based biometric solution. In: Yao, Yiyu, Sun, Ron, Poggio, Tomaso, Liu, Jiming, Zhong, Ning, Huang, Jimmy (eds.) BI 2010. LNCS, vol. 6334, pp. 145–155. Springer, Heidelberg (2010)

9. Abdullah, M., Subari, K., Loong, J., Ahmad, N.: Analysis of effective channel placement for an EEG-based biometric system. In: 2010 IEEE EMBS Conference on Biomedical Engineering and Sciences (IECBES), pp. 303–306, 30 Nov-2 Dec 2010

10. Mohammadi, G., Shoushtari, P., Molaee Ardekani, B., Shamsollahi, M.B.: Person Identification by Using AR Model for EEG Signals. In: Proceedings of World Academy of Science, Engineering and Technology, vol. 11, pp. 281–285 (2006)

11. Nakanishi, I., Baba, S., Miyamoto, C.: EEG based biometric authentication using new spectral features. ISPACS **2009**, 651–654 (2009)

12. Brigham, K., Kumar, B.V.: Subject identification from electroencephalogram (EEG) signals during imagined speech. In: Proceedings of the IEEE Fourth International Conference on Biometrics: Theory, Applications and Systems (BTAS'10) (2010)

13. Riera, A., Soria-Frisch, A., Caparrini, M., Grau, C., Ruffini, G.: Unobtrusive biometric system based on electroencephalogram analysis. EURASIP J. Adv. Signal Process.

14. Marcel, S., del R. Millan, J.: Person authentication using brainwaves (EEG) and maximum a posteriori model adaptation. IEEE Trans. Pattern Anal. Mach. Intell. **29**(4), 743–748 (2007)

15. La Rocca, D., Campisi, P., Scarano, G.: EEG biometrics for individual recognition in resting state with closed eyes. In: Proceedings of the International Conference of the Biometrics Special Interest Group (BIOSIG), pp. 1–12 (2012)

16. Su, F., Xia, L., Cai, A., Wu, Y., Ma, J.: EEG-based personal identification: from proof-of-concept to a practical system. In: 20th International Conference on Pattern Recognition (ICPR 2010), pp. 3728–3731 (2010)

17. Debener, S., Minow, F., Emkes, R., Gandras, K., de Vos, M.: How about taking a low-cost, small, and wireless EEG for a walk? Psychophysiol. **49**(11), 1617–1621 (2012)

18. Barry, R., Clarke, A., Johnstone, S., Magee, C., Rushby, J.: EEG differences between eyes-closed and eyes-open resting conditions. Clin. Neurophysiol. **118**, 2765–2773 (2007)
19. Schwartz, M., Andrasik, F.: Biofeedback: A Practitioner's Guide. Guilford Press, New York (2003)
20. McFarland, D., McCane, L., David, S., Wolpaw, J.: Spatial filter selection for EEG-based communication. Electroencephalogr. Clin. Neurophysiol. **103**(3), 386–394 (1997)
21. Kay, S.: Modern Spectral Estimation: Theory and Applications. Prentice-Hall, Englewood Cliffs, NJ (1988)
22. Ross, A.A., Nandakumar, K., Jain, A.K.: Handbook of Multibiometrics, vol. 6. Springer Science+ Business Media, New York (2006)

Health Informatics

TEBRA: Automatic Task Assistance for Persons with Cognitive Disabilities in Brushing Teeth

Christian Peters[✉], Thomas Hermann, and Sven Wachsmuth

CITEC, Bielefeld University, Bielefeld, Germany
{cpeters,thermann,swachsmu}@techfak.uni-bielefeld.de

Abstract. We introduce a novel Assistive Technology for Cognition. The TEBRA system assists persons with cognitive disabilities in brushing teeth by prompting the user. We develop the system based on an analysis of the task using qualitative data analysis. We recognize different substeps using a recognition component based on a Bayesian network. The user's progress in the task is monitored using a Finite State Machine and a dynamic timing model which allows for different velocities of task execution. We evaluate the TEBRA system in a first study with regular users and analyze the timing behavior of the system in detail. We found that the system is able to provide appropriate prompts in terms of timing and modality to assist a user through the complex task of brushing teeth.

Keywords: Cognitive assistive technology · Task analysis · Bayesian network · Dynamic timing model · User study

1 Introduction

Assistive Technology for Cognition (ATC) aims at developing systems which support persons with cognitive disabilities in the execution of activities of daily living (ADLs). These persons mostly have problems in accomplishing ADLs on their own and need assistance by a human caregiver to perform such tasks successfully. ATC can provide assistance by automatically prompting the users, and keep elderly people or persons with cognitive disabilities further in their own homes which leads to an increase of independence of the persons and a relief of caregiver burden.

The execution of ADLs is very complex: ADLs like washing hands, brushing teeth, or preparing meal contain several substeps which can be combined in a flexible manner. Furthermore, the execution of substeps differs significantly between persons based on their individual abilities. A key paradigm in the development of automatic prompting systems for complex ADLs is to deliver prompts to the user when necessary in order to foster the user's independence. A prompt is necessary if the user forgets a step or gets stuck in a substep during task execution. Users with cognitive disabilities, but also regular users, show a huge variance in spatial and temporal execution of the task: one user may take the

© Springer-Verlag Berlin Heidelberg 2014
M. Fernández-Chimeno et al. (Eds.): BIOSTEC 2013, CCIS 452, pp. 333–353, 2014.
DOI: 10.1007/978-3-662-44485-6_23

brush with the left hand while another user takes the right hand and performs completely different movements at different velocities. An ATC system needs to deal with the spatial and temporal variances to deliver appropriate prompts in terms of timing and modality.

In this paper, we introduce a novel ATC system: the TEBRA (**TE**eth **BR**ushing **A**ssistance) system which assists in *brushing teeth* as an important ADL. We develop the TEBRA system based on a systematic analysis of the task. We apply Interaction Unit (IU) analysis proposed in [1] as a method for qualitative data analysis. The results of IU analysis are utilized for different design decisions. Results are (1) the decomposition of the task into substeps we aim to recognize, (2) the extraction of environmental conditions associated with substeps and task progress, (3) preconditions and effect of substeps.

In order to deal with the variance in spatial execution, we abstract from the recognition of specific movements by tracking objects or the user's hands. Instead, we classify substeps based on environmental configurations in a hierarchical recognition component. We cope with the temporal variance in task execution by using a Finite State Machine and a dynamic timing model allowing for different user velocities. We learn the timing parameters for different velocities of users (fast, medium and slow) and switch the parameters dynamically during a trial based on the velocity of the user. We choose appropriate system prompts using a search in an ordering constraint graph (OCG). An OCG models temporal relations between substeps in terms of preconditions and effects obtained in the IU analysis.

We evaluate the first prototype of the TEBRA system in a study with regular users. Evaluating our system with regular users is suitable. Regular persons show individual ways of task execution which may not coincide with the system's framework of action. The system prompts the user who in turn has to react to the prompts and adapt his/her behavior to successfully execute the task from a system's point of view. Hence, we provoke similar phenomena in terms of prompting and reaction behavior with both regular users and persons with cognitive disabilities. We consider the target group in the development of the TEBRA system because IU analysis is conducted on videos of persons with cognitive disabilities in a residential home setting. The aim of the study with regular users is two-fold: firstly, we evaluate the technical correctness of the system with regard to recognition of substeps, monitoring the user's progress and timing of prompts. Secondly, we determine whether the prompts are appropriate in terms of duration and modality.

The remainder of the paper is structured as follows: Sect. 2 gives an overview of relevant related work. In Sect. 3, we give an overview of the TEBRA system. Section 4 describes IU analysis and the integration of the results in the system design. The main components of the TEBRA system are described in detail in Sect. 5. Section 6 shows the results of the user study, followed by a conclusion in Sect. 7.

2   Related Work

Assistive Technology for Cognition (ATC) is developed for special user groups like elderly or persons with cognitive disabilities in a number of different ADLs: the COACH system [2] assists persons with dementia in the handwashing task, Archipel [3] supports persons with cognitive disabilities in meal preparation. ADLs are complex tasks in terms of spatial execution which makes the recognition of behaviors challenging. Much work is done on recognizing behaviors based on movement trajectories of objects or the user's hands [4–6]. However, user behaviors with similar appearance are hard to distinguish based on movement trajectories only. In this work, we classify behaviors based on environmental states of objects involved in the brushing task.

Users perform behaviors at different velocities due to individual abilities. In the TEBRA system, we explicitly model timing parameters of user behaviors. We use a Finite State Machine and a dynamic timing model to allow for different velocities of the user's movements. In the *Autominder* system [7], persons with memory impairments are assisted in scheduling daily activities. *Autominder* models durations of events and reasons on temporal constraints to provide appropriate reminders. The PEAT system [8] schedules user's activities by applying reactive planning to restructure the schedule when events take more time than expected.

Monitoring the user's progress in the task is a key aspect to provide appropriate prompting. The COACH system uses a Partially Observable Markov Decision Process (POMDP). A belief state models the user's abilities and monitors the progress in the task. Whenever the belief state changes persistently due to sensor observations, the belief state is updated and a system action (prompt/no prompt) is triggered.

The *Activity Compass* assists disoriented users finding a destination by using a hierarchical Markov model to track the location of the user [9]. The specification process of probabilistic models like POMDPs is very hard in terms of determining the probabilities of dependent variables in the model. In the TEBRA system, we monitor the user's progress by utilizing a set of environmental variables which we deterministically update based on the occurrence of user behaviors. We find appropriate prompts using a search procedure on an ordering constraint graph (OCG). An OCG models temporal relations between substeps in terms of a partial ordering of substeps. We don't model every possible way of executing the task as done in the Archipel system [10] which uses a full hierarchical representation of the task. Instead, we use the OCG to model the constraints under which the execution of user behaviors is appropriate.

Most ATC systems are modeled using common-sense knowledge without further analyzing the task. Here, we apply a structured approach of retrieving relevant information on which we develop the TEBRA system. We use Interaction Unit (IU) analysis proposed in [1] as a method for qualitative data analysis to obtain relevant information about the brushing task. IU analysis was used in a similar context in [11] in order to facilitate the specification process of an automatic prompting system using a POMDP.

3   TEBRA Overview

Figure 1 depicts an overview of the TEBRA system. We built a washstand setup which we equipped with sensor technology. The sensor data is passed to a hierarchical recognition component which computes the current most probable user behavior. The main problem in the recognition component is the huge spatial variance in task execution. We tackle the problem by abstracting from tracking the user's hands or objects. Instead, we infer the user's behavior based on an environmental configuration which is expressed by states of objects manipulated during a behavior: we preprocess the sensor data into discrete features representing object states which are fed into a Bayesian network classification scheme. A Bayesian filtering step outputs a belief (conditional probability distribution) over the user behaviors.

The most probable user behavior is calculated each time new sensor data is obtained. A temporal integration mechanism accumulates the user behaviors over time and provides the duration of the behavior in seconds.

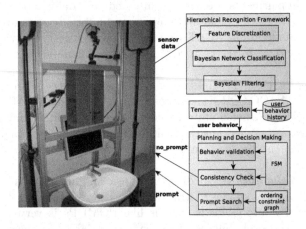

Fig. 1. Overview of the TEBRA system.

A key aspect is the huge variance in temporal execution of the task. One user, for example, is able to perform a substep very quickly while another user takes much longer. We use a Finite State Machine (FSM) to model timing of behaviors. The states of the FSM model different phases during a user behavior: we validate a user behavior and perform a consistency check on the validated behavior with respect to the user's progress in task execution. If the validated behavior is consistent, the system won't prompt, but will instead start a new iteration cycle receiving sensor data. If the validated behavior is inconsistent, the system will search for a consistent behavior to prompt: we use an online search procedure in an ordering constraint graph. An ordering constraint graph models partial orderings between user behaviors in the brushing task and is computed offline using a partial order planner. The consistent behavior found is displayed to the user by an audio-visual prompt delivered via a TFT display at the setup.

In the design of the TEBRA system, we focus on an iterative design process using a method of qualitative data analysis which is described in the following section.

4   Design Process

Developing an ATC system for an everyday task like brushing teeth is a challenging problem. The task consists of several substeps which can be combined in a flexible way during task execution. The analysis of the task and substeps as well as the possible ways of task execution are important steps in the development of an assistance system. We conduct Interaction Unit (IU) analysis which models interaction by describing the conjunction of cognitive and environmental pre- and postconditions for individual actions as described in [1]. We apply IU analysis on 23 videos recorded at our cooperation partner *Haus Bersaba*, a residential home where persons with cognitive disabilities such as dementia, Autistic Spectrum Disorder, intellectual disabilities, etc. permanently live. Each video shows a single trial of a user brushing teeth while being observed and supported by a caregiver.

We are interested in three aspects of IU analysis: (1) decomposition of the task into substeps which we will call *user behaviors* in the following, (2) environmental conditions associated with user behaviors and task progress, (3) preconditions and effects of behaviors. Table 1 shows the results of the IU analysis for brushing teeth.

The brushing task is decomposed into seven user behaviors as described in column *UB*: *paste_on_brush*, *fill_mug*, *rinse_mouth*, *brush_teeth*, *clean_mug*, *clean_brush* and *use_towel*. Each user behavior is further subdivided into single steps described in column *UB steps*. *rinse_mouth* for example consists of three steps: mug is moved to the face, the user rinses the mouth and moves the mug away from the face. Column *Current environment* shows the environmental configuration as a precondition of single user behavior steps. Performing the step changes the environmental state, for example in the first step of *paste_on_brush*: the toothpaste tube is on the counter and taking the tube changes the toothpaste location to 'in hand'.

We utilize the environmental configuration obtained in IU analysis. We extract environmental states in terms of discrete variables as depicted in Table 2. We distinguish between *behavior* and *progress* variables: we apply *behavior* variables to recognize user behaviors in a hierarchical recognition component. The *progress* variables are hard to observe directly due to reasons of robustness: for example, it is very error-prone to visually detect whether the *brush_condition* is *dirty* or *clean*. A specialized sensor at the brushing head is not desirable due to hygienic reasons. We utilize *progress* variables to monitor the user's progress during the task.

We abstract from the recognition of single behavior steps as given in column *UB steps* in Table 1. Instead, we infer the user's behavior based on the *behavior* variables which express states of objects manipulated during a behavior. From column *Current environment*, we extract five *behavior* variables describing important

Table 1. Results of the IU analysis for brushing teeth. Column "UB" describes the different substeps involved in the brushing task. Column "UB steps" lists the ideal steps to execute the according substep. Column "Current environment" shows the environmental configuration in terms of states of objects involved in a particular step. TT = toothpaste tube.

UB	Current environment	UB steps
paste_on_brush	TT on counter	take TT from counter
	TT closed in hand	alter TT to open
	brush on counter	take brush from counter
	brush and TT in hand	spread paste on brush
	TT is open	alter TT to closed
	TT closed in hand	give TT to counter
	TT on counter	
	brush in hand	
fill_mug	mug empty	give mug to tap
	mug at tap, tap off	alter tap to on
	mug at tap, tap on	alter tap to off
	mug filled	
rinse_mouth	mug filled	give mug to face
	mug at face	rinse
	mug else	give mug to counter
	mug counter	
brush_teeth	brush with paste in hand	give brush to face
	brush at face	brush all teeth
	brush at face, teeth clean	take brush from face
	brush not at face	
clean_mug	mug dirty at counter	give mug to tap
	mug dirty at tap, tap off	alter tap to on
	mug dirty at tap, tap on	clean mug
	mug clean at tap, tap on	alter tap to off
	mug clean at tap, tap off	give mug to counter
	mug clean at counter	
clean_brush	brush dirty	give brush to tap
	brush dirty at tap, tap off	alter tap to on
	brush dirty at tap, tap on	clean brush
	brush clean at tap, tap on	alter tap to off
	brush clean at tap, tap off	give brush to counter
	brush clean at counter	
use_towel	towel at hook, mouth wet	give towel to face
	towel at face, mouth wet	dry mouth
	towel at face, mouth dry	give towel to hook
	towel at hook	

Table 2. *Behavior* and *progress* variables extracted from the environmental configuration in Table 1.

Type	State variable	Values
behavior	mug_position	counter, tap, face, else, no_hyp
	towel_position	hook, face, else, no_hyp
	paste_movement	no, yes
	brush_movement	no, yes_sink, yes_face
	tap_condition	off, on
progress	mug_content	empty, water
	mug_condition	dirty, clean
	mouth_condition	dry, wet, foam
	brush_content	no_paste, paste
	brush_condition	dirty, clean
	teeth_condition	dirty, clean

Table 3. Preconditions and effects of user behaviors extracted from the environmental configuration in Table 1.

User behavior	Preconditions	Effects
paste_on_brush	brush_content=no_paste	brush_content=paste
	teeth_condition=dirty	brush_condition=dirty
fill_mug	mug_content=empty	mug_content=water
clean_mug	mug_content=empty	mug_condition=clean
	mug_condition=dirty	
	teeth_condition=clean	
rinse_mouth_clean	mug_content=water	mug_condition=dirty
	mouth_condition=foam	mouth_condition=wet
	teeth_condition=clean	mug_content=empty
rinse_mouth_wet	mug_content=water	mug_condition=dirty
	mouth_condition=dry	mouth_condition=wet
brush_teeth	brush_content=paste	teeth_condition=clean
	teeth_condition=dirty	brush_content=no_paste
	mouth_condition=wet	mouth_condition=foam
		brush_condition=dirty
clean_brush	brush_condition=dirty	brush_condition=clean
	teeth_condition=clean	brush_content=no_paste
use_towel	mouth_condition=wet	mouth_condition=dry
	teeth_condition=clean	

object states: *mug_position*, *towel_position*, *paste_movement*, *brush_movement* and *tap_condition*. The upper part of Table 2 shows the five variables and their according discrete values. For *brush_movement*, we have the states *no*, *yes_sink* and *yes_face*. The latter ones are important to discriminate between the user behaviors *paste_on_brush* and *brush_teeth* based on the movement of the brush. The values of the variables *mug_position* and *towel_position* are the different regions identified in column *Current environment* where the mug and towel appear during task execution. *No_hyp* is used if no hypothesis about the mug/towel position is available.

We utilize *progress* variables to monitor the user's progress in the task. At each time in task execution, the user's progress is modeled by the set of six *progress* variables which we will denote *progress state space* in the following. The lower part of Table 2 shows the variables of the progress state space and their according discrete values.

The occurrence of a user behavior during the execution of the task leads to an update of the progress state space: we define necessary preconditions and effects of user behaviors in terms of *progress* variables. When a user behavior occurs, we check whether the preconditions are met and, if so, update the progress state space with the effects of the current behavior. Table 3 shows the preconditions and effects for user behaviors in terms of *progress* variables extracted during IU analysis. We distinguish between *rinse_mouth_wet* and *rinse_mouth_clean* because the steps have different meanings during task execution: the IU analysis is based on videos recorded at *Haus Bersaba*. The videos showed that wetting the mouth with water using the mug (before brushing the teeth) is a common step as part of the regular daily routine. This step is described as *rinse_mouth_wet* whereas cleaning the mouth after the brushing task is *rinse_mouth_clean*.

The results of the IU analysis described in this section are integrated into the TEBRA system: the decomposition of the task into user behaviors and the *behavior* variables are utilized in the hierarchical recognition component. The progress state space and the preconditions and effects of user behaviors are integrated into the planning and decision making component. Both components are described in more detail in the following section.

5   System Components

We built a washstand setup as depicted on the left in Fig. 1 which we equipped with a set of sensors for environmental perception. We use a combination of unobtrusive sensors installed in the environment and tools. We don't attach any wearable sensors to the user directly, because we don't want to disturb in task execution. The washstand setup is equipped with two 2D cameras observing the scene from an overhead and a frontal perspective. A flow sensor installed at the water pipe measures whether the water flow is on or off. The toothbrush is equipped with a 9-dof[1] sensor module including accelerometer, gyroscope and magnetometer in 3 axis each. We extract a set of features from the sensors which

---

[1] Degree of freedom.

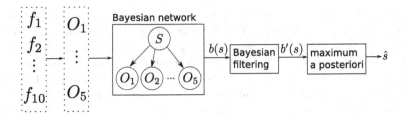

Fig. 2. Overview of the hierarchical recognition component. See text for a detailed description.

we feed into the hierarchical recognition component described in the following subsection.

5.1   Hierarchical Recognition Component

The IU analysis decomposes a task into different user behaviors. Each user behavior is further subdivided into single steps which are described in terms of environmental states. Hence, the IU analysis structures the task into a hierarchy of user behaviors and combines semantic information about the user behavior with environmental states. In our approach, we make use of the hierarchical structure obtained in the IU analysis: we use a two-layered framework for user behavior recognition modeling the hierarchical structure as shown in Fig. 2. A detailed description of the recognition component can be found in [12]. We extract ten features ($f_1...f_{10}$) from the sensory information: from the camera images, we extract the position of the mug, towel and paste using a detector based on color distribution which provides the center positions of the object's bounding boxes. The flow sensor returns a binary feature indicating water flow. For brush movement detection, it is sufficient to extract only the 3-dimensional gyroscope data which measures the angular velocity of the change in orientation in x, y and z direction.

The features $f_1...f_{10}$ are discretized into the *behavior* variables $O_1...O_5$ given in the upper part of Table 2. $O_1...O_5$ are obtained from the IU analysis and encode the environmental configuration of user behaviors. $O_1...O_5$ correspond to the variables *mug_position*, *towel_position*, *paste_movement*, *brush_movement* and *tap_condition*. The position of the mug and towel as well as the movement of the toothpaste are calculated by comparing the bounding box hypotheses of the object detectors to pre-defined image regions like *counter* or *tap*. The movement of the brush is computed using the gyroscope data obtained by the sensor module in the toothbrush. The condition of the tap is set according to the flow sensor.

In our framework, we aim to recognize the user behaviors obtained from the IU analysis in Table 1. We subsume the user behaviors *fill_mug* and *clean_mug* to a common user behavior *rinse_mug* because the *behavior* variables as well as the according states are nearly identical for both user behaviors. In a regular trial, user behaviors don't follow exactly on each other, but mostly alternate with transition behaviors, for example the user's hand approaches or leaves a manipulated object. We consider these transition behaviors by adding a user behavior *nothing* which we treat as any other user behavior in our recognition model.

In this work, we use a Bayesian network (BN) to classify user behaviors based on the observations $O_1...O_5$. A BN is ideally suited to model the structural relations between user behaviors denoted by the random variable S and *behavior* variables $O_1...O_5$. We use a BN with a *Naive Bayes* structure as depicted in Fig. 2. Each observation variable O_i is conditionally independent given the user behavior. The BN with *Naive Bayes* structure has the ability to deal with small training sets since the probability of each O_i depends only on the user behavior S. This is important in our work, because some user behaviors like *clean_brush* are rare compared to other behaviors and the acquisition of data in our scenario is very hard. The result of the BN classification scheme is a belief $b(s)$, a probability distribution over user behaviors. The BN is prone to faulty observations which happen occasionally in the discretization of features into observation variables. Faulty observations lead to rapid changes in the belief b from one time step to the next. This is not desirable in our scenario, because transitions between user behaviors are rather smooth due to the nature of the task. Hence, we extend our component with a transition model which takes into account the belief of the preceding time step. This results in a Bayesian filtering approach similar to the forward algorithm in a Hidden Markov Model. The belief b is updated to a consecutive belief b' using

$$b'(s') = \frac{O(s', \mathbf{o}) \cdot \sum_{s \in S} T(s', s) \cdot b(s)}{C} \tag{1}$$

with the normalization term $C = \sum_{s' \in S} O(s', \mathbf{o}) \cdot \sum_{s \in S} T(s', s) \cdot b(s)$. $O(s', \mathbf{o}) = \prod_{i=1}^{5} P(O_i | s')$ is the probability of making observation $\mathbf{o}$ when the user behavior is s'. The observation model $O(s', \mathbf{o})$ and transition model $T(s', s)$ are learned on manually annotated training data using Maximum Likelihood (ML) estimation. The transition model in Eq. 1 leads to smooth state transitions between user behaviors because single faulty observations can't rapidly change the entire belief from one time step to the next. The maximum a posteriori hypothesis of b' - the user behavior s' with the highest probability - is fed into the temporal integration mechanism.

5.2   Planning and Decision Making

The planning and decision making component determines whether to prompt the user during task execution and chooses a prompt as appropriate. The main paradigm in our system is that a prompt should only be given to the user when necessary in order to foster the independence of the user. A prompt is necessary when the user gets stuck in task execution or performs a step which is not appropriate at that time. In order to check whether a user behavior is appropriate, we maintain the user's progress in the task utilizing the progress state space described in Sect. 4. We update the progress state space based on the occurrence of user behaviors: when a behavior is recognized by the system, the progress state space is deterministically updated with the effects of the user behavior. For example, if *paste_on_brush* is recognized, the variables *brush_content* and *brush_condition* of the progress state space are set to *paste* and *dirty*, respectively, according to the effects in Table 3. All other variables remain unchanged. Monitoring the user's progress in the task is a very challenging problem due to the huge temporal variance in behavior execution. For example, one user may successfully perform *paste_on_brush* much slower compared to another user. Furthermore, the execution time of a single person may vary from day to day, especially for persons with cognitive disabilities. We cope with the huge intra- and inter-personal variance in a generalized timing model as described in the following subsection.

Timing Model. The hierarchical recognition component works in a framewise manner: the sensors in our setup are synchronized to a rate of 15 Hz. For each set of sensor data, the recognition component calculates the behavior s which has the maximum a posteriori probability in $b(s)$. The planning component works on a coarser scale: durations of user behaviors are measured in seconds. Hence, we apply a temporal integration mechanism between recognition and planning component: we maintain a local history, a list of user behaviors to which the behavior s is added. For each s, we calculate the occurrence o_s of s in the history. As long as $o_s \geq 0.8$, the duration of the current behavior is measured in seconds. If $o_s < 0.8$, we reset the current behavior time to 0. With a threshold of 0.8, we allow for misperceptions in the hierarchical recognition component without resetting the current behavior duration in case of single sensor errors. The temporal integration mechanism provides the duration of the current user behavior in seconds which is fed into a Finite State Machine (FSM). The FSM depicted in Fig. 3 models the timing behavior in our planning component. A FSM is suitable to model the different phases during a behavior: we validate a user behavior over a certain period of time (*validation* state) before we check the consistency (*check consistency* state). We refer to this time as validation time t_v^s based on the average duration of the behavior. We compare the duration t_s of the current behavior s to t_v^s. If $t_s \geq t_v^s$, the FSM transits to state *consistency_check*. The consistency of s is checked by comparing the preconditions of s with the progress state space. If the preconditions are not fulfilled, s is inconsistent which means that *ub* is not an appropriate behavior at that time. Hence, a prompt is delivered to the user. We describe the selection process of a prompt in detail in Sect. 5.2. If the preconditions are fulfilled, s is consistent. If the consistency check occurs too early, the user might feel patronized by the system. If the check is too late, the behavior effects might have been erroneously occurred already although the behavior is inconsistent. If *ub* is consistent, the FSM transits to state *pre_effect*. The effects of a user behavior occur after a minimum duration of the behavior which we call effect time t_e^s. State *pre_effect* denotes that the duration of s is too short for the effects to occur because t_e^s is not reached. If $t_e^s \geq t_s$, we update the progress state space by applying the effects of s. The FSM transits to state *post_effect*. For any user behavior (except for *nothing*),

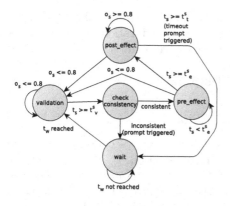
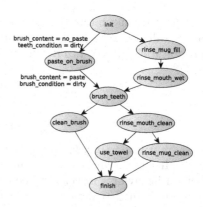

Fig. 3. Finite State Machine used to model the different phases in user behavior timing.

Fig. 4. Ordering constraint graph depicting partial orderings of user behaviors in the brushing task. We depict the preconditions and effects of *paste_on_brush*, exemplarily.

a timeout t_t^s may occur in the *post_effect* state. A timeout t_t^s denotes that the user might be stuck in task execution. If the duration $t_s \geq t_t^s$, a timeout prompt will be selected and delivered to the user. After a prompt, the FSM transits to a state *wait* for a fixed time $t_w = 5\mathrm{s}$ in order to wait for the user to receive the prompt and react properly.

We maintain a different set of timing parameters t_v^s, t_e^s and t_t^s for each user behavior in order to cope with the huge variance in temporal execution of individual behaviors. For example, the duration of *rinse_mouth* is usually much shorter compared to *brush_teeth*. Hence, the effect time t_e^s and timeout t_t^s of the behaviors are completely different. The validation time t_v^s can be set higher for longer behaviors to avoid a misdetection of the behavior due to perception errors. In addition to the different durations of user behaviors, we cope with different velocities of users by maintaining a set of timing parameters for three different user velocities: $t_{v_i}^s$, $t_{e_i}^s$ and $t_{t_i}^s$ where $i = \{f, m, s\}$ corresponding to *fast*, *medium* and *slow* execution velocity chosen manually by the authors. The parameters are estimated using an unsupervised learning mechanism: we apply a k-means algorithm to user behavior durations which we recorded in 18 test trials. We cluster the durations of each user behavior into $k = 3$ classes corresponding to *fast*, *medium* and *slow* execution velocity. We fit a Gaussian distribution $\mathcal{N}_i^s(\mu, \sigma^2)$ over the members of each cluster. We calculate the timing parameters $t_{v_i}^s$, $t_{e_i}^s$ and $t_{t_i}^s$ using the inverse cumulative distribution function (invCDF) of $\mathcal{N}_i^s(\mu, \sigma^2)$. For a given probability p, invCDF returns the duration at which the cumulative distribution function (CDF) is p. We calculate the validation time $t_{v_i}^s$ based on the average length of s in velocity model i:

$$t_{v_i}^s = h_v \cdot \mu_i^s \tag{2}$$

where μ_i^s is the mean duration of s in velocity model i and $h_v = 0.3$. $t_{v_i}^s$ denotes that we validate behavior s in velocity model i for a duration of 0.3 times the mean duration of s in i. The effect time and timeout are set with respect to invCDF:

$$t_{e_i}^s = \text{invCDF}(p_e) \qquad (3)$$

$$t_{t_i}^s = h_t \cdot \text{invCDF}(p_t) \qquad (4)$$

with $p_e = 0.3$, $p_t = 0.9$, $h_t = 2.5$, $\text{invCDF}(x) = \mu_i^s + \sigma_i^s \cdot (-1) \cdot \sqrt{2} \cdot \text{erfcInv}(2x)$ and erfcInv is the inverse complementary error function. $t_{e_i}^s$ denotes that the effects of behavior s in velocity model i occur after a duration of $\text{invCDF}(p_e)$. We chose a small parameter $p_e = 0.3$ in the calculation of the effect time of a behavior due to the following reason: missing a successful behavior and an update of the progress state space due to a large effect time is not desirable because it leads to an incorrect progress state space. Setting the effects of a behavior earlier than necessary is not critical since we already validated the user behavior. The description of $t_{t_i}^s$ is analogue to $t_{e_i}^s$. We have a total of $N = 72$ timing parameters: for each of the eight user behaviors given in Table 3, we have three velocities with three timing parameters each. We calculate the parameters based on four meta-parameters manually set to $h_v = 0.3$, $h_t = 2.5$ $p_e = 0.3$ and $p_t = 0.9$ used in Eqs. 2, 3 and 4.

In order to cope with the intra-personal variance in temporal execution of user behaviors, we use the learned parameters in a dynamic timing model: when a user behavior s switches to a successor behavior s', we determine the duration of s. We categorize the duration into one of the velocity classes *fast*, *medium* and *slow*. We maintain a discrete probability distribution over the velocity classes. We determine the current user velocity by applying a winner-takes-all method on the probability distribution which chooses the velocity occurring most frequently during the trial so far. The timing parameters of the FSM are set according to the user's velocity.

The FSM models the timing behavior of the TEBRA system. The selection mechanism of an appropriate prompt is described in the following subsection.

Prompt Selection. The prompt selection mechanism determines the content and modality of prompts given to the user. The content of a prompt is a hint on a user behavior and denotes an appropriate next step in task execution. The modality describes the sensory channel on which the prompt is delivered to the user. In the TEBRA system, we apply two types of visual prompts accompanying a verbal command: (1) pictograms and (2) real-time videos. The real-time videos are prerecorded showing a human actor performing the desired behavior. In a preliminary study using a Wizard of Oz paradigm with persons with cognitive disabilities [13], some users didn't react to audio prompts and were not able to proceed in task execution. Hence, we implemented an escalation hierarchy to provide prompts with different levels of information to the user: the type of prompt preferred by the system is a pictogram prompt. If the user doesn't react to the prompt, we escalate in the prompting hierarchy and deliver a video prompt where the behavior is depicted in more detail. We conducted interviews with the caregivers at *Haus Bersaba* about appropriate prompting modalities. The caregivers named pictogram and video prompts paired with a verbal command as the preferred modalities. Amongst the choices were pure audio prompts, visual prompts showing objects to use next and cartoon-like prompts. We have a total of 15 prompts: for each user behavior, we have a pictogram and a video prompt plus an additional pictogram showing that the user has reached the final state of the task.

The prompt selection mechanism is triggered in two cases: firstly, when a timeout of a user behavior occurs. Secondly, when the current user behavior is inconsistent with the progress state space. Inconsistent means that there are open preconditions of the current user behavior which are not fulfilled in the progress state space. In both situations, an appropriate prompt to the user addresses the following requirements: the prompt needs to (1) be consistent with the current progress state space, (2) push forward the user's progress in the task, and (3) provide at least one open precondition of the current behavior as an effect. The latter requirement arises because we want to assist the user in his way of executing the task as far as possible. Hence, we aim to find a prompt which supplies the open precondition and allows the user to re-perform the desired behavior after correctly performing the prompted behavior. For example, assume the user has successfully performed *brush_teeth* and performs *use_towel* afterwards. *use_towel* is inconsistent because precondition *mouth_cond=wet* is not fulfilled (*mouth_cond=foam*). In this situation, two prompts are appropriate: *clean_brush* and *rinse_mouth_clean*. The prompt selection mechanism would then decide for *rinse_mouth_clean* because it provides the open precondition *mouth_cond=wet* as an effect. If the user performs *rinse_mouth_clean*, he/she can go on performing *use_towel* as desired which would not be the case with *clean_brush*.

Our prompt selection mechanism performs a search on an ordering constraint graph (OCG) to find an appropriate prompt. An OCG is a visualization of a set of ordering constraints which are calculated using a partial order planner: given an initial state I, a goal state G and a set of STRIPS-like actions A with preconditions and effects, the partial order planner calculates a plan to transit from the initial to the goal state. A plan consists of a set of actions A, a set of ordering constraints O (action a before b), a set of causal links C (action a provides condition x for b) and a set of variable bindings B (variable $v = c$ where c is a constant).

In this work, we use a partial order planner to obtain an OCG for the task of brushing teeth. The user behaviors and according preconditions and effects identified in the IU analysis as given in Table 3 form the set of possible actions A. The initial and goal states are extracted from the IU analysis in Table 1 in terms of progress variables: $G = [mug\_content = empty, mug\_condition = clean, mouth\_condition = dry, brush\_content = no\_paste, brush\_condition = clean, teeth\_condition = clean]$ The initial state differs only in the variable *teeth_condition = dirty*. From the ordering constraints O and the causal links C, we manually construct a directed OCG as depicted in Fig. 4. An arrow in the OCG describes that the source behavior provides necessary preconditions for the target behavior. For example, *rinse_mug_fill* provides the effect *mug_content* = *water* which is a precondition of *rinse_mouth_wet*. The OCG depicts no strict execution plan of the task which the user has to follow, but models the ordering between behaviors in the overall task: the behavior sequence *rinse_mug_fill, paste_on_brush, rinse_mug_fill* e.g. is consistent with respect to the partial ordering given in Fig. 4.

We search for an appropriate prompt in the OCG as described in Algorithm 1: We determine the open preconditions of the inconsistent user behavior *s*. We process each open precondition as described in Algorithm 2: we search for a user behavior *s'* which is a predecessor of *s* in OCG and provides the open precondition. If *s'* exists, we check the consistency with regard to the progress state space. If *s'* is consistent, *s'* is an appropriate prompt. If *s'* is also inconsistent due to open preconditions, we recursively call *selectPrompt* with *s'* in order to find a behavior resolving the open preconditions of *s'*. By recursively calling *selectPrompt*, we resolve chains of open preconditions over several user behaviors by iterating backwards through the OCG. If no predecessor of *s* providing the open precondition is found, we search for a consistent behavior by iterating backwards through the OCG starting at the *finish* node. By starting at the finish node, we ensure to find a consistent behavior which is most closely to the desired goal state. In case of a timeout, the prompt selection mechanism directly searches for a consistent user behavior starting at the *finish* node.

We evaluate the technical correctness of the planning component in a user study described in the following section.

Algorithm 1. Select appropriate prompt.

```
 1: function SELECTPROMPT(s)
 2:     o ← GETOPENPRECONDITIONS(s)
 3:     for all o do
 4:         s' ← PROCESSPRECONDITION(s,o[i])
 5:         add s' to list_s' prompts
 6:     end for
 7:     if |list_s'| ≥ 2 then
 8:         prompt ← GETCLOSESTTOGOAL(list_s')
 9:         return s'
10:     else
11:         return list_s'[0]
12:     end if
13: end function
```

Algorithm 2. Process precondition.

```
     function PROCESSPRECONDITION(s,o)
 2:     s' ← FINDSUPPLYUB(s, o)
        if s' is empty then
 4:         ŝ ← FINDCONSISTENTPRED(finish)
            return ŝ
 6:     else
            CHECKCONSISTENCY(s')
 8:         if s' is consistent then
                return s'
10:         else
                return SELECTPROMPT(s')
12:         end if
        end if
14: end function
```

6   User Study and Results

We evaluate the first prototype of the TEBRA system in a study with 26 trials. Each trial is a single brushing task performed by a regular user. A study with regular users is suitable since regular persons show individual ways in the execution of the task which may not coincide with the system's framework of action. The system prompts the user who has to adapt to the prompts to successfully execute the task from a system's point of view. Hence, we provoke similar phenomena in terms of prompting and reaction behavior in a study with regular users compared to a study with persons with cognitive disabilities. However, we aim to conduct a study with persons with cognitive disabilities in the future. The goal of the study described here is two-fold: we aim to evaluate (1) the technical correctness of the system and (2) the user's reaction to system prompts. The reaction of regular users to system prompts is a measure whether the prompts are semantically reasonable to a minimum degree: if regular users have problems understanding the prompts, they might most likely be inappropriate for persons with cognitive disabilities. The 26 trials were performed by 13 users. Each user performed a single trial in each of two different scenarios: in the *free* scenario, users received the instruction to brush teeth as they would regularly do. The system generates prompts if necessary according to the user's task execution. In the *collaborative* scenario, the user is instructed to perform the brushing task in collaboration with the system: the user ought to follow the prompts whenever they are appropriate.

6.1   Technical Correctness and System Improvement

The technical correctness of the TEBRA system is highly dependent on the performance of the recognition component and the planning and decision making component. In the recognition component, we can't distinguish between *rinse_mouth_clean* and *rinse_mouth_wet* because the *behavior* variables are nearly identical for both behaviors. In the planning component, we need to distinguish between *rinse_mouth_clean* and *rinse_mouth_wet* since the behaviors are different in terms of preconditions and effects as given in Table 3. *Rinse_mouth_wet* describes taking water before brushing the teeth and *rinse_mouth_clean* is performed after the brushing step. We apply a systematic heuristic: when *rinse_mouth* is classified by the recognition component, it will be set to *rinse_mouth_wet* if *brush_teeth* has already been recognized in the trial because *brush_teeth* is the only behavior which provides the open preconditions of behavior *rinse_mouth_wet*. Hence, *brush_teeth* serves as a logical border between *rinse_mouth_clean* and *rinse_mouth_wet* during a trial. If we *brush_teeth* has not been recognized in a trial, *rinse_mouth* will be set to *rinse_mouth_clean*. We apply the same heuristic for *rinse_mug_fill* (before *brush_teeth*) and *rinse_mug_clean* (after *brush_teeth*) which are subsumed to a common behavior *rinse_mug* in the recognition component. Table 4 shows the classification rates of the user behaviors: The classification rates of *rinse_mug_fill*, *paste_on_brush*, *clean_brush* and *use_towel* are very good with 86 %, 99.2 %, 85.4 % and 87.1 %, respectively. However, the rate of *rinse_mouth_clean* is very low with 42 %. The heuristic is highly dependent on the recognition of *brush_teeth* which has a classification rate of 70 %. The recognition of *brush_teeth* is challenging: the gyroscope in the brush sensor module measures the angular velocity of the change in orientation. The changes are integrated over time to obtain the absolute orientation on which the behavior variable *brush_movement* is set. In the integration process, small errors are accumulated. For a behavior like *brush_teeth* which usually has a long

Table 4. Classification rates of user behaviors. RMgC - rinse_mug_clean, RMgF - rinse_mug_fill, UT - use_towel, PB - paste_on_brush, RMC - rinse_mouth_clean, RMW - rinse_mouth_wet, BT - brush_teeth, CB - clean_brush, N - nothing.

	RMW	RMC	RMgF	RMgC	BT	PB	CB	UT	N
RMW	**79.5**	0.0	0.0	0.0	0.0	0.6	0.0	0.0	19.9
RMC	32.5	**42.0**	2.4	1.6	0.0	0.1	9.9	0.0	11.5
RMgF	1.9	0.8	**86.0**	4.4	0.0	0.0	3.9	0.0	3.0
RMgC	0.0	6.2	0.0	**78.8**	0.0	0.0	0.5	8.0	6.6
BT	0.6	0.0	0.1	0.0	**70.0**	25.6	0.3	1.0	2.4
PB	0.0	0.0	0.0	0.0	0.5	**99.2**	0.0	0.0	0.3
CB	0.8	0.0	0.2	1.8	1.9	6.6	**85.4**	0.1	3.2
UT	0.0	0.0	0.0	0.2	0.1	0.6	0.0	**87.1**	12.1
N	3.9	2.4	1.3	1.4	2.8	22.8	2.0	9.8	**53.7**

duration compared to other behaviors, the accumulation of errors leads to mis-classifications: *brush_teeth* was mixed up with *paste_on_brush* in 25.6 % of the cases. However, the average classification rate of 75.7 % over all user behaviors is a very good result with regard to the huge variance in task execution.

6.2   System Performance

The performance of the TEBRA system in the *collaborative* scenario is excellent as depicted in column *FSR(%)* in Table 5:

Table 5. Coll - collaborative scenario, #Prompts - number of prompts, avg #Prompts - average nr of prompts, SC - ratio of semantically correct prompts, C - ratio of correct reactions to a prompt, CSC - ratio of correct reaction to a semantically correct prompt, dur - minimum (maximum) duration, FSR - final state reached.

	#Prompts	avg #Prompts	SC(%)	C(%)	CSC(%)	dur	FSR(%)
free	87	6.7	59	10	10	63 (184)	8
coll	117	9	66	75	85	142 (292)	100

Each of the 13 users reached the final state in this scenario where users ought to follow the prompts when appropriate. In the *free* scenario, only a single user reached the final state: regular users have an individual way of executing the task which may not coincide with the system's framework of action. In order to reach the final state, users have to adapt to the system prompts. In the *free* scenario, users were not explicitly encouraged to react to prompts, but were instructed to brush their teeth as they would regularly do. Since all users were capable

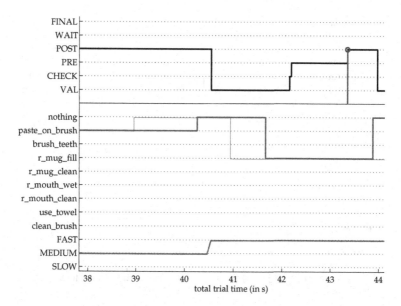

Fig. 5. Section of a trial of user 5 showing the state of the Finite State Machine (black line), the estimate of the user's behavior according to the recognition component (blue line), the estimate of the velocity (thick red line), and the ground truth annotation of behaviors (thin red line). The vertical blue line denotes the update of the state space by applying the effects of the current user behavior which is *rinse_mug_fill* here (Color figure online).

of brushing their teeth independently, all users except one didn't follow system prompts which leads to the rate of 8 %.

However, the excellent results in the *collaborative* scenario show that the system is able to assist a user in trials which differ significantly in duration: the minimum (maximum) duration in seconds are 63 (184) seconds in the *free* and 142 (292) in the *collaborative* scenario. The trials not only differ in the overall durations, but also in the durations of single user behaviors. We cope with the different behavior durations using the dynamic timing model as described in Sect. 5.2. Exemplarily, we show the advantage of the dynamic timing model in two situations.

Figure 5 visualizes the state of the FSM (black line), the estimate of the users behavior according to the recognition component (blue line), the estimate of the users velocity (thick red line), and the ground truth annotation of behaviors (thin red line). The visualization covers an interval of about six seconds in a trial of user 5. User 5 finishes *paste_on_brush* at about 40.3 s. Due to the duration of *paste_on_brush* and the velocities of the preceding behaviors, the velocity model is updated from medium to fast at 40.5 s. At 41.8 s, the user starts *rinse_mug_fill* which is performed for 2.2 s. Due to velocity model fast, the effects of the behavior occur after 1.6 s which is depicted by the vertical blue line at 43.4 s. With the model for medium velocity, *rinse_mug_fill* would not have been recognized

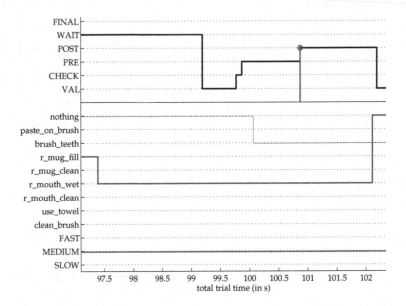

Fig. 6. Section of a trial of user 2. For a description of the lines, see Fig. 5.

correctly since the effect time of 3.3 s would not have been reached. Hence, the effects of *rinse_mug_fill* would not have been applied to the progress state space leading to erroneous prompts in the remainder of task execution.

A second situation is depicted in Fig. 6 showing a section of a trial of user 2. The velocity model of the user is medium. *Nothing* and *brush_teeth* are erroneously classified as *rinse_mouth_wet* for 4.7 s. Using the velocity model fast, which is the initial model of a trial, a timeout would have been reached after 4.4s. Hence, an erroneous prompt would have been issued based on the perception error. By adapting the velocity model to medium during the course of the trial, the dynamic timing model avoided the delivery of an erroneous prompt.

6.3 Appropriate Prompting

An important measure of the performance of our system is the number of prompts which are semantically correct. Semantically correct means that the type of prompt is appropriate with regard to the user's progress in the task so far: we determine the semantical correctness by using a ground truth annotation of the behaviors in the task which was done by the first author of the paper. In the *collaborative* scenario, 66 % of the total number of 117 prompts were semantically correct as depicted in Table 5. The 34 % of semantically incorrect prompts contain follow-up prompts arising from perception errors: when a user behavior was successfully performed, but not recognized correctly, the system delivers follow-up prompts which are semantically incorrect. If we only regard the first prompts of the system in each trial, the number of semantically correct prompts increases to 92 % which is a very good rate with regard to the complexity of the task.

The users' reactions to prompts indicate whether the prompts are meaningful to the user. The reaction of the user is correct when he/she updates the behavior to what was prompted by the system. We found 75 % of correct reactions in the *collaborative* scenario including semantically correct and incorrect prompts. Correct reactions to semantically correct prompts are much higher with 85 %. A major challenge in enhancing the TEBRA system will be the increase of semantically correct prompts which depends highly on the performance of the hierarchical recognition component. The system performance is very good measured in terms of correct reactions to prompts which are appropriate for the user. Additionally to the appropriateness of prompts in terms of timing and meaning, we asked the users to judge the modality of the prompts in terms of prompt duration and understandability in a questionnaire. Prompt duration and understandability were evaluated with a score of 5.5 and 6.1 of 7, respectively. A value of 1 denotes insufficient and 7 denotes perfectly good. This result underlines that the prompts are understandable at least to regular persons. We will evaluate in a future study whether persons with cognitive disabilities are also able to follow the prompts. We are optimistic, because the prompting modalities were selected as a result of interviews with caregivers of *Haus Bersaba* who found the modalities of prompting to be appropriate for the target group.

The results show that the TEBRA system is able to deal with the huge variance in spatial and temporal execution of the task: the system mostly gives semantically correct prompts and is able to assist users through the entire task of brushing teeth.

7 Conclusions

In this paper, we describe a novel Assistive Technology for Cognition: the TEBRA system aims to assist users with cognitive disabilities in the complex task of brushing teeth. We use a structured approach based on IU analysis by utilizing the results in the development of the TEBRA system. We tackle the huge variance in spatial and temporal execution of the task: in the hierarchical recognition component, we abstract from tracking objects or the user's hands. Instead, we infer behaviors based on the environmental state of objects. We deal with the temporal variance in task execution in our planning component: we apply a Finite State Machine and a dynamic timing model to allow for different velocities of users. We showed in a study with regular users that the TEBRA system is able to monitor the user's progress in the task and provide appropriate prompts to the user if necessary. The user's reactions show that prompting modalities are meaningful to a minimum degree with regard to regular users. Due to the structured approach using IU analysis, the TEBRA system can be easily extended to different tasks.

Future work includes technical enhancements of the system: we aim to recognize user behaviors more robustly by improving the classification mechanisms.

We will enhance the dynamic timing model by learning the parameters on a larger set of training data. We aim to evaluate the TEBRA system with persons with cognitive disabilities in the near future.

Acknowledgements. The authors would like to thank the inhabitants and caregivers of Haus Bersaba for their cooperation. This work was funded by the German Research Foundation (DFG), Excellence Cluster 277 "Cognitive Interaction Technology". Special thanks to Simon Schulz from the CITEC Central Lab Facilities for integrating the sensor module into the toothbrush.

References

1. Ryu, H., Monk, A.: Interaction unit analysis: a new interaction design framework. Hum. Comput. Interact. **24**, 367–407 (2009)
2. Hoey, J., Poupart, P., Bertoldi, A.V., Craig, T., Boutilier, C., Mihailidis, A.: Automated handwashing assistance for persons with dementia using video and a partially observable markov decision process. Comput. Vis. Image Underst. **114**, 503–519 (2010)
3. Bauchet, J., Giroux, S., Pigot, H., Lussier-Desrochers, D., Lachapelle, Y.: Pervasive assistance in smart homes for people with intellectual disabilities: a case study on meal preparation. Assistive Robot. Mechatron. **9**, 42–54 (2008)
4. Moore, D., Essa, I., Hayes, M.: Object spaces: context management for human activity recognition. In: International Conference on Audio-Visual Biometric Person Authentication, AVBPA'99 (1999)
5. Nguyen, N., Phung, D., Venkatesh, S., Bui, H.: Learning and detecting activities from movement trajectories using the hierarchical hidden markov model. In: International Conference on Computer Vision and Pattern Recognition, CVPR'05, pp. 955–960 (2005)
6. Pusiol, G., Patino, L., Bremond, F., Thonnat, M., Suresh, S.: Optimizing trajectories clustering for activity recognition. In: International Workshop on Machine Learning for Vision-based Motion Analysis, MLVMA'08, (2008)
7. Pollack, M.E., Brown, L.E., Colbry, D., McCarthy, C.E., Orosz, C., Peintner, B., Ramakrishnan, S., Tsamardinos, I.: Autominder: an intelligent cognitive orthotic system for people with memory impairment. Robot. Auton. Syst. **44**, 273–282 (2003)
8. Najjar, M., Courtemanche, F., Hamam, H., Dion, A., Bauchet, J.: Intelligent recognition of activities of daily living for assisting memory and/or cognitively impaired elders in smart homes. Ambient Comput. Intell. **1**, 46–62 (2009)
9. Patterson, D.J., et al.: Opportunity knocks: a system to provide cognitive assistance with transportation services. In: Mynatt, E.D., Siio, I. (eds.) UbiComp 2004. LNCS, vol. 3205, pp. 433–450. Springer, Heidelberg (2004)
10. Giroux, S., Bauchet, J., Pigot, H., Lussier-Desrochers, D., Lachappelle, Y.: Pervasive behavior tracking for cognitive assistance. In: PETRA'08 International Conference on PErvasive Technologies Related to Assistive Environments, pp. 1–7 (2008)
11. Hoey, J., Plötz, T., Jackson, D., Monk, A., Pham, C., Olivier, P.: Rapid specification and automated generation of prompting systems to assist people with dementia. Pervasive Mob. Comput. **7**, 299–318 (2011)

12. Peters, C., Hermann, T., Wachsmuth, S.: User behavior recognition for an automatic prompting system - a structured approach based on task analysis. In: International Conference on Pattern Recognition Applications and Methods, ICPRAM'12, pp. 162–171 (2012)
13. Peters, C., Hermann, T., Wachsmuth, S.: Prototyping of an automatic prompting system for a residential home. In: RESNA/ICTA 2011 Conference Proceedings (online) (2011)

Integration of Smart Home Health Data
in the Clinical Decision Making Process

Axel Helmer[1(✉)], Frerk Müller[1], Okko Lohmann[1], Andreas Thiel[1],
Friedrich Kretschmer[2], Marco Eichelberg[1], and Andreas Hein[1]

[1] R&D Division Health, OFFIS Institute for Information Technology,
Escherweg 2, 26121 Oldenburg, Germany
`axel.helmer.job@gmail.com`, {`frerk.mueller,`
`andreas.thiel,marco.eichelberg,andreas.hein`}`@offis.de`,
`okko.lohmann@web.de`
[2] NIH/NEI/NNRL, 6 Center Drive, Bethesda, MD 20892-0610, USA
`frierich@kretschmer.de`

Abstract. Patients suffering from COPD benefit from the performance of any
kind of physical activity. The 3D layer context (3DLC) model characterizes data
from smart home environments in relation to their relevance for the clinical
decision making process. We have used this model to show how data from an
ambient activity system in the domestic environment can be used to provide a
more informed and thereby better treatment management for COPD patients.
We set up an experiment to calculate an individual intensity relation between
household activities and telerehabilitation training on a bicycle ergometer.
We have extracted features from the power data of devices, which are used
during the performance of two example every day activities to calculate the
energy expenditure for the performance of these activities.

Keywords: Telemedicine · Physiological modeling · Knowledge management ·
Electronic health records

1 Introduction

1.1 Background

Chronic Obstructive Pulmonary Disease (COPD) is a collective term for different
diseases affecting the respiratory system. The World Health Organization estimates that
COPD affects 210 million people worldwide [1]. The illness is the third leading cause
of death in the United States, where the yearly direct/indirect costs are estimated with
29.9/49.5 billion USD.

National and international clinical guidelines, which summarize large randomized
controlled trials (RCT), show that the performance of rehabilitation training with relative high intensity provides many benefits for COPD patients e.g. an improved
exercise tolerance, less exacerbations and an improvement in the quality of life [2, 3].
Typically, a patient will begin the rehabilitation after he/she had an exacerbation, which
often leads to a stationary hospital stay. After the patient has been stabilized, a number
of clinical assessments such as a physical exercise tolerance test will be performed to

© Springer-Verlag Berlin Heidelberg 2014
M. Fernández-Chimeno et al. (Eds.): BIOSTEC 2013, CCIS 452, pp. 354–366, 2014.
DOI: 10.1007/978-3-662-44485-6_24

determine the individual functional capacity. This data is the basis for the medical staff to create a training schedule, which is then used to perform a supervised ambulatory or inpatient training in a rehabilitation clinic.

The current versions of the relevant clinical guidelines emphasize that the training has to be continued at home to preserve the positive effects of the clinical rehabilitation. Several systems were developed to implement a supervised or automatically controlled COPD related telerehabilitation training at home [4, 5].

The goal of the clinical or home-based rehabilitation training is that the patient performs a specific amount of physical activity over time. This amount is defined by frequency, duration, and intensity of performed activities and can be measured as energy expenditure. However, the rehabilitation training with its high intensity is only one specific activity of many that a patient will perform in his/her everyday life. Studies show that also activities with moderate intensity like walking or household activities are able to preserve the benefits that were reached during the clinical rehabilitation [6]. This data could also be relevant for follow-up examinations. For example, a trend that shows a reduction in the performance of physical activity could indicate that the health state of a patient becomes worse. This could be a hint towards an upcoming exacerbation or indicate the need of a medication change.

The detection of household activities for COPD patients is a good example that shows how data from the domestic domain can be used by clinicians to derive more informed and potentially better diagnoses or prognoses. One general problem that prevents the usage of this information is the difficulty to decide which of this collectable information is of real relevance to the clinical decision making process.

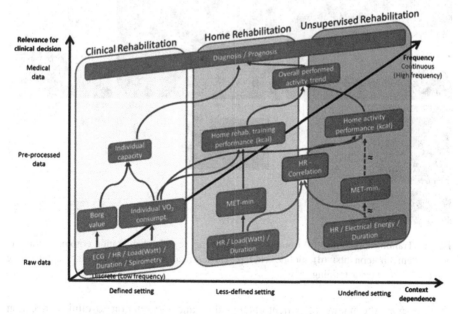

Fig. 1. The three dimensional layer context (3DLC) model (Color figure online).

The currently increasing number of assistive systems at home and approaches to connect user-centered IT systems like Personal Health Records (PHRs) with professional Electronic Health Records (EHRs) reinforces the need for a structured approach to clarify this question.

1.2   Related Work

The professional and the domestic domain have been strictly separated in the past when it comes to data sharing. Little research has been done concerning the combination of data from both domains. Most of the work concentrates on the clinical decision making process and data quality in the professional environment e.g. for clinical trials, but does not regard measurements that were obtained by the patient him/herself [7–10]. Electronic and Personal Health Records are IT systems where the professional and the domestic domains meet. Häyrinen et al. have conducted a systematic review on the definition, structure, content use and impact of EHRs. They state that further studies on the EHR content are needed; especially on patient self-documentation [11]. Tang et al. discussed the dependence of patient generated PHR content and clinical decision making in [12]. They recognized the problem and put it in a nutshell as follows:

> *"The reliability of patient-entered data depends on the nature of the information per se, the patient's general and health literacy, and the specific motivations for recording the data."*

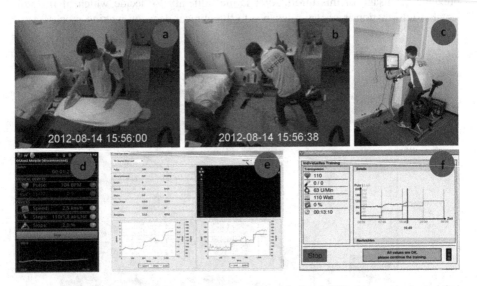

Fig. 2. Training modalities: (a) ironing, (b) vacuuming, (c) telerehabilitation ergometer training and system components: (d) mobile vital parameter recording, (e) monitoring and training control, (f) ergometer training view.

However, the nature of patient-entered data and the relation to clinical decision making were not further characterized.

The field of activity detection can be divided into approaches based on body-worn sensors like accelerometers or heart rate sensors and ambient sensors like cameras or motion sensors. Activity detection with body-worn sensors is well researched; commercial products are available, used in many studies and showing satisfying results [13–15]. However, obvious problems with these sensors are that patients constantly have to wear an electrical device, remember to put it on and to charge the batteries [16]. Systems for activity detection with ambient sensors are currently under research in the field of ambient intelligence. They use statically installed motion sensors [17–19], microphones [20], light sensors [21], and cameras. The main disadvantage for most of these systems is that they can be intrusive and depend on a lot of sensors that have to be installed in the user's environment. This probably leads to acceptance problems and high installation costs. Frenken et al. introduced a system that detects activities of daily living [22]. They use one single sensor that measures the power consumption of electrical devices that are used during these activities.

None of the mentioned ambient systems is able to derive the intensity or the energy expenditure of the performed activities.

1.3 Aim and Scope

The combination of clinical data with patient obtained information from the domestic environment is a general but well-known problem for clinicians. The emerging use of new health-related systems in patient's homes adds a new technical dimension to this problem. This complicates the decision making process, but also holds the potential to make more informed and better decisions.

The aim of this work is to show how new assistive systems can be included in the practice of medical decision making. We applied our prior developed three dimensional layer context (3DLC) model to the data of the COPD rehabilitation process. The model gives a structured approach for the combination of clinical and domestic data. As a proof of concept we combined data from home based telerehabilitation trainings for COPD patients with an activity detection system based on the power consumption of electrical devices. We used this detection system in an experiment to develop a comprehensive method to estimate the energy expenditure for household activities.

2 Methods

2.1 Three Dimensional Layer Context Model (3DLC)

The 3DLC model was first published in [23]. It distinguishes among three continuous dimensions (*frequency, context dependence, and relevance to clinical decision* making) to characterize data from different domains (see Fig. 1). These three dimensions will be described in more detailed in the following paragraphs.

The *context dependence* dimension reflects the influence of the environment on the data acquisition. A laboratory is used to minimize or stabilize influences of the environment that could have a possible impact on the acquired data. Such a very well-understood and controlled setting is termed a *defined setting* in the *context dependence* dimension of the 3DLC. If a normalized clinical test/protocol is performed outside of a laboratory then this is termed a *less-defined* setting. The rest of our everyday activity, which is possibly performed without the intention of capturing medical data, is termed a *undefined setting* in 3DLC.

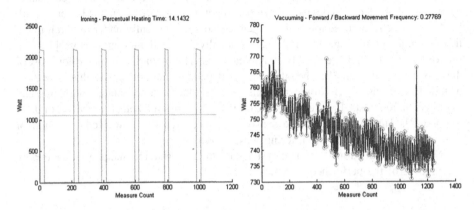

Fig. 3. Intensity feature extraction (cyan) from the power curve of an iron (left) and a vacuum cleaner (right) (Color figure online).

Frequency reflects the occurrence in which one test or dataset is being performed or received. A higher (ideally: *continuous*) frequency is desirable in most situations, to gain a more fine-grained picture of the observed item. However, many tests in the medical domain (e.g. x-ray) can only be performed punctually (discrete).

Relevance to clinical decision making separates the abstraction and importance of data into three layers. The most valuable data to make decisions in the medical domain e.g. for diagnoses is other medical facts in form of *clinical knowledge*. When this knowledge is not sufficient to make a diagnosis the clinician has to perform further tests and is normally interested in the results in form of a trend or some other kind of *pre-processed* data. This information is based on *raw data* that often represents a physical measurement and is typically not directly relevant for the decision making process.

2.2   Experiment Design

The Experiment aimed to show how data from the domestic domain can be useful for medical decisions. Therefore, we wanted to obtain the individual relation between the rehabilitation training at home and two different household activities.

As COPD patients were not available and the experiment is a proof of concept for the applicability of the 3DLC method, it was conducted with healthy test persons. The participants performed two household activities (ironing, see Fig. 2a., and vacuuming, see Fig. 2b) and one step test on a bicycle ergometer (see Fig. 2c) in the home lab of the OFFIS institute. Both household activities were performed for five minutes with low intensity and for five minutes with high intensity. The participants rested for 3 min between the two intensities to recover themselves. The step test on the bicycle ergometer consisted of four steps with a length of 7:30 min each. The starting load was 30 W and increased each step by 40 W, so that the overall length of the training was 30 min with a maximum load of 150 W.

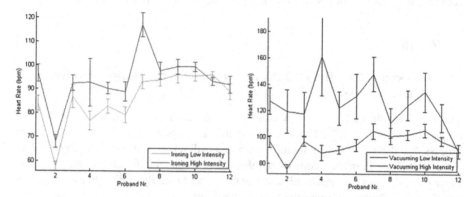

Fig. 4. Heart rate and standard deviation during the performance of household activities per test person (Color figure online).

To perform the tests and to collect the data, three software components were developed: The first component runs on an Android mobile phone (see Fig. 2d) and collects the vital sign measurements during the household activity tests. The second component system was used to create training plans and to monitor the training with the bicycle ergometer (see Fig. 2e). The third component was mainly developed during OSAmI project [5]. It runs on the training device for bicycle ergometer training and controls the load of the device depending on training plan (see Fig. 2f).

All systems have an integrated user management to allow multi-user access and are capable of using several vital sign sensors. For this experiment the Polar Wearlink+ was chosen. A video was recorded during household activities to synchronize the data from the different systems in time before analysis.

2.3   Feature Extraction

We use the individual energy consumption of the iron and the vacuum cleaner to extract features that can be used to determine the intensity of the performed activity.

Figure 3 shows the power curve of the iron (left) and the vacuum cleaner (right) during the use in the experiment. The periods where the iron heats up can be clearly recognized. If a test person performs ironing with higher intensity the iron has to heat up more of the material (e.g. a shirt) that is intended be freed from wrinkles. This results in longer heating periods over the time of the trial. Therefore, we calculated the duration of the heating periods in which the power consumption lied above a certain threshold (cyan line in left in Fig. 3).

The power consumption of the vacuum cleaner changes, with the load of its motor, which depends on how much air is drawn into the opening. This amount differs during the forward and backward movements of the suction head over the floor. So, the flickering of the power curve of the vacuum cleaner at the right of Fig. 3 reflects these movements, which was validated with help of the video that was taken during the experiment. We determine the frequency of the forward and backward movements by counting the peaks of the power curve (cyan circles on the right of Fig. 3).

3 Results

3.1 Inclusion of Data from the Domestic Domain in the Medical Decision Making Process

The target of physicians in our COPD example scenario is to make a decision or prognosis (see top of Fig. 1) for a patient that is as good as it can be. The typical process for the COPD rehabilitation with three different stations was described in Sect. 1.1. These domains (*clinical rehabilitation, home rehabilitation, and unsupervised rehabilitation*) are reflected in the 3DLC model (see three blue framed boxes in Fig. 1) where they span along the three dimensions.

Clinical rehabilitation is the most defined setting, where a patient is strongly supervised and external influences are avoided as much as possible during the data acquisition. The frequency is very low because the patient cannot perform tests in the clinic more than once or twice a year due to the effort that this would take from her/him and the medical staff and also for cost reasons. The *home rehabiliation* can (and should) be perfomed with a higher frequency, but the setting is less defined than in the clinic. The patient performs a normalized training that is defined by a clinician and may also be supervised. *Unsupervised rehabilitation* takes place in the patient's everyday life and reflects her/his normal behaviour, which can be a very active or passive lifestyle. It is clear that these activities are being performed with high frequency.

The contents of the dark blue box in the lower left of Fig. 1 display the raw data that is being obtained during the physical exercise tolerance test, when that functional capacity of the patient is estimated. Typically this is done by a stepwise increase of the training load e.g. on a bicycle ergometer. Parts of the data that can be recorded during this test are the *heart rate* during the test, *load* of the training modality and duration of the different test-steps. Further rather complex sensors like *electrocardiography* and *spirometry* may be used in the clinical setting. The spirometry data is very important in case of COPD because it reflects the oxygen consumption (VO2) under different loads, which is different for each patient. The VO2 value can used to precisely compute energy expenditure during physical activities. The *Borg value* [24] is provided by the patient and expresses the individual perceived exertion of a physical activity. Physicians use VO2 and Borg to estimate the *individual capacity* (lower and centre left in Fig. 1) and finally create an individual training plan for one patient.

The partially normalized home rehabilitation uses such a training plan to perform training in the domestic environment by using a device (also typically a bicycle ergometer) for telerehabilitation. Typically such a system provides a subset of the data that is also measured during the physical exercise test (see bottom in centre frame in Fig. 1), but without the complicated and expensive sensors like the ones used for the spirometry. Load and duration are known for the specific activity of ergometer training.

This information can be used to calculate the so called *MET minutes* (see bottom in centre and middle frame in Fig. 1). MET stands for Metabolic Equivalent of Task which is a measure to express the energy cost of physical activities. It is based on the oxygen consumption of the muscles and expresses the energy consumption as a factor of the mean resting metabolic rate for a specific activity. This data can be used over a longer time period and a number of trainings by an appropriate IT system like a PHR to calculate a *home rehabilitation training performance trend* for a patient.

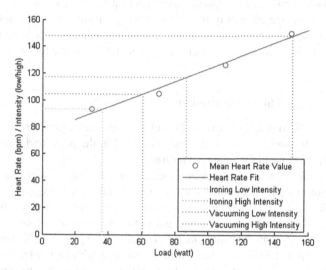

Fig. 5. Relation of the household activities ironing and vacuuming with low and high intensity to the fitted load values of a step test on an bicycle ergometer.

The third domain summarizes unstructured activity in an undefined (domestic) setting, where unsupervised rehabilitation takes place. Currently, the typical method to estimate activities in this domain is a patient diary, where performed activities are documented. We use an *electrical power consumption* sensor to detect performed activities, their duration and their intensity at home (bottom in right frame in Fig. 1). The corresponding MET values can be looked up in a catalogue and can then be used to estimate the energy expenditure and for the calculation of the *home activity performance* of a patient (centre in right frame in Fig. 1). This estimation of the energy expenditure for household activities only with the activity and duration is imprecise (dashed line right in Fig. 1) because the MET concept does not take any individual or physical parameters into account, except of gender and weight.

The energy expenditure estimation of rehabilitation training and household activities can be improved by including the individual VO2 value of the formerly obtained exercise tolerance test. The two trends are being combined to an *overall performed activity trend* that is of high relevance for the clinical decision making process in which a physician estimates the health state of a patient. This trend can be compared with the former defined individual capacity to make better diagnoses and prognoses, e.g. to predict exacerbations before they occur.

The estimation of the energy expenditure for the detected household activities can be further improved by not only including the MET minutes and individual VO2 but also the intensity of the performed activity.

In our proof of concept experiment, we calculate two individual correlations between the recorded heart rate (HR) and the energy expenditure during the rehabilitation training and during the household activities (bottom two red arrows in Fig. 2). This enables us to relate the detected activity with their duration and intensity to the telerehabilitation training. Since HR reflects the impact of an activity on the metabolism, it can be used to estimate which specific amount of household activity substitutes one complete rehabilitation training session. In other words, the correlation with the rehabilitation training over HR can be used to estimate the energy expenditure of household activities.

3.2 Energy Expenditure Determination

12 healthy test persons aged between 27 and 39 years participated in the experiment that took place between August and September 2012 in the home and assessment labs of the OFFIS Institute in Oldenburg, Germany.

Figure 4 shows the standard deviation and mean HR of the test persons during the performance of the household activities ironing (left) and vacuuming (right). Except for one trial, HR was lower when the test persons were instructed to perform an activity with low intensity. Compared to the low intensities, the overall HR was 9.4 % higher during ironing with high intensity (mean low int. 86.0 bpm $\pm$ 11.2, mean high int. 94.1 bpm $\pm$ 11.0) and 31.2 % higher during vacuuming with high intensity (mean low int. 95.2 $\pm$ 8.2, mean high int. 124.9 $\pm$ 18.2). HR rises during the step test with each increment of the load in each of the four steps (mean HR in bpm of step 1: 97.19 $\pm$ 7.1, step 2: 109.44 $\pm$ 8.2, step 3: 126.1 $\pm$ 13.4, step 4: 144.37 $\pm$ 16.4) (Fig. 5).

To determine the intensity of ironing, the total heating time of the iron was extracted as feature from the power curve. Except for two trials, the heating time was lower when test persons were instructed to iron with low intensity. The iron heated in mean 16.02 % of the time during trials with low intensity and 16.95 % of the time when intensity was high, which corresponds to a difference between low and high of 6.53 % in heating time.

To determine the intensity of vacuuming we extracted the frequency of forward/backward movements from the power curve. Except for one trial, the frequency was lower during trials that should be performed with low intensity. The frequency during trials with low intensity was 0.40 $\pm$ 0.069 movements per second and for high intensity 0.65 $\pm$ 0.124 movements per second, which corresponds to a 62.5 % higher frequency.

The target of the next step was to calculate the energy (E) for household activities. Therefore, we first calculated for each household activity ($h = \{ironing, vacuuming\}$) the linear correlation between the detected intensity ($I_h(E) = \{low, high\}$) and the recorded heart rate (HR_h). This results in a simple linear model that enables us to calculate the heart rate for an activity and intensity:

$$R_h(I_h(E)) = HR_h = c_h + i * f_K^h \tag{1}$$

The same procedure was applied to the four intensities ($l = \{30,70,110,150\}$) of the step test on the bicycle ergometer:

$$R_e(l) = HR_e = c_e + l * f_e \tag{2}$$

These formulas are then solved for l:

$$R_h(I_h(S)) = R_e(l) \tag{3}$$

$$l = (c_h + i * f_h - c_e)/f_e$$

The principle of this linkage over the HR is shown in Fig. 4, where the household activities have been set in relation with the bicycle ergometer load for one patient. With usage of the measured power data we can now estimate the corresponding load values for an activity and intensity. Ironing with low intensity corresponds to 36.39 W (green dashed lines), ironing with high intensity to 86.58 W (magenta dashed lines), vacuuming with low intensity to 60.36 W (blue dashed lines) and vacuuming with high intensity to 150.8 W (black dashed lines).

4 Discussion

Since the 3DLC model characterizes data on a very abstract level, it is usable when new applications or technical improvements take place and data from different domains have to be merged. However, the model has to be used with concrete examples to show its worth. Our proof of concept experiment was conducted without real COPD patients, but is a detailed blueprint that shows how activity data from the domestic domain could be included in the clinical decision making process to improve medical diagnoses.

The experiment results show that the heart rate correlates with the given instructions about the intensity with which an activity should be performed. This intensity is also reflected by the features that were extracted from the power sensor. It can be said that the difference in HR and feature values is expressed stronger during vacuuming in comparison to ironing. This can be explained with the kinetics of the movements, which demands or allows the use of the whole body during vacuuming, in comparison to ironing, where only the upper body is used. The smaller difference during the measured heating time of ironing reflects the smaller difference in HR. Even if the detection is not perfect, the data shows that the intensity (high/low) in which a household activity has been performed can be robustly detected by usage of an unobtrusive ambient power sensor.

The mapping between the intensities of household activities and the bicycle ergometer over heart rate is to our knowledge the first attempt to bring these different modalities together. The usage of a simple fit with one parameter as a model for energy expenditure is not sufficient for all practical needs. For example, the model would predict negative values when it extrapolates the energy expenditure under certain circumstances. The most important factor for a precise prediction of the energy expenditure is the individual oxygen consumption of a patient. To determine this value the patient has to perform a load test with a cost intensive breath by breath gas analysis. Hence, we are currently working on a more complex model that reflects the physiology of the human body in greater detail. It takes the individual VO_2 consumption and also environmental factors like temperature into account and should, thereby, enable a more precise prediction.

5 Conclusions

The use of the 3DLC model for the case study of an enhanced energy expenditure determination for COPD patients shows a way how data from the domestic domain could be used to improve the clinical decision making process. We substantiated this abstract path with an experiment that was conducted to create an intensity relation between the telerehabilitation training on a bicycle ergometer and the household activities ironing and vacuuming. We showed that intensities of the activities can be distinguished simply from the power consumption of electrical devices that are used during the performance of such an activity. We extracted heating time for ironing and the frequency of forward/backwards movements for vacuuming as features from the power curves. These features proved to be sufficient measures to distinguish between two intensities in which the activities were performed. Finally we used them for a correlation with the ergometer training to estimate the energy expenditure for house-hold activities with an ambient power sensor.

Acknowledgements. This work was funded in part by the German Ministry of Education and Research within the research project "OSAMI-D" (grant 01 IS 08003) and in part by the Ministry for Science and Culture of Lower Saxony within the Research Network "Design of Environments for Ageing" (grant VWZN 2701).

References

1. Bousquet, J., Khaltaev, N.: Global surveillance, prevention and control of chronic respiratory diseases - a comprehensive approach. In: Bousquet, J., Khaltaev, N. (eds.) Word Health Organization. World Health Organization, Geneva (2007)
2. Abholz, H.H., Gillissen, A., Magnussen, H., Schott, G., Schultz, K., Ukena, D., Worth, H.: Nationale Versorgungsleitlinie - Chronisch Obstruktive Lungenerkrankung – COPD, German, February 2010. http://www.copd.versorgungsleitlinien.de/
3. Rodriguez-Roisin, R., Vestbo, J.: Global strategy for the diagnosis, management, and prevention of chronic obstructive pulmonary disease, Report published by the Global Initiative for Chronic Obstructive Lung Disease, Inc., February 2011. http://www.goldcopd. org/uploads/users/files/GOLD_Report_2011_Feb21.pdf
4. Busch, C., Baumbach, C., Willemsen, D., Nee, O., Gorath, T., Hein, A., Scheffold, T.: engSupervised training with wireless monitoring of ecg, blood pressure and oxygen-saturation in cardiac patients. engJ Telemed Telecare **15**(3), 112–114 (2009). doi:10.1258/ jtt.2009.003002

5. Lipprandt, M., Eichelberg, M., Thronicke, W., Kruger, J., Druke, I., Willemsen, D., Busch, C., Fiehe, C., Zeeb, E., Hein, A.: OSAMI-D: an open service platform for healthcare monitoring applications. In: Proceedings of the 2nd Conference on Human System Interactions HSI '09, Catania, pp. 139–145 (2009)

6. Grams, L., Tegtbur, U., Kueck, M., Guetzlaff, E., Marschollek, M., Kerling, A.: Energieumsatzmessungen unter kontrollierten Bedingungen - Vergleich von Accelerometer, Multisensorsystem und mobiler Spiroergometrie. Deutsche Zeitschrift fuer Sportmedizin **6**, 160–165 (2011)

7. Pauker, S.G., Kassirer, J.P.: The threshold approach to clinical decision making. New Engl. J. Med. **302**(20), 1109–1117 (1980)

8. Kuperman, G.J., Bobb, A., Payne, T.H., Avery, A.J., Gandhi, T.K., Burns, G., Classen, D. C., Bates, D.W.: Medication-related clinical decision support in computerized provider order entry systems: a review. J. Am. Med. Inf. Assoc. **14**(1), 29–40 (2007). http://www.nejm.org/doi/full/10.1056/NEJM198005153022003

9. Williams, G.W.: engThe other side of clinical trial monitoring; assuring data quality and procedural adherence. engJ. Soc. Clin. Trials **3**(6), 530–537 (2006). doi:10.1177/174077450 6073104

10. Carson, E., Cramp, D., Morgan, A., Roudsari, A.: Clinical decision support, systems methodology, and telemedicine: their role in the management of chronic disease. IEEE Trans. Inf. Technol. Biomed. **2**(2), 80–88 (1998)

11. Haeyrinen, K., Saranto, K., Nykaenen, P.: Definition, structure, content, use and impacts of electronic health records: a review of the research literature. Int. J. Med. Inf. **77**(5), 291–304 (2008). http://www.sciencedirect.com/science/article/pii/S1386505607001682

12. Tang, P.C., Ash, J.S., Bates, D.W., Overhage, J.M., Sands, D.Z.: Personal health records: definitions benefits and strategies for overcoming barriers to adoption. J. Am. Med. Inf. Assoc. JAMIA **13**(2), 121–126 (2006). doi:10.1197/jamia.M2025

13. Mattila, J., Ding, H., Mattila, E., Sarela, A.: Mobile tools for home-based cardiac rehabilitation based on heart rate and movement activity analysis. In: Conference Proceedings of IEEE Engineering in Medicine and Biology Society, vol. 2009, pp. 6448–6452 (2009)

14. Chen, Y.-P., Yang, J.-Y., Liou, S.-N., Lee, G.-Y., Wang, J.-S.: Online classifier construction algorithm for human activity detection using a tri-axial accelerometer. Appl. Math. Comput. **205**(2), 849–860 (2008). Special Issue on Advanced Intelligent Computing Theory and Methodology in Applied Mathematics and Computation. http://www.sciencedirect.com/science/article/pii/S0096300308003640

15. Bauldoff, G.S., Ryan-Wenger, N.A., Diaz, P.T.: Wrist actigraphy validation of exercise movement in copd. W. J. Nurs. Res. **29**(7), 789–802 (2007). http://wjn.sagepub.com/content/29/7/789.abstract

16. Scanaill, C.N., Carew, S., Barralon, P., Noury, N., Lyons, D., Lyons, G.M.: engA review of approaches to mobility telemonitoring of the elderly in their living environment. engAnn. Biomed. Eng. **34**(4), 547–563 (2006). doi:10.1007/s10439-005-9068-2

17. Barger, T.S., Brown, D.E., Alwan, M.: Health-status monitoring through analysis of behavioral patterns. IEEE Trans. Syst. Man Cybern. Part A Syst. Hum. **35**(1), 22–27 (2005)

18. Virone, G., Noury, N., Demongeot, J.: A system for automatic measurement of circadian activity deviations in telemedicine. Trans. Biomed. Eng. **49**(12), 1463–1469 (2002)

19. Virone, G., Alwan, M., Dalal, S., Kell, S.W., Turner, B., Stankovic, J.A., Felder, R.: Behavioral patterns of older adults in assisted living. Trans. Inf. Technol. Biomed. **12**(3), 387–398 (2008)

20. Chen, J., Zhang, J., Kam, A.H., Shue, L.: An automatic acoustic bathroom monitoring system. In: Proceedings of the IEEE International Symposium Circuits and Systems ISCAS 2005, pp. 1750–1753 (2005)
21. Monekosso, D.N., Remagnino, P.: Monitoring behavior with an array of sensors. Comput. Intell. **23**(4), 420–438 (2007). doi:10.1111/j.1467-8640.2007.00314.x
22. Frenken, T., Wilken, O., Hein, A.: Technical approaches to unobtrusive geriatric assessments in domestic environments. In: Proceedings of the 5th Workshop on Behaviour Monitoring and Interpretation, BMI'10, Karlsruhe, Germany, 21 September 2010
23. Helmer, A., Lipprandt, M., Frenken, T., Eichelberg, M., Hein, A.: 3DLC: a comprehensive model for personal health records supporting new types of medical applications. J. Healthc. Eng. **2**, 321–336 (2011). ISSN 2030-2295. doi:10.1260/2040-2295.2.3.321
24. Borg, G.: Perceived exertion as an indicator of somatic stress. Scand. J. Rehabil. Med. **2**(2), 92–98 (1970)

Supporting Contextualisation of ABAC Attributes Through a Generic XACML Request Handling Mechanism

Brecht Claerhout[1], Kristof De Schepper[1(✉)],
David Perez-Rey[2], and Anca Bucur[3]

[1] Custodix, Kortrijksesteenweg 214, 9830 Sint-Martens-Latem, Belgium
{brecht.claerhout,kristof.deschepper}@custodix.com
[2] Grupo de Informática Biomédica, Dept. Inteligencia Artificial,
Facultad de Informática, Universidad Politécnica de Madrid,
Campus de Montegancedo S/N, 28660 Boadilla del Monte, Madrid, Spain
dperez@infomed.dia.fi.upm.es
[3] Phillips Research, Healthcare Information Management,
High Tech Campus 34, 5656 AE Eindhoven, The Netherlands
anca.bucur@phillips.com

Abstract. When building cross-organisation data sharing environments in the clinical domain, one is confronted with high security demands. Although at the present time, a broad range of security technology is available, typically not all desired functionality can be easily met. One of these requirements is managing access over dynamically instantiated contexts in collaborative environments. This requirement was encountered during the EU funded projects INTEGRATE and EURECA. This paper presents a solution which enriches XACML with context awareness without changing the policy language itself. Furthermore, it is shown that the presented mechanism (XACML request modification) can also be used for uniformly addressing other security challenges.

Keywords: Contextual attributes · Contextual roles · ABAC · XACML · Context aware policies · Extension mechanism · INTEGRATE project · EURECA project · Collaborative environments

1 Introduction

As ICT technology is advancing and the volume of available clinical data is accumulating, new ways to put this information into use for advancing medical science are devised. Examples include scenarios aiming to re-use EHR data for improving clinical trial processes (e.g. for patient recruitment, trial protocol feasibility studies [1]) and various practical applications in data mining (e.g. for building clinical decision support systems or disease models [2]).

All these new applications have in common that they rely on increased exchange of (sensitive) information across organizations and consequently also across country borders. It goes without saying that this evolution introduces new challenges in data protection. Compliance with data protection regulation becomes increasingly difficult

© Springer-Verlag Berlin Heidelberg 2014
M. Fernández-Chimeno et al. (Eds.): BIOSTEC 2013, CCIS 452, pp. 367–377, 2014.
DOI: 10.1007/978-3-662-44485-6_25

as these solutions scale up. Governance (both on an administrative and technical level) becomes a major issue.

At the technical level this means that data protection policies which have been defined at the administrative level are to be uniformly implemented (enforced) over all involved IT systems. One need that thus clearly arises is the uniform management of access policies in heterogeneous distributed environments.

The work presented in this paper has been performed in the context of implementing such a uniform management framework for two EU research projects: INTEGRATE [3] and EURECA [4]. INTEGRATE focusses on building various tools (a.o. a patient screening tool, a cohort selection tool, clinical data analysis tool) for the Breast International Group (BIG), a large international consortium of breast cancer treatment hospitals and research sites [5]. The EURECA project is dealing with a wide range of novel applications based on secondary use of EHR data [6]. Both projects are building IT solutions for unlocking large amounts of clinical data in a distributed environment. The projects share the same architectural approach (loosely coupled, service oriented architecture) and part of the implementation, including the security framework which is conceived as a generic identity and access management for distributed environments treating sensitive personal data.

This paper proposes a solution to the specific issue of contextualization of access control policies and illustrates a generic mechanism useful for implementing several security features with XACML.

2   XACML

Attribute-Based Access Control (ABAC) presents an access control model inherently capable of meeting many of the 'modern' access control demands (e.g. data dependent access policies, environment dependent policies, etc.). In ABAC, attributes that are associated with a user, action or resource serve as inputs to the decision of whether a given user may access a given resource in a particular way.

The eXtensible Access Control Markup Language (XACML) [7] implements ABAC. It is an XML based declarative access control policy language defining both a policy, decision request and decision response language. The work presented in this paper (and as such the examples) is based on XACML 2.0 which has been an OASIS standard since 2005. Recently (January 2013) XACML 3.0 was approved as a standard, however the mechanisms explained in this paper remain relevant.

3   Contextualization of Attributes

3.1   Introduction

XACML provides a rich feature set for defining and processing access control policies out of the box and as such is a good technical choice for environments faced with high security demands. However, not all functionality encountered in such environments can be easily implemented with XACML.

One of the functionalities sought after in the mentioned projects is the ability to define a uniform policy - uniform in a 'context' - which is applicable in different instances of that context. To clarify this, an example is given from the INTEGRATE project. In this project, tools are shared among researchers working on different clinical trials. One wants to be able to define a single uniform policy over all trials (trial context), which should be properly evaluated in each specific trial (context instance). Such a general policy would for example state that an investigator in a trial should be able to access all data from his own patients within the trial and that a trial chairman should be allowed to see all data objects in a trial.

One approach to implement this form of contextualization in XACML would be to shift the responsibilities to the policy authoring tools. This basically means that for each context instance separate policy files would be generated (e.g. from templates describing the default context policies). This approach suffers from the inherent issues associated with all 'auto-generating' solutions (be it for configurations or source code). Every change requires many policies to be rewritten (regenerated) and possibly redistributed. Furthermore, synchronization becomes an issue when one wants to allow exceptions in auto-generated policies ('manual' additions). For these reasons, this solution is not the most favorable for large scale environments.

Another option is to demand that PEPs make separate requests for each context. Putting this burden on the individual PEPs creates multiple points of failure (and hampers auditability).

Instead, in the presented approach the responsibility of ensuring that requests are restricted to one context has been offloaded into a separate component. This 'request handler' processes incoming XACML multi-context requests and splits them into separate requests which can be handled by any standard XACML PDP.

Architecturally speaking this component is part of the 'context handler' as defined by the OASIS XACML specification (which functionality is out of scope of the XA-CML specification). The component was implemented as a transparent XACML request handler, which can be pipelined together with other similar XACML requests handlers (dealing with other functionality) and put in front of a generic XACML PDP. Section 4 illustrates the implementation of this pipeline into an XACML based authorization service.

3.2   Handler Operation

The contextualization extension handler proposed splits up an incoming XACML request (containing possible multiple contexts and context instances in the attributes) in multiple single-context requests. More specifically, for each different context instance that is included in the original XACML request, a separate new context-specific XA-CML request is generated. These context-specific requests are sent separately to the PDP engine for evaluation. By splitting up the requests, the PDP engine can provide an access decision for each context instance using manageable context-specific policies. The XACML responses containing the context-specific access decisions are sent back to the contextualization extension handler where a global (no context specific) XACML request is generated. This request is then sent back to the PDP engine which evaluates the request, making use of defined global policies. The access decision is finally returned to the extension handler and is passed back to the access control requestor as final outcome of the original request.

Incoming XACML Request. To contextualize attributes of a subject and mark the resources he/she wishes to access that are context-specific, a special attribute syntax is defined in the incoming XACML request that is recognizable by the handler. The pseudocode below shows how a contextual role attribute of a user is included in a request as an XACML subject attribute with attribute id "role" and a special formatted attribute string value containing the role, the context and the context instance:

```
<att id="role">
[context role]@[context]:[context instance]
</att>
```

For including the global context roles of a user in the request, this special formatting is not needed, because these roles are relevant for each of the contexts. In this case, the attribute value is only the role itself.

To illustrate the inclusion of contextual roles in more detail, the example given in Sect. 3.1 is further elaborated. Consider a user called "John Doe" whom has different roles within different trials. He has a global role of "clinical staff" (i.e. in the global context), the role of "investigator" in trial A (context trial A) and the role of "principal investigator" in another trial B (context trial B). These three roles of John are provided as subject attributes:

```
<Request>
  <Subject>
    <att id="name">John Doe</att>
    <att id="role">investigator@trial:A</att>
    <att id="role">principal investigator@trial:B</att>
    <att id="role">clinical staff</att>
  </Subject>
  ...
</Request>
```

To indicate that a resource is context-specific, an extra predefined attribute is included in the corresponding XACML resource object with the attribute id "contextInstance" and a special formatted attribute string value containing a reference to the context and context instance:

```
<att id="contextInstance">
  [context]:[context instance]
</att>
```

Note that a resource can belong to more than one instance of the same context. In the case where John Doe wants to access three resources named EHR001 in the trial context of trial A and trial B (multiple context instances), EHR002 in the trial context of trial B and EHR003 in the global context, the resource attributes in the request would look as follows (in pseudocode):

```
. . .
<Resource>
  <att id="resID">EHR001</att>
  <att id="contextInstance">trial:A</att>
  <att id="contextInstance">trial:B</att>
  <att id="type">crf</att>
</Resource>
<Resource>
  <att id="resID">EHR002</att>
  <att id="contextInstance">trial:B</att>
  <att id="type">adm</att>
</Resource>
<Resource>
  <att id="resID">EHR003</att>
  <att id="type">doc</att>
</Resource>
. . .
```

Context-Specific Requests. When a new XACML request enters the contextualization handler, an inventory is made of all context and context instance included in this request. For each different context instance found, a new context-specific XACML request is generated.

For the context-specific role attributes in the example, this means that the role attribute of the original request is copied to the corresponding context-specific request. During this copying the context instance part is stripped from the role attribute value (this enables the generation of context-specific policies, see further). Non context-specific attributes are copied to each of the requests (so that they are available during contextual policy evaluation in each of the context instances).

Each context-specific resource object and the underlying resource attributes defined in the incoming request are copied to one or more of the corresponding context-specific requests, depending on the context instance(s) that is included in the context attribute value(s) of this resource (a resource can have more than one context/context instance). The context resource attribute(s) itself is omitted during the copying, because it is not relevant for evaluation. Non context-specific resources are not copied to a request, these will be used later in the global request (see further).

Finally in each generated context-specific request the corresponding context and context instance are added as environment attributes. This allows easy definition of policies concerning certain contexts or context instances. Other environment and action attributes in the original request are again copied to all context-specific requests.

Figure 1 illustrates the result of splitting the incoming request attributes of the given examples.

Context-Specific Policies. After the request is split in multiple context-specific requests, these requests are sent to the PDP engine for evaluation. Because of this split up, the management and creation of context-aware policies is simplified. Generating policies for a context and/or context instance can be easily done in various ways, e.g. through the XACML "target" specification. The XACML pseudocode below shows how to target a trial context (targeting all trial instances):

Split request @ trail A   **Split request @ trail B**

```
...                                    ...
<Subject>                              <Subject>
<att id="name">John Doe</att>          <att id="name">John Doe</att>
<att id="role">investigator@trial</att> <att id="role">principalinvestigator@trial</att>
<att id="role">clinical staff</att>    <att id="role">clinical staff</att>
</Subject>                             </Subject>
...                                    ...
<Res>                                  <Res>
<att id="resID">EHR001</att>           <att id="resID">EHR001</att>
<att id="type">crf</att>               <att id="type">crf</att>
</Res>                                  </Res>
...                                    <Res>
<Env>                                  <att id="resID">EHR002</att>
<att id="context">trial</att>          <att id="type">adm</att>
<att id="contextInstance">trial:A</att> </Res>
</Env>                                  ...
                                       <Env>
                                       <att id="context">trial</att>
                                       <att id="contextInstance">trial:B</att>
```

Fig. 1. Splitted context-specific XACML requests (pseudo-code).

```
<Target>
  <Environments><Environment>
    <EnvMatch MatchId="string-equal">
      <AttValue>trial</AttValue>
      <EnvAttDesignator AttId="context"/>
    </EnvMatch>
  </Environment></Environments>
</Target>
```

The following XACML pseudocode illustrates how to write a policy that targets a context instance trial A:

```
<Target>
  <Environments><Environment>
    <EnvMatch MatchId="string-equal">
      <AttValue>trial:A</AttValue>
      <EnvAttDesignator AttId="contextInstance"/>
    </EnvMatch>
  </Environment></Environments>
</Target>
```

The next XACML policy excerpt illustrates how a policy (or rule) could be written dealing with the access rights tied to a specific contextual attribute across instances (the role from the example in this case).

```
<Target>
  <Subjects><Subject>
    <SubjectMatch MatchId="string-equal">
      <AttValue>principalinvestigator@trial</AttValue>
      <SubjAttDesignator AttId="role"/>
    </SubjectMatch>
  </Subject></Subjects>
</Target>
```

Note that the attribute annotation "@[context]" (in the example, the "trail" context) facilitates policy authoring and review. It indicates that this is a contextual attribute and will thus be evaluated on a "per instance basis".

The examples illustrate that the presented mechanism for enhancing XACML with contextual attributes allows policies to be written in structured and manageable way.

Global Request. The access control decisions for the different context-specific requests are sent back to the contextual extension handler. In the handler a new XACML request (global request) is generated based on the original access control request. This global request additionally contains the contextual access decision results for each resource.

To include the access results of the context-specific resources in this global request, one or more resource attributes "contextResult" are added for each resource. The attribute value has a special formatted string:

```
<att id="contextResult">
  [Resource AC Decision]@[context]
</att>
```

The following shows the generated request for the above example; this request is sent again to the PDP for evaluation:

```
...
<Subject>
  <att id="name">John Doe</att>
  <att id="role">clinical staff</att>
</Subject>
...
<Res>
  <att id="resID">EHR001</att>
  <att id="contextResult">permit@trial</att>
  <att id="contextResult">deny@trial</att>
  <att id="type">crf</att>
</Res>
<Res>
  <att id="resID">EHR002</att>
  <att id="contextResult">permit@trial</att>
  <att id="type">adm</att>
</Res>
<Res>
  <att id="resID">EHR003</att>
  <att id="type">doc</att>
</Res>
...
</Env>
...
```

This second evaluation, which includes the "contextResult" responses, allows "global" policies to be written that rely on the result(s) of the context-specific resources or override them (i.e. there is total freedom in combination algorithm). The following shows how a rule could incorporate such a contextual result:

```
. . .
<Rule Effect="Permit">
    <Condition>
        <Apply FunctionId="any-of">
            <Apply FunctionId="string-equal"/>
            <AttValue>permit@trial</AttValue>
            <SubjAttDesignator
AttId="contextResult"/>
        </Apply>
    </Condition>
</Rule>
. . .
```

Note that in the special case where a resource belongs to more than one context instance of the same context, both a "permit" and "deny" for the same context could be present. Although this seems contradictory at first, it allows decision combination to be specified in the XACML policies and thus gives total freedom to policy writers.

Finally, the resulting XACML response, containing the access result for each resource is sent back to the contextualization extension handler. The handler in its turn will return this response as outcome to "calling" handler.

4   Pipelined Request Handlers

4.1   Authorization Server

As explained in the previous section, the contextualization of XACML policies was implemented though a request handler mechanism (modifying XACML requests) rather than by putting constraints on the policy constructs or introducing specific demands for the PEPs when constructing access request. The request handler itself was integrated into a simple authorization service. Figure 2 shows its high level architecture.

In this service, request handlers can be loaded into a pipeline which sits between the authorization service endpoint and the (internal) XACML Policy Decision Point (PDP). The service endpoint takes care of communication and authorization of incoming (authorization) requests.

When an access request is made towards the service, it enters the pipeline of handlers. Each handler gets as input an XACML request from the service endpoint or another handler; subsequently transforms this request; and finally feeds the transformed request to the next handler or the PDP (last stage in the pipeline). Access control responses follow the same route back.

Each stage in the pipeline is responsible for dealing with its own XACML enhancement and should ensure transparency of operation. The final stage in the pipeline is a standard XACML policy decision engine.

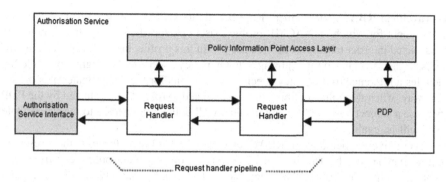

Fig. 2. High level architecture request handler pipeline.

4.2 Attribute Mapping/Modification

It is clear from the previous that handlers are ideal for modifying or complementing security attributes in XACML access control requests. Two such extension handlers have been built: one for translate security attributes between security domains, another one as an alternative for implementing hierarchical Role Based Access Control (RBAC).

Security Domain Mapping. The first handler is used to translate security attributes between different security domains. It has been used in a distributed environment where different applications in different organization are required to comply with an overall access policy (defining the minimum access restriction required for achieving compliance). In order to enforce the overarching access policy, a central authorization service ('master PDP') was introduced in the architecture.

An access request in an application thus requires requests to be made both at a local decision point (where locally defined policies are loaded) and the master PDP. As can be expected in a cross-organizational environment, different organizations use a different security vocabulary (e.g. identity attributes with different roles). Rather than to enforce each of the applications to create a separate request for the master PDP in the vocabulary used in the overarching governance domain, an attribute translation handler was introduced.

That way, applications can just forward their access requests to the master PDP in their 'local' vocabulary. The attribute translation handler, active at the central authorization service, would automatically modify all attributes in the local vocabulary to the attribute vocabulary used by the master PDP. This modification encompasses removal, translation and introduction of new attributes (through inference).

In the presented implementation (re-implementation of [8] as a pipelined component), the mapping between the different security domains and the central one was ontology-based, allowing for an easy way to model complex mappings. Such mappings typically include relationships like for example classification of data-types of local applications to (one or more) confidentiality classes know in the general domain. The centralized translation allowed for convenient management of the security vocabulary mappings (separate from the authorization infrastructure implementation) and facilitated auditing.

Hierarchical RBAC. Hierarchical Role Based Access Control (RBAC) is a common requirement for which an XACML profile exists that specifies how policies should be constructed in order to support this feature [9]. In this profile, hierarchical evaluation of access rights is achieved by requiring that any policy specifying the rights for a role 'A' includes a reference to the policy specifying the rights for the subordinate roles of 'A'. This way, when the policy defining the rights linked to role 'A' is evaluated by the PDP (because a subject has role 'A'), automatically all policies connected to the subordinate roles will be evaluated.

Alternatively, hierarchical RBAC can also be enforced through the presented request handler mechanism instead of by relying on this fixed policy construct (referencing to policies of subordinate roles). The handler simply needs to expand incoming hierarchical role attributes, so that all subordinate roles are explicitly added to the list of attributes in the access request. This also ensures that all relevant policies are evaluated, but shifts management of the hierarchical relationship away from the XACML policy definitions (no hierarchical links need to be maintained between the access right policies).

As a final remark, it should be mentioned that the latter two handlers follow the principle of supplying all required attributes for policy evaluation up-front to the PDP. In both cases, the functionality could also have been implemented through callbacks (to a Policy Information Point) whenever the PDP encounters unknown attributes. In the presented use cases this would however bring no advantage (on the contrary).

5   Summary

The presented work has been performed in the context of building IT security solutions for sharing large amounts of clinical data in a distributed environment. This paper discussed how the specific problem of contextualization of ABAC attributes can be tacked when working with XACML.

A mechanism of pipelined access control request handers was introduced to extend the XACML functionality without touching the standard XACML core specification. The contextualization request handler presented splits up a request containing different contexts to multiple context-specific requests. After evaluation of these requests, a global request is created which in his turn is evaluated by the PDP. The advantage of this extension is that policies can be written specific for one context without making these policies very complex.

Finally, it was shown how the pipelined request hander approach also can help solve other issues, two examples where given: translation of security attributes between security domains and hierarchical RBAC.

Acknowledgements. This work is partially funded by the European Commission under the 7th Framework Programme (FP7-ICT-2009-6-270253) and (FP7-ICT-20011-7-288048).

References

1. Cuggia, M., Besana, P., Glasspool, D.: Comparing semi-automatic systems for recruitment of patients to clinical trials. I. J. Med. Inf. (IJMI) **80**(6), 371–388 (2011)
2. Rüping, S., Anguita, A., Bucur, A., Cirstea, T.C., Jacobs, B., Torge, A.: Improving the implementation of clinical decision support systems. In: EMBC'13, Osaka, Japan, 3–7 July 2013
3. INTEGRATE: Driving Excellence in Integrative Cancer Research. http://www.fp7-integrate.eu
4. EURECA: Enabling information re-Use by linking clinical REsearch and Care. http://www.eurecaproject.eu/
5. Breast International Group (BIG). http://www.breastinternationalgroup.org/
6. Vdovjak, R., Claerhout, B., Bucur, A.: Bridging the gap between clinical research and care - approaches to semantic interoperability, security & privacy. In: HEALTHINF 2012, pp. 281–286 (2012)
7. eXtensible Access Control Markup Language (XACML), Version 2.0, February 2005. http://docs.oasis-open.org/xacml/2.0/access_control-xacml-2.0-core-spec-os.pdf
8. Ciuciu, I., Claerhout, B., Schilders, L., Meersman, R.: Ontology-based matching of security attributes for personal data access in e-health. In: Meersman, R., et al. (eds.) OTM 2011, Part II. LNCS, vol. 7045, pp. 605–616. Springer, Heidelberg (2011)
9. XACML Profile for Role Based Access Control (ABAC), Draft 01, February 2004. http://docs.oasis-open.org/xacml/cd-xacml-rbac-profile-01.pdf

Modality Preference in Multimodal Interaction for Elderly Persons

Cui Jian[1(✉)], Hui Shi[1], Nadine Sasse[2], Carsten Rachuy[1],
Frank Schafmeister[2], Holger Schmidt[3], and Nicole von Steinbüchel[2]

[1] SFB/TR8 Spatial Cognition, Universität Bremen, Bremen, Germany
{ken, shi, rachuy}@informatik.uni-bremen.de
[2] Medical Psychology and Medical Sociology, University Medical Center
Göttingen, Göttingen, Germany
{n.sasse, frank-schafmeister,
nvsteinbuechel}@med.uni-goettingen.de
[3] Neurology, University Medical Center Göttingen, Göttingen, Germany
h.schmidt@med.uni-goettingen.de

Abstract. This paper is focusing on two important aspects: on the one hand, it presents our work on designing, developing and implementing a multimodal interactive guidance system for elderly persons to be used in autonomous navigation within complex building; on the other hand, it summaries and compares the data of a series of empirical studies that have been conducted to evaluate the effectiveness, efficiency and user satisfaction of the elderly-centered multimodal interactive system regarding different multimodal input-possibilities such as speech, gesture via touch-screen and the combination of both under simulated conditions. The overall positive results validated our systematically developed and empirically improved design guidelines, foundations, models and frameworks for supporting multimodal interaction for elderly persons.

Keywords: Multimodal interaction · Elderly-friendly interface · Dialogue management · Human-computer interaction in AAL · Formal methods · Interactive system evaluation

1 Introduction

There has been a rapidly growing interest in research and development of multimodal interaction over the past few decades ([1, 2]). Specifically, multimodal interaction is showing its importance and necessity for more effective interaction compared to single modal interaction (see e.g. [3, 4]). It also holds the potential of further enhancing users' individual as well as overall performance (see e.g. [5]). Moreover, multimodal interfaces can be used to achieve a more natural human-computer communication and increase the robustness of the interaction with complementary information (see e.g. [6]).

However, the typical multimodal interaction mechanisms are usually only suitable for users with sufficient familiarity with information and communication technology, which poses a particular challenge for people with less knowledge about this kind of

© Springer-Verlag Berlin Heidelberg 2014
M. Fernández-Chimeno et al. (Eds.): BIOSTEC 2013, CCIS 452, pp. 378–393, 2014.
DOI: 10.1007/978-3-662-44485-6_26

interaction, especially for the constantly growing group of elderly persons due to the acceleration of population ageing nowadays in almost all industrialized countries ([7]). Therefore, in order to maximize the advantage of multimodal interaction, special focus has been laid on the research of multimodal interaction with respect to the emerging area Ambient Assisted Living and its potential user group: elderly persons or persons with special needs (see e.g. [8–10]).

Adding to this body of literature, our work is concentrating on multimodal inter-action in AAL context for elderly persons by taking ageing-related characteristics into account. It can be divided into two important aspects: (a) the design, development and implementation of multimodal interaction for elderly persons; and (b) the empirical evaluation of a minutely developed and systematically improved elderly-friendly multimodal modal interactive system with elderly persons.

For (a) two fundamental aspects are proposed for supporting our system design and development: a list of elaborated design guidelines regarding traditional design prin-ciples of conventional interactive systems and the most common elderly-centered characteristics corresponding to ageing-related decline of sensory, perceptual, motor and cognitive abilities of elderly persons ([11]); and a formal language supported unified dialogue modelling and management approach, which combines a finite state based generalized dialogue model and a classic agent based management theory, and therefore can support a flexible and context-sensitive, yet formally tractable and controllable multimodal interaction ([12]). According to the two development foundations, a mul-timodal interactive guidance system was then especially designed and implemented.

For (b) a series of empirical studies were conducted with groups of elderly persons to practically evaluate the multimodal interactive system with respect to its multiple input modalities as well as to enable a continuously improved development based on the data of each empirical study. Specifically, a touch-screen graphical user interface was implemented and tested in a pre-study in [11]; a spoken language interface and its dialogic interaction were tested in [12]; with the test data, the touch-screen interface and the spoken language interface are accordingly improved, tested and compared with each other in a follow-up study in [13]; then the combination of the two modalities are evaluated in the next study and the results are described in [14].

Therefore, in order to perform a detailed comparison of all the input modalities, namely the touch-screen, the spoken language and the combination of both, as well as the assessment of the complete multimodal interactive system concerning its effec-tiveness of task performance, efficiency of interaction and user satisfaction about the system, the data of all the experimental studies were summarized and analyzed, then the results are described and discussed in this paper.

The rest of the paper is structured as follows: Sect. 2 briefly introduces the proposed and improved design and development foundation of our work and presents the min-utely implemented multimodal interactive system for elderly persons; Sect. 3 describes the experimental studies on evaluating the complete interactive system while focusing on comparing the multiple input modalities; Sect. 4 summarizes and analyzes the empirical data and discussed about the results according to an adapted version of a classic evaluation framework for conventional interactive systems; finally, Sect. 5 gives a conclusion of the reported work and outlines the direction of our future research focus.

2 System Design, Development and Implementation

This section first introduces our theoretical foundation for designing and developing multimodal interaction for elderly persons, then according to the design and development foundation a multimodal interactive guidance system for elderly persons is implemented and presented.

2.1 The Foundation of System Design and Development

The theoretical foundation of our work consists of two aspects: a set of guidelines to support the design and development of elderly-friendly multimodal interaction; and a unified dialogue modelling approach with a formal method based framework for dialogue management.

Design Guidelines of Multimodal Interaction for Elderly Persons. During the ageing process, elderly persons often suffer from decline of sensory, perceptual, motor and cognitive abilities, especially for the seven most common human abilities, as shown in Fig. 1.

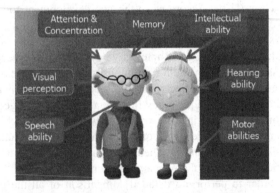

Fig. 1. The seven most common ageing-related human abilities.

Specifically, **visual perception** declines for most people with ageing; physically the size of the visual field is decreasing and the peripheral vision can be lost; It is more difficult to focus on objects up close and to see fine details, including rich colors and complex shapes that make images hard or even impossible to identify; rapidly moving objects are either causing too much distraction, or become less noticeable ([15]); **speech ability** declines while ageing in the way of being less efficient for pronouncing complex words or longer sentences, probably due to reduced motor control of tongue or lips ([16]); [17] also confirmed that, elderly-centered adaptation of speech-enabled interactive components can improve the interaction quality to a satisfactory level; **attention and concentration** drop while ageing, elderly persons either become more easily distracted by details and noise, or find other things harder to notice when concentrating on one thing ([18]); they show great difficulty with situations where divided attention is needed ([19]); **memory functions** decline differently. Short term memory holds fewer items with age and working memory becomes less efficient ([20]). Semantic information is normally preserved in long term memory ([21]); **intellectual ability** does not decline much during the normal ageing process, yet [22] believed that crystallized intelligence can assist elderly persons to perform better in a stable well-known interface environment; **hearing ability** declines to 75 % between the age of 75 and 79 ([23]). High pitched sounds are hard to perceive; complex sentences are difficult to follow ([24]); **motor abilities** decline generally due to loss of physical activities while ageing. Complex fine motor activities are more difficult to perform, e.g. to grab small or irregular targets ([25]); conventional input devices such as a computer mouse are less preferred by elderly persons as good hand-eye coordination is required ([26]).

According to the above empirical findings and much more other research work on effects of ageing using computer based systems (see e.g. [27–29]), it is necessary to consider age-related characteristics while developing interactive systems for elderly persons. Therefore, based on the common design principles for conventional interactive systems and the ageing-related characteristics regarding the seven most common human abilities, a set of guidelines for designing and developing multimodal interactive system for elderly persons was proposed in [11]. These guidelines have been implemented into the first versions of our interactive systems, evaluated by our previous empirical studies with elderly persons, and then accordingly improved on the basis of the evaluation results. The final set of improved design guidelines were summarized in [13] and have been used as the first fundamental aspect of our work ever since, especially for the development of the final version of our multimodal interactive system.

Formal Language Supported Unified Dialogue Modelling and Management. One of the most essential issues of developing an interactive system is the interaction management, i.e., how the interaction flow is controlled in the dialogues between users and the system. In most of the related work on dialogue modelling and management, the following two methods can be deemed as the basic and important ones among others:

- The generalized dialogue models, which can abstract dialogue models by describing discourse patterns as illocutionary acts in the classic recursive transition networks, without reference to any direct surface indicators ([30]);
- The information state update based management theories, which focus on the modelling of discourse context as the attitudinal state of an intelligent agent and show a powerful way to handle dynamic information for a context sensitive dialogue management ([31]).

However, these two methods have their own drawbacks. On the one hand, the generalized dialogue models are based on finite state transition models, which are criticized for their inflexibility of dealing with dynamic information; on the other hand, the information state update models have the problem of controlling their complexity for state manage and model extension.

Therefore, a unified dialogue modelling approach is proposed ([11]), which extends a generalized dialogue model with the information state update based components, such that finite state transitions can only be triggered by fulfilled conditions and followed by updated information state with a set of predefined information state update rules (see the left part of Fig. 2).

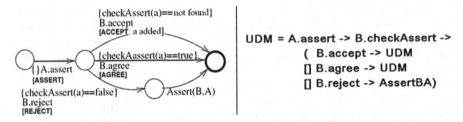

Fig. 2. A simple unified dialogue model with its CSP specification on the illocutionary level.

In order to support the development and implementation of unified dialogue models and their integration into practical interactive systems, the formal dialogue development framework (abbreviated as FormDia) was developed and proposed in [32].

Figure 3 illustrates the structure of the FormDia framework, which consists of six important components according to the development process of a unified dialogue model based dialogue manager:

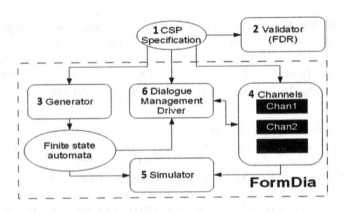

Fig. 3. The structure of the FormDia framework.

1. **CSP Specification:** every unified dialogue model is based on a generalized dialogue model, whose illocutionary structure can therefore be specified as a machine readable Communication Sequential Processes (CSP [33]) program (see an example CSP specification of a simple unified dialogue model in the right part of Fig. 2)

2. **Validator:** the CSP specified program can then be validated by the Failures-Divergence Refinement tool, (FDR [34]), which is a model checking tool for validating and verifying the properties of CSP specifications.
3. **Generator:** according to the validated CSP specification, machine readable finite state automata can be generated by the Generator.
4. **Channels:** with the finite state automata, channels regarding all the generated states can be defined with related information state update rules. These channels are at first black boxes, which will be filled with deterministic behavior of concrete components according to their application contexts.
5. **Simulator:** with the finite state automata and the defined communication channels, dialogue scenarios can be simulated via a graphical interface, which visualizes dialogue states as a directed graph and provides a set of utilities to trigger dialogue events and updating of dialogue states for testing and verification.
6. **Dialogue Management Driver:** after validation and verification, the unified dialogue model can be integrated into a practical interactive dialogue system via the dialogue management driver.

The unified dialogue modelling approach with the formal language based dialogue management framework FormDia is detailed in [11] and serves as the second fundamental aspect of our work. A unified dialogue model for the multimodal interaction for elderly persons was developed. With the FormDia framework, this model is implemented and integrated into our interactive system for elderly persons, then evaluated via empirical studies and improved accordingly (see e.g. [12, 13]).

2.2 System Description

According to the two introduced development foundations, we developed MIGSEP, a general Multimodal Interactive Guidance System for Elderly Persons. MIGSEP runs on a portable touch-screen tablet PC and will serve as the interactive media to be used by an elderly or handicapped person seated in an autonomous electronic wheelchair (see Fig. 4) that can carry its user to desired locations within complex environments autonomously.

Fig. 4. An autonomous electronic wheelchair ([35]) to be interacted via MIGSEP with an elderly or handicapped person.

The Architecture of MIGSEP. The architecture of MIGSEP is illustrated in Fig. 5. The Unified Dialogue Manager, which was developed according to the introduced unified dialogue model and the FormDia framework, functions as the central processing unit of the MIGSEP system and supports a flexible and context-sensible yet formally tractable multimodal interaction management. An Input Manager receives and interprets all incoming messages from the GUI Action Recognizer for GUI input events, the Speech Recognizer for natural language instructions and the Sensing Manager for other possible sensory data. An Output Manager on the other hand, handles all outgoing commands and distributes them to the View Presenter for presenting visual feedbacks, the Speech Synthesizer to generate and utter natural language responses and the Action Actuator to perform necessary motor actions, such as sending a driving command to the autonomous electronic wheelchair. The Knowledge Manager, which is closely connected with the unified Dialogue Manager, uses a Database to keep the static data of certain environments and the Context to process the dynamic information exchanged with the users during the interaction.

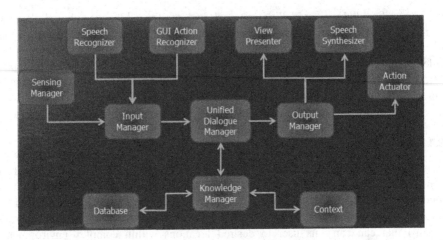

Fig. 5. The architecture of MIGSEP.

The communication between the components of MIGSEP uses a uniform XML-protocol and each component can be treated as an open black box which can be accordingly modified or extended for specific use, without directly affecting other components in the MIGSEP architecture. It provides a general open platform for both theoretical research and empirical studies on single- or multimodal interaction that can relate to different application domains and scenarios.

Multimodal Interaction with MIGSEP. The current instance of MIGSEP has been set in the application domain of a simulated hospital environment. Figure 6 shows the configuration of the MIGSEP system where a user can use spoken-language, gesture via touch-screen or the combination of both to interact with MIGSEP.

Fig. 6. The current instance of MIGSEP.

This system consists of a tablet PC, on which the MIGSEP instance is running and the interface is displayed, a button device for triggering a "press to talk" signal, and a green lamp to signalize the "being pressed and ready to talk" state, The MIGSEP interface itself can be divided into the following two areas:

- **Function-area** contains the function button "start" on the top left for going to the start state, the function button "toilet" showing the most basic need of an elderly person, and the text area for displaying the system responses;
- **Choice-area** displays information entities as single cards that can be clicked, with a scrollbar indicating the position of the current displayed cards and a context sensitive colored bar showing the current concerned context if necessary.

Figure 7 shows an example of interaction between MIGSEP and a user who would like to go to the registration room of the endocrinology department.

Sys:	\<shows 3 cards, green for persons, yellow for rooms and blue for departments\> "Good day, I can help you find the desired location to go to."
User:	"I want to go to the endocrinology departement."
Sys:	\<shows a blue bar texted with endocrinology on the top, and one brown card, one green card and one yellow card in the middle\> "Do you want me to show all the persons or the rooms in the endocrinology?"
User:	\<press the first card\>
Sys:	\<enlarges the first brown card, shows yes/no button\> "Do you want to go to the registration room of endocrinology?"
User:	\<press the yes button\>
Sys:	\<resizes the first brown card and then shows the 3 cards in the start menu again\> "OK, I have saved this goal. Where else do you want to go?"

Fig. 7. A sample interaction between a user and MIGSEP.

3 Experimental Studies

In order to evaluate how well elderly persons can be assisted by MIGSEP system regarding different input modalities, i.e. gesture via touch-screen, speech, or the combination of both (abbreviated as combi), a series of experimental studies were conducted with the elderly persons in the fixed age range and the same experiment settings, which will be introduced in this section.

3.1   Participants

Altogether 31 elderly persons (m/f: 18/13, mean age of 70.7, standard deviation 3.1), all German native speakers, participated in the study. Every participant had to finish the mini-mental state examination (MMSE), a screening test to measure cognitive mental ability (cf. [36]). The participants are having the score of 29.0 averagely (std. = .84), which indicates that they have slight decline in cognitive abilities.

3.2   Stimuli and Apparatus

As shown in Fig. 6, visual stimuli were presented via a graphical user interface on the screen of a portable tablet PC and a green lamp, which is on if the speech input is activated; audio stimuli were generated by the MIGSEP and played via two loud-speakers at a well-perceivable volume. All tasks were given as keywords on the pages of a calendar-like system.

Three kinds of input possibilities were used to interact with MIGSEP: (1) speech input, activated if a button was being pressed and the green lamp was on, and deactivated if the button was released; (2) gesture via touch-screen, which is directly performable on the touch-screen display; (3) the combi modality, which allows participants to freely choose between speech and gesture via touch-screen as input modality.

The same data set contains virtual information about personnel, rooms and departments in a common hospital, was used in the experiment.

During the experiment each participant was accompanied by the same investigator, who gave an introduction to the system as well as predesigned instructions to interact with the system.

An automatic internal logger was implemented inside MIGSEP and used to record all the real-time system internal data, while an audio recorder program kept the whole audio data during the dialogic interaction process.

An evaluation questionnaire, focusing on the user satisfaction with MIGSEP regarding the subjective assessment using the combi modality, was especially designed for this study. It contains 6 questions concerning the quality of using the combi modality compared to the single modalities, each of which concerns with the feasibility, the advantages, the usability, the appropriateness and the preference. This questionnaire was answered by each participant via a five point Likert scale.

3.3   Procedure

At the beginning a brief introduction was given to the participants, so that they could get the basic idea and an overview of the experiment.

Then in order to minimize the learning or bias effect with respect to the use of one modality, we introduced a cross-over procedure with altogether three test runs, in which 16 participants out of 31 had to first use the gesture via touch-screen and then the speech input, and then the combi modality, while the other 15 first used speech and then the gesture via touch-screen, and finally the combi modality. Before each test run each participant was given comprehensive instructions and enough time to get familiar with each to be tested modality and its interaction with MIGSEP. During each test run each participant had to perform 11 tasks, each of which contained incomplete yet sufficient information about a destination the participant should select. For example a task can be to drive to "room 2603", to "Sonja Friedrich", or to "room 1206 or room 2206 with the name OCT-Diagnostics". For each modality the tasks were different at the content level, yet similar at the complexity level. Each task was fulfilled or ended, if the goal was selected or the participant gave up trying after six minutes.

After all tasks were run through, each participant was asked to fill in the questionnaire for evaluation.

4   Results and Discussion

According to the classic evaluation framework Paradise ([37]), the performance of an interactive system is determined by the effectiveness of task success, the efficiency of interaction and the user satisfaction about the system. Therefore, these three criteria are also used to analyze the data of the multimodal interaction between elderly persons and MIGSEP, while focusing on comparing the effects of using speech, gesture via touch-screen and combi input modalities.

4.1   Regarding Effectiveness of Task Performing

In the classic Paradise framework, kappa coefficient is used to measure the effectiveness of task performing of an interactive system. However, in order to calculate the kappa coefficient, a confusion matrix is needed and the original way of constructing a confusion matrix can only be applied to an interactive system with a single modality, which is not suitable for the data collected during multimodal interaction. Therefore, in order to construct the needed confusion matrix, we proposed the concept of an attribute value tree (AVT) in [14] to replace the attribute value matrix in the classic Paradise framework, where an AVT contains all information to be exchanged during the multimodal interaction between MIGSEP and elderly persons and therefore can function similarly to the original attribute value matrix, yet with the ability of dealing with multimodal interaction data.

As shown in Fig. 8, an AVT already contains all the information exchanged by using either gesture via touch-screen or speech input and thus, the 11 AVTs used in [14] for assisting in evaluation the combi modality can also be used for evaluating the two single modalities.

Thus, the confusion matrix can then be constructed for all tasks and all modalities. For example Table 1 shows a confusion matrix for the task "go to Dr. Prescher", where

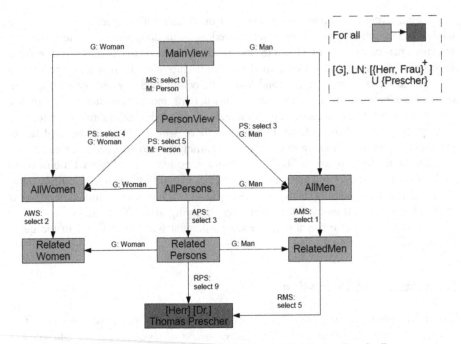

Fig. 8. An Attribute Value Tree for the task "go to Dr. Prescher".

Table 1. The confusion matrix for the task "go to Dr. Prescher".

Data	MS		LN		...	M		sum
	M	N	M	N		M	N	
MS	23							23
LN			33	4				37
...								...
M						34	1	35

"M" and "N" denote whether the actual data match with the expected attribute values in the AVTs. There were 23 correctly selected actions in the MetaSelect (MS) state; the spoken language command regarding the last name (LN) was misrecognized by the system for 4 times; and the MetaType (M) "Person" was wrongly selected for one time. Note that, because of the width of the text, not all attributes of this confusion matrix can be shown in this example.

The data for all confusion matrices were merged and final confusion matrices of the 11 performed tasks for all the three modalities were created. Given the final confusion matrices, the Kappa coefficients were calculated with $\kappa = \frac{P(A)-P(E)}{1-P(E)}$ ([37]), where $P(A) = \frac{\sum_{i=1}^{n} M(i,M)}{T}$ is the proportion of times that the actual data agree with the attribute values, and $P(E) = \sum_{i=1}^{n} (\frac{M(i)}{T})^2$ is the proportion of times that the actual data

Table 2. The kappa coefficients of task performing regarding the three modalities.

	Speech	Touch-screen	Combi
κ	0.74	0.88	0.91
Std.	0.13	0.1	0.04

are expected to be agreed on by chance, where $M(i, M)$ is the value of the matched cell of row i, $M(i)$ the sum of the cells of row i, and T the sum of all cells.

As shown in Table 2, the three overall kappa coefficients show a sufficiently successful interaction between MIGSEP and the participants. The κ of speech input is 0.74, which is a bit lower than the gesture via touch-screen with $κ = 0.88$, this is mainly caused by the automatic recognition errors. However, κ of Combi modality is even higher, indicating that with the combi modality the ASR errors were reduced considerably. The standard deviation 0.04 also implies that task performing using the combi modality is much more stable than the other two single modalities.

4.2   Regarding Efficiency of Interaction

Table 3 presents the results calculated from the quantitative data automatically recorded during the interaction using all the three input modalities, with respect to the most important factors for the efficiency analysis of an interactive system: the user and system turns in one dialogue for performing a task, the time used to complete a task and the number of speech recognition errors.

Table 3. Quantitative results regarding interaction efficiency for each participant and each task.

	Touch-screen		Speech		Combi	
	Mean	Std.	Mean	Std.	Mean	Std.
User turns	15.5	4.1	4.3	1.7	7.4	3.6
Sys turns	15.4	3.9	4.3	1.6	7.4	3.6
Elapsed time (s)	88.9	40.2	57.6	24.2	48.7	20
ASR error times	–	–	0.8	0.7	0.3	0.4

With only half of the user/system turns and about 40 s less per participant per task, the interaction using combi modality shows a much better efficiency in all factors while comparing with gesture via touch-screen.

Although only averagely 4.3 user turns and 4.3 system turns were needed for each task using speech input, the error times of the automatic speech recognition (ASR) are much higher than using combi modality, which caused the 9 more seconds than using the combi modality for each participant while performing each task. This is also reflected in the lower standard deviation for combi modality, indicating that less extreme speech recognition problems occurred with combi modality than pure speech input.

4.3   Regarding User Satisfaction About the Modalities

The results of the questionnaire regarding the subjective comparison between the combi modality and the single modalities were analyzed and summarized in Table 4.

Table 4. The subjective comparison between the combi modality and single modalities.

	Mean	Standard deviation
Better than single modality?	4.4	1.1
Easier solving tasks?	4	1.3
Showing advantages?	4.5	1
Usable to use combi-modality?	4.1	1.5
Prefer to use combi-modality?	4.4	1.3
Not confusing?	4.5	0.9
Overall	4.3	1

A very good overall user satisfaction with the combi modality is displayed via the score of 4.3 out of 5. Specifically, the elderly participants found the combi modality better and easier for performing tasks than the single modality with the score of 4.4 and 4.0; they could see the advantages of using combi modality and regard it as usable with the score of 4.5 and 4.1; they would prefer to use the combi modality with the score of 4.4; and they didn't find using the combi modality confusing with the score of 4.5. However, the scores of easier solving tasks and the usability of the combining modality were a bit lower than the others and the corresponding standard deviations were also a bit higher. Given a further insight into the data, this is mainly due to two elderly participants, who only used gesture via touch-screen even though both the modalities are enabled, and had made unpleasant impression of using only that, and therefore gave comparably lower score in the questionnaire.

5   Conclusions and Future Work

This paper reported our work on multimodal interaction for elderly persons regarding the following two important aspects:

- The brief presentation of our theoretical and technical foundation for supporting designing, developing and implementing elderly-centered multimodal interactive systems;
- The summary and analysis of empirical data of a series of empirical studies for evaluating a practical multimodal interactive guidance system for elderly persons to be used in navigation scenarios in complex buildings, in which three modalities, speech, gesture via touch-screen and the combination of both, were compared.

In general, the overall positive results showed high effectiveness of task performing, high efficiency of interaction and high user satisfaction with the implemented system, which validated our systematically developed and empirically improved design guidelines, foundations, models and frameworks for supporting multimodal interaction for elderly persons. The comparison between the input modalities also showed that the combination of gesture via touch-screen and speech input performs much better, more efficiently and is much more preferred by the elderly participants.

The presented work continued the pursuit of our final goal of building effective, efficient, adaptive and robust multimodal interactive systems and frameworks for elderly persons in Ambient Assisted Living environments. Corpus-based supervised and reinforcement learning techniques will be applied to support and improve the formal language driven unified dialogue modelling and management approach. Further application domain of autonomous navigation for elderly or handicapped persons on more spatially-related interaction is being investigated with qualitative spatial reasoning based frameworks. Special focus will also be placed on multimodal interaction within a smart home environment with a fully equipped technological infrastructure for the elderly or people with disabilities.

Acknowledgements. We gratefully acknowledge the support of the Deutsche Forschungs-gemeinschaft (DFG) through the Collaborative Research Center SFB/TR8, the department of Medical Psychology and Medical Sociology and the department of Neurology of the University Medical Center Göttingen.

References

1. Oviatt, S.: Multimodal Interfaces. In: Jacko, J.A., Sears, A. (eds.) The Human-Computer Interaction Handbook, pp. 286–304. L. Erlbaum Associates Inc., Hillsdale (2002)
2. Jaimes, A., Sebe, N.: Multimodal human-computer interaction: a survey. In: Sebe, N., Lew, M., Huang, T. (eds.) Computational Vision and Image Understanding, pp. 116–134. Elsevier Science Inc., New York (2007)
3. Goldin-Meadow, S.: Hearing Gesture: How Our Hands Help Us Think. Harvard University Press, Cambridge (2005)
4. Kendon, A.: Gesture: Visible Action as Utterance. Cambridge University Press, Cambridge (2004)
5. Vitense, H.S., Jacko, J.A., Emery, V.K.: Multimodal feedback: establishing a performance baseline for improved access by individuals with visual impairments. In: Proceedings of the Fifth International ACM Conference on Assistive Technologies (Assets'02), pp. 49–56. ACM, New York (2002)
6. Reeves, L.M., Lai, J., Larson, J.A., Oviatt, S., Balaji, T.S., Buisine, S., Collings, P., Cohen, P., Kraal, B., Martin, J.-C., McTear, M., Raman, T.V., Stanney, K.M., Su, H., Wang, Q.Y.: Guidelines for multimodal user interface design. Commun. ACM – Multimodal Interfaces that Flex, Adapt, and Persist **47**(1), 57–59 (2004). ACM, New York
7. Lutz, W., Sanderson, W., Scherbov, S.: The coming acceleration of global population ageing. Nature **451**, 716–719 (2008). International Institute for Applied Systems Analysis, Laxenburg
8. D'Andrea, A., D'Ulizia, A., Ferri, F., Grifoni, P.: A multimodal pervasive framework for ambient assisted living. In: Proceedings of the 2nd International Conference on PErvasive Technologies Related to Assistive Environments (PETRA'09), pp. 9–13. ACM, New York (2009)

9. Margetis, G., Antona, M., Ntoa, S., Stephanidis, C.: Towards accessibility in ambient intelligence environments. In: Paternò, F., de Ruyter, B., Markopoulos, P., Santoro, C., van Loenen, E., Luyten, K. (eds.) AmI 2012. LNCS, vol. 7683, pp. 328–337. Springer, Heidelberg (2012)

10. Anastasiou, D., Jian, C., Zhekova, D.: Speech and gesture interaction in an ambient assisted living lab. In: Proceedings of the 1st Workshop on Speech and Multimodal Interaction in Assistive Environments, pp. 18–27. Association for Computational Linguistics (ACL), Stroudsburg (2012)

11. Jian, C., Scharfmeister, F., Rachuy, C., Sasse, N., Shi, H., Schmidt, H., von Steinbüchel-Rheinwll, N.: Towards effective, efficient and elderly-friendly multimodal interaction. In: Proceedings of the 4th International Conference on PErvasive Technologies Related to Assistive Environments, pp. 45:1–45:8. ACM, New York (2011)

12. Jian, C., Scharfmeister, F., Rachuy, C., Sasse, N., Shi, H., Schmidt, H., von Steinbüchel-Rheinwll, N.: Evaluating a spoken language interface of a multimodal interactive guidance system for elderly persons. In: Proceedings of the Fifth International Conference on Health Informatics, pp. 87–96. SciTepress, Vilamoura (2012)

13. Jian, C., Shi, H., Schafmeister, F., Rachuy, C., Sasse, N., Schmidt, H., Hoemberg, V., von Steinbüchel, N.: Touch and speech: multimodal interaction for elderly persons. In: Gabriel, J., Schier, J., Van Huffel, S., Conchon, E., Correia, C., Fred, A., Gamboa, H. (eds.) BIOSTEC 2012. CCIS, vol. 357, pp. 385–400. Springer, Heidelberg (2013)

14. Jian, C., Shi, H., Scharfmeister, F., Rachuy, C., Sasse, N., Schmidt, H., von Steinbuechel, N.: Better choice? combing speech and touch in multimodal interaction for elderly persons. In: Proceedings of the 6th International Conference on Health Informatics. SciTepress, Barcelona (2013)

15. Fozard, J.L.: Vision and hearing in aging. In: Birren, J., Sloane, R., Cohen, G.D. (eds.) Handbook of Metal Health and Aging, vol. 3, pp. 18–21. Academic Press, New York (1990)

16. Mackay, D., Abrams, L.: Language, memory and aging. In: Birren, J.E., Schaie, K.W. (eds.) Handbook of the psychology of Aging, vol. 4, pp. 251–265. Academic Press, San Diego (1996)

17. Moeller, S., Goedde, F., Wolters, M.: Corpus analysis of spoken smart-home interactions with older users. In: Calzolari, N., Choukri, K., Maegaard, B., Mariani, J., Odjik, J., Piperidis, S., Tapias, D. (eds.) Proceedings of the Sixth International Conference on Language Resources Association, pp. 735–740. ELRA (2008)

18. Kotary, L., Hoyer, W.J.: Age and the ability to inhibit distractor information in visual selective attention. Exp. Aging Res. (Experimental Aging Research) 21(2), 159–171 (1995)

19. McDowd, J.M., Craik, F.: Effects of aging and task difficulty on divided attention performance. J. Exp. Psychol. Hum. Percept. Perfor. (American Psychological Association, Inc) 14(2), 267–280 (1988)

20. Salthouse, T.A.: The aging of working memory. Neuropsychology (American Psychological Association, Inc) 8(4), 535–543 (1994)

21. Craik, F., Jennings, J.: Human memory. In: Craik, F., Salthouse, T.A. (eds.) The Handbook of Aging and Cognition, pp. 51–110. Erlbaum, Hillsdale (1992)

22. Hawthorn, D.: Possible implications of ageing for interface designer. Interact. Comput. 12, 507–528 (2000)

23. Kline, D.W., Scialfa, C.T.: Sensory and perceptual functioning: basic research and human factors implications. In: Fisk, A.D., Rogers, W.A. (eds.) Handbook of Human Factors and the Older Adult, pp. 27–54. Academic Press, San Diego (1996)

24. Schieber, F.: Aging and the senses. In: Birren, J.E., Sloane, R.B., Cohen, G.D. (eds.) Handbook of Mental Health and Aging, vol. 2. Academic Press, San Diego (1992)

25. Charness, N., Bosman, E.: Human Factors and Design. In: Birren, J.E., Schaie, K.W. (eds.) Handbook of the Psychology of Aging, vol. 3, pp. 446–463. Academic Press, San Diego (1990)
26. Smith, M.W., Sharit, J., Czaja, S.J.: Aging, motor control, and the performance of computer mouse tasks. Hum. Factors **41**(3), 389–396 (1999)
27. Czaja, S.J., Sharit, J.: The influence of age and experience on the performance of a data entry task. In: Proceedings of the Human Factors and Ergonomics Society 41st Annual Meeting, pp. 144–147 (1997)
28. Sharit, J., Czaja, S.J., Nair, S., Lee, C.C.: Effects of age, speech rate, and environmental support in using telephone voice menu system. Hum. Factors **45**, 234–252 (2003)
29. Ziefle, M., Bay, S.: How older adults meet complexity: aging effects on the usability of different mobile phones. Behav. Inf. Technol. **24**(5), 375–389 (2005)
30. Ross, R.J., Bateman, J., Shi, H.: Using generalised dialogue models to constrain information state based dialogue systems. In: The Symposium on Dialogue Modelling and Generation, Amsterdam, Netherlands (2005)
31. Traum, D., Larsson, S.: The information state approach to dialogue management. In: van Kuppevelt, J., Smith, R. (eds.) Current and New Directions in Discourse and Dialogue, pp. 325–354. Kluwer, Dordrecht (2003)
32. Shi, H., Bateman, J.: Developing human-robot dialogue management formally. In: Proceedings of Symposium on Dialogue Modelling and Generation, Amsterdam, Netherlands (2005)
33. Roscoe, A.W.: The Theory and Practice of Concurrency. Prentice Hall, New York (1997)
34. Broadfoot, P., Roscoe, B.: Tutorial on FDR and its applications. In: Havelund, K., Penix, J., Visser, W. (eds.) SPIN 2000. LNCS, vol. 1885, p. 322. Springer, Heidelberg (2000)
35. Mandel, C., Huebner, K., Vierhuff, T.: Towards an autonomous wheelchair: cognitive aspects in service robotics. In: Proceedings of Towards Autonomous Robotic Systems (TAROS 2005), pp. 165–172 (2005)
36. Folstein, M., Folstein, S., Mchugh, P.: "Mini-mental state", a practical method for grading the cognitive state of patients for clinician. J. Psychiatr. Res. **12**(3), 189–198 (1975)
37. Walker, M.A., Litman, D.J., Kamm, C.A., Kamm, A.A., Abella, A.: Paradise: a framework for evaluating spoken dialogue agents. In: Proceedings of the Eighth Conference on European Chapter of Association for Computational Linguistics, NJ, USA, pp. 271–280 (1997)

Centralised Electronic Health Records Research Across Health Organisation Types

Samantha S.R. Crossfield$^{(\boxtimes)}$ and Susan E. Clamp

Leeds Institute for Health Sciences, University of Leeds, Leeds, UK
{s.crossfield,s.clamp}@leeds.ac.uk

Abstract. Where health and care provision is divided across organisation types, such as child health and palliative care, it is difficult for researchers to access comprehensive healthcare data. Integrated electronic health records offer an opportunity for research into care delivered within and across organisations. In this paper a new centralised model for accessing such data is justified using the critical success factors of an established research data provider. This validates a model that will facilitate integrated health research to inform the evidence base and systems used in clinical practice across organisations.

Keywords: Research database · Shared health records · EPR · Patient record access · ResearchOne · Multiple health providers · Clinical system · Database validation

1 Introduction

1.1 Summary

The multitude of health and social care organisations that may be involved in a patient's care is frequently under-considered by research projects and in translating results into evidence-based care. Electronic health record (EHR) data is increasingly being used in research due to their widespread use in clinical practice for gathering detailed and structured data. EHRs are often non-shareable, used in only the organisation that generated them, such as a general practice or a hospital ward [1]. Such isolated records cannot comprehensively represent to the research community the health of patients who receive care in multiple settings. Alternative structures with EHR data sharing between clinicians at different health organisations can improve clinical practice and reduce errors [2, 3]. Research on records from across organisation types has enhanced healthcare through investigating clinical practice through a whole systems approach [4]. Research using integrated EHRs in England has nonetheless remained infrequent due to accessibility issues.

1.2 The Silo Issue in Health and Social Care and Research

Health and social care in many countries including the US, UK and China is delivered through multiple organisation types working with independence. These care providers

© Springer-Verlag Berlin Heidelberg 2014
M. Fernández-Chimeno et al. (Eds.): BIOSTEC 2013, CCIS 452, pp. 394–406, 2014.
DOI: 10.1007/978-3-662-44485-6_27

struggle with an outcome of this specialisation which is often termed 'silo working' wherein service deliverers with different aims and professional languages gather information on separate aspects of patient care [5, 6] and store these in silos: unlinked records in closed databases. Such siloed systems often have different structures or formats that cannot be easily linked. As such these records may not be shared with other healthcare organisations of the same, or different, type. As a consequence each organisation holds partial patient records rather than the entirety of the patient's medical history. If a patient has a general practitioner (primary physician) then they often receive summary discharge letters from other organisations, though this unstructured and delayed information flow is one-way. With closed systems it becomes difficult to share timely and pertinent information, such as diagnoses, allergies, medication and professional insights with other healthcare providers that are also intervening with and monitoring the health of the patient. This results in issues of duplication and missing data. Patient information held in such silos provides less support to patients that cross healthcare organisation types and reduces the capacity to perform longitudinal assessments [7].

This silo issue is also of relevance for the research community who consider patient health in an array of fields including health informatics, epidemiology, health economics, clinical care and medicine. Traditional data collection involves invasive, timely and resource-intensive methods such as conducting interviews and questionnaires. The increasingly routine use of EHRs in clinical practice, for example among 76 % of European general practitioners, makes EHRs an efficient source of large cohort research data [8]. This is particularly applicable in the UK where 97 % of general practices use EHRs [9]. The capacity for large EHR cohorts facilitates research on low frequency incidences or diagnoses. For example, this enabled identification of the correlation between emergency department waiting times and outcomes of mortality and readmission, by using the records of over 14.5 million emergency department attendances [10].

The silo issue impacts the timeliness and security of EHR research. Patient data dispersal in EHRs across multiple organisation types often necessitates the involvement of identifiable data for undertaking data linkage. This brings security issues and the time taken to gather and link siloed data reduces the timeliness of cross-organisation research. Further time on the part of the researchers and the data providers is often required to update the research dataset. The ethical issues surrounding the identification of relevant patients and in developing a fully informed consent mechanism remain. Nevertheless such research has successful results and was crucial in resolving the disputed link between Autism and the Measles, Mumps and Rubella vaccine [11]. Researchers, in using non-shared EHR data, face the same constraints as clinicians in not being able to view the full patient pathway in a timely, cost-effective and secure, audited manner.

QResearch has an established ten year record in supporting electronic health records research. QResearch is a UK-based not-for-profit general practice EHR research database. Over 650 practices contribute data [12]. It was developed with the aim of consolidating de-identified, siloed EHR data from a large representative cohort of general practices to provide data for ethical research purposes [13]. It has facilitated research into the development of tools for identifying patient risk of, for example, developing cardiovascular disease (CVD) and diabetes [14, 15]. Despite the importance of risk assessment in health promotion, evidence suggests that risk tools are under-used. A study of clinical practice showed that 66 % of clinicians who identify the need to perform a global CVD risk assessment fail to follow the guidelines by employing such a tool, valuing subjective assessment alone [16]. As patient risk crosses organi-sation boundaries it may be that the relevance of these tools is constrained by the external validity of using non-shared records data from one organisation type.

As the UK population both pay for and receive lifelong care from the National Health Service (NHS), there is a focus on efficient illness prevention. The longitudinal information already existing in EHRs is seen as a resourceful means to facilitate research to help meet this goal. The NHS Quality, Innovation, Productivity and Pre-vention program encourages the innovative use of existing resources [17] and the Prime Minister, in a speech made in London in December 2011, specified 'opening up' NHS data to research as being crucial to supporting health, the economy, and the life sci-ences. Subsequently in May 2012 the government's Department of Health (DH) issued a call for more efficient EHR research. The call linked this to a move towards sharing information and delivering care across organisation boundaries in response to the increasingly complex needs of a diverse, multi-morbidity population [18]. Cross-organisation EHR research will be required in addressing this need for efficient research that is relevant across multiple organisation types.

1.3   The Integrated Solution

There are well established alternatives to non-integrated clinical practice. Kaiser Per-manente provides comprehensive care packages to 8 million patients in America. Their shared standards result from collaboration between all care providers [19]. In the UK, SystmOne is a centrally hosted clinical system provided by TPP (The Phoenix Part-nership) that enables record sharing between many of the health and care organisation types in the NHS (Fig. 1). Since 1999 its centralised database (cloud) has contained one integrated EHR per patient. From this, data is shared with the patient and across the health organisations that use SystmOne, where access rights are legitimate. Through SystmOne, 26 million patients have a shareable record (Table 1) that facilitates inte-grated care [20]. Both Kaiser Permanente and SystmOne exemplify long-standing alternatives that reduce the 'silo effect' in healthcare.

Integrated EHRs assist in cross-organisation type care management that efficiently utilises resources. Integration at Kaiser Permanente organisations contributes to the number of bed stays being 3.5 times fewer than in the NHS for 11 leading causes [21]. Benefits from EHR sharing are indicated by patient management improvements in cases that involve professionals from primary and secondary care sectors, such as are frequent in the treatment of long-term conditions. In such cases, secondary care consultants more frequently use their EHR where the patient's general practitioner is also registered with SystmOne as the record contains updates since the previous outpatient appointment and includes advice from other specialists, recent medications and blood test results [20]. This replaces reliance upon summary letters, patient awareness or being able to tele-phone other organisations [22]. Comprehensive information enables all organisations to review and communicate regarding medication, ongoing treatments and appointment

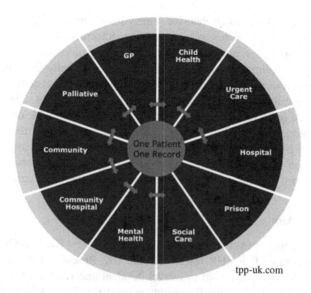

tpp-uk.com

Fig. 1. Healthcare organisation types in which SystmOne is used and between which information can be shared with the exception of prison data. SystmOne can provide hospitals with the clinical record, patient administration, bed management, e-prescribing and e-discharge.

Table 1. Approximate count of organisations and patients with a relationship recorded on SystmOne for ten organisation types. Other organisations using SystmOne include Speech and Occupational Therapies, Community and Social Services, Dietetics, Palliative Care, School Nurses and Endocrinology.

Health Care Organisation Type	Patients with data on SystmOne	Count of organisations using SystmOne
General practice	22 million	2100
Child health	6 million	50
District nursing	5 million	1300
Out of hours	3 million	120
Health visitor	2 million	190
Physiotherapy	1 million	60
Acute hospital	1 million	20
Podiatry	1 million	40
Community primary care clinic	1 million	360
Minor injuries/Accident and emergency	1 million	50

non-attendances [22]. Shared records also assist in prompt medicines reconciliation between care settings, which has been shown to identify errors in 38 % of prescriptions [23]. Through such means clinical systems integration delivers the benefits of an "electronic highway" envisioned by the NHS National Programme for IT [24].

The benefits that more comprehensive, timely information bring to clinical practice could also be brought to the research community. Information sharing among Kaiser Permenante organisations supports research that considers care provision across organisation types. Clinical practice has altered internationally in response to links uncovered, using Kaiser Permanente data, between hospital admissions and drugs such as rofecoxib being issued in ambulatory (primary) care settings [4, 25]. Using shared EHRs in research replaces linkage exercises that involve identifiable data and result in biased, incomplete datasets [26]. Shared EHRs enable research on the otherwise lost communications between healthcare organisations, such as referral trails. Research on cross-organisation type records can validate siloed research in a cost-effective, timely manner and inform clinical practice that occurs in these multiple settings.

1.4 Research Aim

The aim of this paper is to determine the capacity of a new ResearchOne database to facilitate cross-organisation EHR research in the UK. ResearchOne is a not-for-profit organisation with ethical approval to extract de-identified EHR data from the centralised SystmOne database into the ResearchOne database. This enables secure, audited access to anonymous records data for the research community. This access is with the purpose of developing new clinical understanding through research to improve patient care. This model must be investigated in order to justify that the ResearchOne database may bring benefits to research in the way that SystmOne does for clinical practice.

2 Method

The method was designed to assess the capacity of the ResearchOne database to support EHR research and to justify its potential benefits to integrated records research using English health data. Information regarding SystmOne and ResearchOne were determined from the ResearchOne Database Protocol and through interviews [27]. QResearch is specifically designed for EHR research in the UK and was taken as an academically established 'standard'. A search of Web of Science, PubMed and Google Scholar identified articles published on research that used QResearch data. These and the QResearch Protocol were reviewed. The key features of QResearch were taken as critical success factors, as justified in Table 2, against which the model presented by the ResearchOne database was appraised. These factors are the headings in the following section. From this the ResearchOne database could be validated with the potential to perform to the existing standard for a research database of NHS data, in order that it can facilitate cross-organisation research.

3 Results

3.1 Data Consolidation

QResearch facilitates research on EHR data consolidated from over 650 non-integrated general practice (GP) databases [12]. The ResearchOne database can similarly hold

Table 2. Features of the QResearch general practice EHR research database, with reasoning behind their necessity.

Critical factor	Reason
Data consolidation	The database assists researchers in accessing data that has been consolidated from many health organisations and so reduces the invasion, time and cost for clinicians and researchers, who must otherwise perform repeated extracts
Large cohort of research EHRs	Larger sample sizes bring both power and validity to research outcomes, enabling more research questions to be addressed [28]
De-identified EHR data	De-identification of EHR data protects privacy and permits research access without a public health mandate or consent, which could not be feasibly and non-invasively acquired for a significantly large cohort [29, 30]
Representative coverage	The external validity of a research outcome depends upon it being derived from a representative sample of the population
Ethical research practice	Success relies upon the database being securely developed and used for ethical purposes

EHR data contributed by multiple practices and so it meets this critical success factor. Moreover it can hold data from other organisation types, as it mirrors the infrastructure of SystmOne. SystmOne integrates data from multiple organisations into one centralised record per patient and so no consolidation is required in order to extract data in SystmOne from multiple organisation settings into the ResearchOne database.

Data linkages to other sources undertaken by QResearch are also feasible with ResearchOne. QResearch links GP EHR information to socio-economic, Hospital Episode Statistics (HES), disease-specific registry and death registration data [14, 31]. ResearchOne has national ethical and governance approval to undertake or request such linkage and consolidation [27]. An NHS National Institute for Health Research funded study, Improving Prevention of Vascular Events in Primary Care, has successfully piloted the capacity to link ResearchOne data to HES and Myocardial Ischaemia National Audit Project (MINAP) data.

3.2 Large Cohort of Research EHR Data

General practice involvement in QResearch has grown steadily to over 650, surpassing the original aim of 500 practices [12, 13]. SystmOne hosts patient information for over 26 million patients across England from more than 4500 organisations that may participate in ResearchOne. This includes 2100 general practices, 170 community services and 80 palliative care organisations (Table 1) that can contribute their 'real-world' data. Prison data recorded in SystmOne cannot be extracted into the ResearchOne database. TPP already hosts this data in SystmOne and has the data management skills and capacity to hold such a large cohort of records from multiple organisation settings in the ResearchOne database.

3.3 De-Identified EHR Data

Both the ResearchOne database and QResearch have a nationally approved governance framework under which they can hold de-identified data. Neither database can contain free text with potentially identifiable data, nor, for example, full dates of birth or death. Furthermore, given the comprehensiveness of cross-organisation type records, the ResearchOne database excludes diagnostic cases that are present in fewer than five records. QResearch requires consent from each practice in order to extract data from the practice's database. While SystmOne is centrally hosted, ResearchOne follows this practice in requesting consent from contributing organisations, and also provides the opportunity for patients to 'opt out' from providing the non-identifiable data. Consent is electronically audited through SystmOne, the centralisation of which ensures that any changes will automatically update the ResearchOne database within seven days.

3.4 Representative Coverage

QResearch practices are "spread throughout the UK", offering representative GP coverage [13]. ResearchOne has the capacity to provide an England-wide representation of cross-organisation type health and social care, through the more than 4500 invited organisations. England is divided into 433 lower tiers of local government – Local Authorities - of which over 85 % have patient representation on SystmOne. There are more than 26 million patients who have contributed to 300 million years of patient records, geographically distributed across England (Fig. 2). Over 12 million patients have registered with more than one care organisation on SystmOne at some point, and 365,000 patients have received care from five or more organisations using SystmOne. As such these records are cross-organisational. SystmOne holds 5 billion

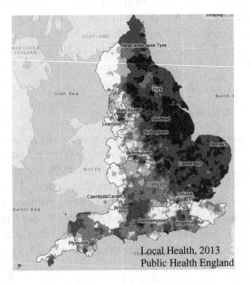

Fig. 2. A chloropleth map of SystmOne patient coverage.

diagnostic codes, inputted by clinicians whose specialties range from ante-natal to geriatric, rehabilitation to neuropathology. This coverage crosses community, primary and secondary care. The capacity for representation also covers the indices for rurality and deprivation defined by the UK Economic and Social Data Service [32].

3.5 Ethical Research Practice

QResearch and ResearchOne are specifically designed for ethical research access with the aim of improving healthcare. The frameworks for both QResearch and the ResearchOne database have been developed with ethical and governance approval from the relevant national bodies [27, 31]. The protocol for ResearchOne was developed with input from patients, clinicians, researchers and information governance experts. Any change in either protocol is reviewed by the national boards and a database committee of patients and clinical professionals along with experts in informatics, database architecture and governance [27, 31].

A national research ethics committee has approved both the QResearch and ResearchOne database frameworks to review data requests based on the benefit to clinical practice of each project proposal [27, 31]. Research proposals submitted to QResearch are reviewed by a QResearch Scientific Committee and a national multicentre Research Ethics Committee. The ResearchOne database has a nationally approved de-identification procedure and access protocol whereby a ResearchOne Project Committee considers all data requests. This approved de-identification and project acceptance process enables project proposals to be reviewed more promptly by a specialised committee.

Ethical accessibility of both QResearch and the ResearchOne database is supported by their being not-for-profit organisations. With SystmOne data existing centrally, the cost of database maintenance is low for ResearchOne, which reduces the cost further for the research community. Remote access to the secure ResearchOne data warehouse is audited for the purpose of maintaining this ethical practice.

4 Discussion

The results of this investigation show that the ResearchOne database matches the capacity of the existing standard of a research database with EHR data. ResearchOne has been nationally approved to extract de-identified EHR data from consenting health and care organisations. There is potential for the inclusion of data from a large cohort of shared EHRs with representative coverage both geographically and demographically across England. The framework has been designed to ethically support research that delivers benefit to patient health.

4.1 Further Research Capacities

The ResearchOne database has the capacity to perform beyond the current standard in terms of cross-organisation integration. It maintains data from more organisation types in England than the current standard, and so can support more comprehensive and representative research between and within these areas of healthcare. These shared EHRs enable research on data such as communication and referral trails that isolated records cannot, even when linked, portray. The diminished likelihood of integrated records having missing or duplicated data when compared to the standard is also a benefit for research [26]. The ResearchOne database is not consolidated from multiple settings, but rather it is integrated at the point of entry. Through ResearchOne the integrated care records of more than 25 million patients have the capacity to support research, with information from over 4500 health organisations in primary, secondary and social care (Table 1). This gives ResearchOne the capacity to bring the benefits of record-sharing into the research arena.

By centrally extracting data from integrated records, the ResearchOne database moves beyond the current standard of consolidating data from isolated organisations. This extract procedure is more secure and has no potential to incorporate bias through incomplete linkage for research. This is beneficial because Bohensky et al. [26] reviewed linkage sensitivity to range between 74–98 %. Centralised extraction also does not disturb clinical practice and reduces the cost of extraction, with this saving being passed to the research community. The centralisation of SystmOne maintains an up-to-date audit of organisation consent and also enables a patient to opt out of the ResearchOne database by informing just one of their care organisations that uses SystmOne. Such centralised capacity is of relevance in ethically supporting the research community to enhance healthcare.

A further factor that assists in health research is the timeliness of the data available from ResearchOne. Timely data is required in order for research to reflect the evolving field of clinical practice and continual changes in the population and health demographics. EHRs should facilitate timely research [33]. The centralisation of SystmOne ensures that research data could reflect real-time clinical developments, without affecting SystmOne users. The ResearchOne database has a seven day update frequency though items can be extracted more frequently as required for public health monitoring. A further reason for timeliness is that consent withdrawal from the ResearchOne database results in data being removed within seven days. The timeliness of research projects is further enhanced by the streamlined data request process. Timely data provision can occur securely at minimum cost due to the centralised nature of SystmOne, which is a speedier alternative than linking data from multiple sources.

While most projects can benefit from timely access to de-identified real-world records data, some projects require identifiers. This may be to enable linkage with datasets in other sectors, such as education, or information from patient observations and interviews. With UK data, such projects require ethical approval and consent, or permission under Section 251 of the NHS Act 2006. Having de-identified data, neither the ResearchOne database nor QResearch can assist these projects. ResearchOne has experience in providing pseudonymised linkable data from records on SystmOne for such approved projects. This is a more timely mechanism for project teams than extracting data from siloed databases held in different organisations. ResearchOne can similarly support SystmOne users in joining randomised control trials or pilot studies by embedding research in the clinical system. The closeness between ResearchOne and SystmOne enables these further capabilities.

4.2   Next Steps

The capacity of ResearchOne to maintain a large cohort of de-identified EHR data from multiple organisation types depends upon organisation participation. The joining process is simple and extracts do not inconvenience SystmOne users given its centralised nature. Information about joining ResearchOne is available on SystmOne and the ResearchOne website has further information, including current project details. Organisations providing data can be assured that the ResearchOne database is maintained under the same security principles as SystmOne in an NHS-accredited data centre [27]. The aim of ResearchOne, to bring research outcomes into clinical practice, assures that it is beneficial to contribute to the ethically approved process. The success of QResearch should assist SystmOne users, some of whom may have contributed to QResearch previously, in recognising this beneficial invitation. Already organisations are participating and the count of anonymised records in the database is approaching five million. This indicates the successful realisation of the capacity of ResearchOne. As more organisations join, they will enable one of the potentially largest health databases in the world to be developed.

The aim of ResearchOne includes not only pulling data for research purposes, but pushing outcomes back into health and care. Results may be openly published, and as these can have relevance to clinical care across many organisation types, they may initiate more comprehensive innovations in care delivery. SystmOne will incorporate developments so that the clinical system continually improves the support it provides to over 145,000 users across multiple organisation types. The closeness between ResearchOne and SystmOne closes the distance between care providers and researchers so that issues and outcomes become shared. The ResearchOne Project Committee considers projects from a clinical needs driven perspective, with input from SystmOne users. Therefore a cyclical, evaluative, cloud-based model of clinical practice and research is encapsulated in the SystmOne-ResearchOne infrastructure. This can be envisioned as a global model for the future.

ResearchOne facilitates research both for validation purposes and in novel areas. Explorations will continue to compare ResearchOne data with national statistics to validate its representative coverage. Research performed on other datasets can be validated using the ResearchOne database. The impact of integrated and isolated EHR data on research may also be investigated, to explore the role of ResearchOne. ResearchOne has a structure to support different types of projects from anonymised epidemiological monitoring to randomised control trials. Research projects may use data from single or multiple care organisation types. In this way ResearchOne aims to facilitate outcomes that are of relevance across all organisations that are contributing data.

Cross-organisation records research can be used to extend clinical knowledge from one organisation type to another. Research outcomes previously based on data from one organisation type may not be pertinent to either other organisations, or when considering patients receiving treatment from multiple organisations. Research would indicate whether other organisations can see in their 'real-world records' the information required to utilise research outcomes. As such, the decision support from single-organisation-type research can be built on by extending investigations across organisations.

The global capacity for EHR research is continually increasing. Progression in science, particularly in the fields of security technology and machine learning, will lead to data mining of entire EHRs. The anonymisation of free text through advancements in natural language programming, and the reduction of human involvement in data analysis will open up the capacities of EHRs to ethically support research on real-world data. The number and types of organisations across which SystmOne provides integrated EHRs is rising, particularly in the acute sector, so that ResearchOne has increasing potential to represent comprehensive care in England and globally. With such future developments ResearchOne will increasingly support research that benefits healthcare.

5 Conclusions

The contribution of EHRs to research is increasing, but is hindered by the division of data across healthcare organisations as a result of the 'silo effect' in clinical care. Benefits in integrated care delivery have been evidenced from record-sharing. ResearchOne offers an alternative for research using this real-world shared EHR data. The model of the ResearchOne database has been critiqued using the success factors of QResearch, an established provider of EHR research data in the UK. ResearchOne meets this existing standard and brings further developments to the research community. This is particularly in terms of the timely provision of integrated, cross-organisation type data and in feeding results back into clinical care. This offers a global model for integrated evolution of innovations between clinicians, patients and research.

References

1. ISO/TR 20514: Health Informatics – Electronic Health Record – Definition, Scope, and Context. ISO Technical Report. www.iso.org/iso/ (2004). Accessed 21 Aug 2012
2. Twomey, P., Whittaker, A., Jervis, C.: Implementation of the National Service Framework for Diabetes: Initial opportunities for a PCT to facilitate practice and health community action plans – year one experience. Qual. Prim. Care 12(3), 213–218 (2004)
3. Ammenworth, E., Schnell-Inderst, P., Machan, C., Siebert, U.: The effect of electronic prescribing on medication errors and adverse drug events: A systematic review. J. Am. Med. Inform. Assoc. 15(5), 585–600 (2008)
4. Graham, D.J., Campen, D., Hui, R., Spence, M., et al.: Risk of acute myocardial infarction and sudden cardiac death in patients treated with cyclo-oxygenase 2 selective and non-selective non-steroidal anti-inflammatory drugs: Nested case-control study. The Lancet 365 (9458), 475–481 (2005)
5. Wilson, R., Baines, S., Cornford, J., Martin, M.: 'Trying to do a jigsaw without the picture on the box': Understanding the challenges of care integration in the context of single assessment for older people in England. Int. J. Integr. Care 7, 1–11 (2007)
6. Kawonga, M., Blaauw, D., Fonn, S.: Aligning vertical interventions to health systems: A case study of the HIV monitoring and evaluation system in South Africa. Health Res. Policy Syst. 10(2), 1–13 (2012)

7. Kuperman, G.J.: Health-information exchange: why are we doing it, and what are we doing? J. Am. Inf. Assoc. **18**, 678–682 (2011)

8. Dobrey, A., Haesner, M., Hüsing, T., Korte, W.B., Meyer, I.: Benchmarking ICT use among general practitioners in Europe: final report. Empirica, Germany (2008)

9. Schoen, C., Osborn, R., Squires, D., Doty, M., Rasmussen, P., Pierson, R., et al.: A survey of primary care doctors in 10 countries shows progress in use of health information technology, less in other areas. Health Aff. (2012). doi:10.1377/hlthaff.2012.0884

10. Guttmann, A., Schull, M.J., Vermeulen, M.J., Stukel, T.A.: Association between waiting times and short term mortality and hospital admission after departure from emergency department: Population based cohort study from Ontario, Canada. Br. Med. J. **342**, d2983 (2011)

11. Taylor, B., Miller, E., Farrington, C.P., Petropolous, M.-C.: Autism and measles, mumps, and rubella vaccine: No epidemiological evidence for a causal association. The Lancet **353**, 2026–2029 (1999)

12. Vinogradova, Y., Coupland, C., Hippisley-Cox, J.: Exposure to bisphosphonates and risk of cancer: a protocol for nested case–control studies using the QResearch primary care database. Br. Med. J. Open **2**, e000548 (2012)

13. Hippisley-Cox, J., Stables, D., Pringle, M.: QRESEARCH: A new general practice database for research. Inf. Prim. Care **12**, 49–50 (2004)

14. Hippisley-Cox, J., Coupland, C., Vinogradova, Y., Robson, J., et al.: Predicting cardiovascular risk in England and Wales: Prospective derivation and validation of QRISK2. Br. Med. J. **336** (7659), 1475–1482 (2008)

15. Hippisley-Cox, J., Coupland, C., Robson, J., Sheikh, A., et al.: Predicting risk of type 2 diabetes in England and Wales: prospective derivation and validation of QDScore. Br. Med. J. **338**, b880 (2009)

16. Graham, I.M., Stewart, M., Hertog, M.G.L.: Factors impeding the implementation of cardiovascular prevention guidelines: findings from a survey conducted by the European Society of Cardiology. Eur. J. Preventative Cardiol. **13**(5), 839–845 (2006)

17. Department of Health: The NHS Quality, Innovation, Productivity and Prevention Challenge: An Introduction for Clinicians. Crown Copyright, London (2010)

18. Department of Health: The power of information: putting all of us in control of the health and care information we need. Crown Copyright, London (2012)

19. Light, D., Dixon, M.: Making the NHS more like Kaiser Permanente. Br. Med. J. **328**, 763 (2004)

20. Stoves, J., Connolly, J., Cheung, C.K., Grange, A. et al.: Electronic consultation as an alternative to hospital referral for patients with chronic kidney disease: a novel application for networked electronic health records to improve the accessibility and efficiency of healthcare. Qual. Saf. Health Care **19**(5), e54 PMID: 20554576 (2010)

21. Ham, C., York, N., Sutch, S., Shaw, R.: Hospital bed utilisation in the NHS, Kaiser permanente, and the US medicare programme: analysis of routine data. Br. Med. J. **327**, 1257 (2003)

22. Keen, J., Denby, T.: Partnerships in the digital age. In: Glasby, J., Dickinson, H. (ed.) International Perspectives on Health and Social Care: Partnerships Working in Action, pp. 95–107. Blackwell Publishing Ltd, Sussex (2009). ISBN-10: 1444322591

23. Moore, P., Armitage, G., Wright, J., Dobrzanski, S., et al.: Medicines reconciliation using a shared electronic health care record. J. Patient Saf. **7**(3), 148–154 (2011)

24. Department of Health: Making IT happen: information about the National Programme for IT. www.dh.gov.uk/en/Publicationsandstatistics/ (2003). Accessed 27 July 2012

25. Cheetham, T.C., Levy, G., Niu, F., Bixler, F.: Gastrointestinal safety of nonsteroidal anti-inflammatory drugs and selective cyclooxygenase-2 Inhibitors in patients on warfarin. Ann. Pharmacother. **43**(11), 1765–1773 (2009)

26. Bohensky, M.A., Jolley, D., Sundararajan, V., Evans, S., et al.: Data Linkage: A powerful research tool with potential problems. BMC Health Serv. Res. **10**(1), 346 (2010)

27. Crossfield, S.S.R., Bates, C.J., Parry, J.: ResearchOne Database Protocol. Copyright ResearchOne (2012)

28. Cohen, J.: A power primer. Psychol. Bull. **112**(1), 155–159 (1992)

29. Lowrance, W.: Learning from experience: privacy and the secondary use of data in health research. J. Health Serv. Res. Policy **8**(1), 2–7 (2003)

30. Wellcome Trust: Towards Consensus for Best Practice: Use of patient records from general practice for research. www.wellcome.ac.uk/GPrecords (2009). Accessed 01 Aug 2012

31. Hippisley-Cox, J., Stables, D.: QRESEARCH: An ethical high quality general practice database for research. Copyright QRESEARCH (2011)

32. Economic and Social Data Service: www.ccsr.ac.uk/esds/variables/ (2012). Accessed 28 Aug 2012

33. Powell, J., Buchan, I.: Electronic Health Records Should Support Clinical Research. Journal of Medical Internet Research 7(1), e4 URL: www.jmir.org/2005/1/e4/ (2005)

Improving Partograph Training
and Use in Kenya Using
the Partopen Digital Pen System

Heather Underwood[1(✉)], John Ong'ech[2], Grace Omoni[3],
Sabina Wakasiaka[3], S. Revi Sterling[1], and John K. Bennett[1]

[1] University of Colorado Boulder, Boulder, USA
{heather.underwood,srsterling,jkb}@colorado.edu
[2] Kenyatta National Hospital, Nairobi, Kenya
J.Ongech@knh.or.ke
[3] University of Nairobi School of Nursing, Nairobi, Kenya
{gomoni250,swakasiaka}@gmail.com

Abstract. This paper presents the findings from two studies of the PartoPen system – a digital pen software application that enhances the partograph, a paper-based labor-monitoring tool used in developing regions. Previous studies have shown that correct use of the partograph significantly reduces pregnancy complications; however, partographs are not always correctly completed due to resource and training challenges. The PartoPen addresses these challenges by providing real-time decision support, instructions, and patient-specific reminders. The preliminary studies described in this paper examine how the PartoPen system affects classroom-based partograph training among nursing students at the University of Nairobi, and partograph completion in labor theater use by nurse midwives at Kenyatta National Hospital in Nairobi, Kenya. Initial results indicate that using the PartoPen system enhances student performance on partograph worksheets, and that use of the PartoPen system in labor wards positively affects partograph completion rates and nurses' level of expertise using the partograph form.

Keywords: Partograph · Digital pen · ICTD · Health informatics · Maternal health · Kenya

1 Introduction

In 2010 the World Health Organization (WHO) estimated that 287,000 women die every year due to pregnancy related complications [1]. The vast majority (99 %) of annual maternal deaths occur in developing countries. Many of these deaths can be prevented with skilled care before, during, and after childbirth [1]. In addition, the rate of maternal morbidities, which include fistula, uterine rupture, uterine prolapse, and mental health concerns, is estimated to be between 15 and 20 million cases per year. Treatment for these complications, when available, costs an estimated $6.8 billion per year [2].

© Springer-Verlag Berlin Heidelberg 2014
M. Fernández-Chimeno et al. (Eds.): BIOSTEC 2013, CCIS 452, pp. 407–422, 2014.
DOI: 10.1007/978-3-662-44485-6_28

The WHO advocates the paper partograph as the single most effective tool for monitoring labor and reducing labor complications in developing countries. The partograph facilitates the tracking of maternal condition, fetal condition, and cervical dilation versus time during labor [3]. Used correctly, the partograph can serve as a tool for early detection of serious maternal and fetal complications during labor. Early detection of pregnancy complications, especially in rural clinics, allows transport decisions to be made in time for a woman to reach a regional facility capable of performing emergency obstetric procedures. However, in order to be effective, the partograph must be used correctly. A recent study in Kenya reported that while 88.2 % of the 1057 evaluated patient records contained a partograph, only 23.8 % of the forms had been used correctly [4]. This is not unusual for developing countries where lack of training and continuing education, exacerbated by limited resources, represent serious barriers to effective partograph use [5–7].

The goal of the PartoPen project is to increase the effectiveness of the partograph using an interactive digital pen with custom software, together with partograph forms printed with a background dot pattern that is recognized by the pen [8, 9]. The digital pen uses internal handwriting recognition and paper-based location awareness to interpret the measurements made on the partograph form. These interpreted measurements can then trigger alerts for attending health care providers when conditions arise that require additional observation or intervention. In addition, timers on the digital pen can be triggered when measurements are plotted in order to provide audio reminders to take routine patient measurements at specified time intervals. The PartoPen thus provides a low-cost, and intuitive solution that addresses several of the identified barriers to successful partograph use, including form complexity and data interpretation challenges.

This paper describes two preliminary studies that examined the PartoPen in use in Nairobi, Kenya from June 2012 – August 2012; the first was conducted with ninety-five third and fourth-year nursing students at the University of Nairobi School of Nursing Sciences, and the second, with nurse midwives in the labor wards of Kenyatta National Hospital (KNH) and Pumwani Maternity Hospital (PMH) over a period of one month. The principal findings of these two studies, reported in more detail below, are (1) the PartoPen improved the ability of nursing students to accurately complete partograph worksheets using synthetic maternal data, (2) use of the PartoPen during actual labor increases both the rate of partograph completion, and partograph accuracy, and (3) that the PartoPen was readily accepted and adopted by both students and practitioners.

2   Related Work

There is a large body of research that examines the potential relationships between paper-based systems and digital tools, particularly mobile phones. Mobile phone tools have been designed to simplify data collection [10], improve community health worker performance and effectiveness [11–15], and digitize data from paper forms [16, 17].

Digital pens offer the unique affordances of retaining the physical motion of natural writing, and the simultaneous creation of both a paper and a digital record. Digital pens have been customized for context-specific research tools [18–21] due to their programmability, portability, audio and note synchronization, and their ability to digitize sketches as well as handwritten notes for easy transmission via email. A specific example of how digital pens have been used in a healthcare setting is the TraumaPen. The TraumaPen [22] integrates paper emergency patient intake forms with a digital display component in the exam room to reduce redundancy of verbal data transmission between health care practitioners.

Other research on improving the paper partograph form includes the ePartogram device in development by Jhpiego [23], and the partograph e-Learning tool created by the WHO [24]. Jhpiego is currently testing three ePartogram implementations, which include an Android tablet application, a digital clipboard system, and a custom hardware solution, but at this time, no data has been collected or analysed for any of these models. The WHO e-Learning tool is distributed to facilities like KNH via CD-ROM. However, these CD-ROMs are not provided to every student or directly incorporated into the nursing curriculum. Single copies of the tool are often passed from student to student throughout the academic year, placing the primary responsibility for learning the material upon the students themselves. Less than half of the students who participated in the PartoPen study had used the WHO eLearning tool.

To the best of our knowledge, the PartoPen system is the only standalone digital partograph solution that can be used interchangeably as a training tool *and* in active labor theaters without altering the currently paper-based system or requiring significant additional training for the technology itself.

3 The PartoPen System

The current implementation of the PartoPen system uses the Livescribe 2 GB Echo digital pen, which can capture and synchronize audio and handwritten text, and digitize handwritten notes into searchable and printable PDF documents. These pens use an infrared camera in the tip of the pen that is triggered when a user presses the pen tip to a piece of paper. The camera captures a pre-printed unique dot pattern (see Fig. 1) at a rate of 70 images per second.

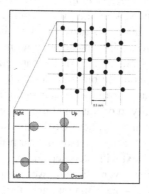

Fig. 1. The Dot Positioning System (DPS) uses printed microdots, as seen above, arranged in specific patterns. The dot pattern allows the digital pen to determine where on the form it is placed. The dot pattern is patented by Anoto AB Group.

Each printed dot represents location information, which the pen interprets and uses to perform location-specific functions, such as playing an audio instruction prompt when an instruction button is tapped, or triggering a decision-support prompt when a birth attendant plots a measurement indicating abnormal labor progression.

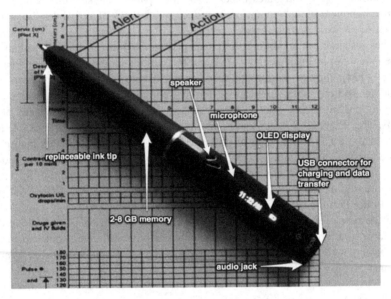

Fig. 2. The digital pen used in the PartoPen system is depicted. The speaker, microphone, OLED display, USB connector, audio jack, memory storage, and replaceable ink tip are identified.

The digital pens also include a speaker, a microphone, a 3.5 mm audio headphone jack, up to 8 GB of memory storage (approximately 800 h of audio recording storage), an OLED display, a rechargeable lithium-ion battery, and a micro-USB connector for charging and data transfer (see Fig. 2). Ink cartridges are low-cost and can be replaced without pen disassembly.

The PartoPen provides partograph training instructions, task-oriented reminders, and context-specific audio feedback in real time. Tapping the pen in different areas on the partograph form provides audio instructions taken directly from the WHO partograph manual, which reinforces birth attendant training. The pen detects abnormal labor progression by analyzing data entered on the partograph form, and provides audio and text-based feedback to encourage birth-attendants to take appropriate action.

The PartoPen is appropriate for use in resource-challenged environments. It does not require network connectivity to operate, and uses a rechargeable lithium ion battery that can be charged using a standard micro-USB cell phone charger. The dot pattern, printed on the partograph forms using a standard laser printer and printer paper, allows the pen to synchronize written text with recorded audio. Most importantly, the PartoPen is low cost, durable, consumes very little power, requires minimal training, and enhances – rather than replaces – the common paper tool in near-ubiquitous use in the developing world.

4 Nursing Student Study

The nursing student study took place at the University of Nairobi (UoN) School of Nursing Sciences in Nairobi, Kenya. The university is closely affiliated with KNH, and the nursing students at UoN perform clinical rotations in the maternity wards at KNH. The goals of the nursing student study were to establish a baseline of common partograph errors based on the type of error (e.g., incorrect values, incorrect form location, or incorrect action based on entered data), determine if using the PartoPen decreases the number of common partograph errors in relation to the established baseline, and approximate the amount of training needed to use the PartoPen and access the majority of the built-in PartoPen functionality.

4.1 Methodology

Participants. Ninety-five nursing students in their third and fourth years of study participated in the study. Local research assistants recruited participants from the population of 148 third and fourth year nursing students at the UoN. All students had previously been taught how to use the partograph to monitor labor during a 10–15 min in-class discussion as part of the nursing curriculum, and during their clinical rotations in the maternity wards.

The 95 student participants were separated by year (i.e., third or fourth year nursing students) and then randomly divided into three groups, resulting in six total groups. Group 1 was the control group, and Groups 2 and 3 were the intervention groups, which focused on the discoverability of the functionality, and the affect on partograph performance, respectively. Group 1 students completed a partograph worksheet task with a PartoPen in "silent logging mode," and received no instructions on how to use the technology. In the "silent logging mode" the digital pen records student answers, and logs when and where on the form student answers would have triggered feedback from a fully functional PartoPen. This control group provided a baseline for students' performance on the partograph worksheet task.

Group 2 completed the same worksheet task, but used a fully functional PartoPen in "use" mode. The PartoPen software in "use" mode for the student pilot has two main components: instructions and decision support. For the nursing student study with nursing students completing a partograph worksheet, the reminders (enabled only for the maternity ward study) were disabled. In addition, playing pre-recorded spoken audio provided the decision support, in contrast to the maternity ward decision support, which was provided by scrolling text across the OLED display.

Group 2 received no training on how to use the technology. In "use" mode, the digital pen logs when errors are made on the form, which was compared to the baseline results recorded from the first class of students. Students in this group received audio feedback from the pen when data was entered incorrectly on the form, and thus, corrected errors were also recorded in this mode. The data collected from this group tested the discoverability and intuitiveness of the PartoPen functionality.

Group 3 received a fully functional PartoPen in "use" mode and a 15-minute introduction and demonstration of the PartoPen system before completing the partograph worksheet task. The digital pen recorded errors, corrections, and all marks made on the partograph form. By comparing the results of Group 3 with the results of Group 2, the authors determined the affect of providing a PartoPen tutorial on partograph performance. Groups 2 and 3 attempt to simulate PartoPen deployments in which students/nurses do and do not receive training prior to using the device. Given that most of the PartoPen functionality is "pushed" to users during normal partograph form completion, we hypothesized that training on the PartoPen system should not significantly alter the results of participants with the same level of prior partograph knowledge – Groups 2 and 3, respectively.

Partograph Worksheet Grading. For all student groups, the partograph worksheets consisted of two patient case studies and two blank partograph forms printed with the dot pattern. The students were asked to record the patient data on the blank partograph forms as if they were actively monitoring that patient during labor.

The principal investigator created a grading scheme based upon the partograph grading schemes currently used to evaluate nursing student partograph completion performance during their clinical rotations. Each measurement category on the partograph (e.g., fetal heart rate, contractions, pulse, etc.) was graded in five sub-categories, and a numerical score was assigned relative to a set number of points specific to the particular case study. The five sub-categories included "measurement present", "mark accurate", "correct symbol", "plotted on correct time line", and "correct spacing". In the grading example shown in Fig. 3, the partograph form sections are listed vertically in the far-left column, and the sub-categories and possible point totals run horizontally along the top of the spreadsheet. The cells that have been grayed-out represent categories that are not applicable to the specific case study.

Three research assistants graded all 95 worksheets according to the grading scheme described above. Each nursing student was assigned an overall worksheet grade based on the total number of points possible for the two case studies they received.

	Marks are there?	(possible)	Marks are accurate value?	(possible)	Symbols correct?	(possible)	Plotted on correct vertical line? (possible)	Spacing?	(possible)
Patient Info	7	7							
FH	9	9	9	9	9	9	0	9	8	8
Liquor	1	2			1	2	0	2	0	1
Moulding	0	2			0	2	0	2	0	1
Cervix	1	2	1	2	1	2	0	2	0	1
Descent	2	3	2	3	2	3	0	3	1	2
Time	1	5	0	5			0	5	0	4
Contractions	9	9	9	9	5	9	0	9	8	8
Oxytocin										
Drops/min										
Drugs given										
Pulse	9	9	9	9	9	9	0	9	8	8
BP	1	2	2	4	1	2	0	2	0	1
Temp	2	3	2	3			0	3	1	2
Protein	1	2	1	2	1	2	0	2	0	1
Acetone	1	2	1	2	1	2	0	2	0	1
Volume	1	2	1	2			0	2	0	1

(Annotations within figure: "Grading categories", "Partograph form sections", "Points possible for each form section for a given case study")

Fig. 3. An example of the grading spreadsheet used for grading the worksheets in the teaching and training study. Partograph form sections are listed vertically in the left hand column, grading sub-categories are listed horizontally across the top of the spreadsheet, and points possible for each category are listed in red. Grayed out cells indicate that these categories were not applicable to the case study being graded.

Focus Groups. After completing the worksheet task, 5–10 students from each group were randomly selected to participate in a focus group discussion. The goals of the focus group discussion were to gain an understanding of current partograph training programs used at the study site, determine students' perceptions about the partograph form as a labor monitoring tool, and to record students' perceptions of the PartoPen system as an in-class training and active labor monitoring tool. The focus group discussions took between 15 and 30 min, and student responses were audio recorded using a digital pen and later transcribed.

4.2 Results

Quantitative Results. Using the grading scheme outlined in Sect. 4.1, scores were calculated as a percentage of total points correct out of the total possible points. We performed an unpaired t-test to determine if there was any significant difference between groups, particularly if Groups 2 and 3 showed any improvement in performance over Group 1 – the control group. Due to time constraints and limited grading resources, only the data from fourth year students is presented here. Group 1 from year four, which used the PartoPen in silent logging mode to complete the worksheet had an average score of 58 %, which means that on average students in this group correctly plotted 58 % of the measurements from both case studies in the worksheet (the highest possible score was 100 %). The average score for Group 2, which used the PartoPen in "use" mode but received no instructions, was 63 %. And the average score for Group 3, which used the PartoPen in "use" mode and received instructions, was 66 %. The difference in the average scores for the worksheet task suggest that use of and training on the PartoPen facilitated more accurate data recording on the partograph forms (Table 1).

Table 1. Average scores on worksheet completion task for fourth year students divided by PartoPen functionality group number. This table illustrates an increase in student performance with increasing PartoPen functionality and training.

Group # and PartoPen mode	Avg. score
Group 1 – silent logging mode	58 %
Group 2 – use mode, no training	63 %
Group 3 – use mode, training	66 %

In each worksheet, students received two patient case studies. All students received the "Mrs. B" case study, and either "Mrs. C" or "Mrs. A." The three case studies represent three possible labor outcomes. Mrs. A's data represents an uncomplicated, timely labor that progresses without medical intervention. Mrs. B's data illustrates a case of prolonged or obstructed labor, which is addressed by the administration of oxytocin – a labor-inducing drug. Finally, Mrs. C's labor progression data illustrates an increasing number of complications, including fetal distress, and ultimately results in a caesarean section. Thirty-four instructional audio prompts are available for all students and all patient case studies. However, only the Group 3 students were informed how to access the instruction prompts by tapping the pen on the text to the left of the graphs on the form. The average scores for each group based on patient case study are shown in Table 2. Using an unpaired t-test, the difference between Group 1 and Group 3 for the patient case study Mrs. C was found to be significant (t (8) = 2.71, p < .05). These data could suggest that for more challenging or complex labor cases, the availability and utilization of the instruction prompts promotes more accurate form completion.

Table 2. Average scores on worksheet completion task for fourth year students divided by patient case study and group number. This table illustrates a significant difference (t (8) = 2.71, p < .05), between Group 1 and Group 3 for the most complex patient case study (Mrs. C).

	Mrs. A	Mrs. B	Mrs. C
Group 1	61.3 %	58.6 %	**52.0 %**
Group 2	63.5 %	62.9 %	62.9 %
Group 3	65.2 %	62.7 %	**72.2 %**

Qualitative Results. Qualitative data from the focus groups examined three factors: students' previous partograph training, students' perceptions about the usefulness and effectiveness of the partograph as a labor monitoring tool, and students' feedback on PartoPen usability.

Students explained that the in-class introduction of the partograph ranged from a 5–15 min explanation by the lecturer. Lecturers reportedly demonstrated the partograph, but did not consistently fill one out completely in class. Students themselves did not generally practice filling out the partograph form. Students gained the majority of their experience using the partograph during their clinical rotations in the maternity wards. Individual experiences using the partograph in maternity wards fluctuated due to the number of nurses in the ward available to facilitate partograph training and the number of patients per day in the ward requiring a partograph (i.e., in active labor).

Students identified the recording of contractions as the most difficult part of the normal partograph form, because of having to remember the different shading styles that indicate contraction duration. Students also reported challenges when plotting the descent of the fetal head, moulding, and liquor (i.e., amniotic fluid). All of the students who participated in the focus groups from Groups 2 and 3 expressed the view that the PartoPen significantly mitigated these challenges and made the difficult form sections easier to fill out. One student commented: "In a classroom setup, it would be good because it will really help when we are first learning [the partograph]. It solidifies the basic things we need to know." Another student said: "At first, when you asked us what action to take when a measurement was made across the alert line, we were silent. But now, after we used it, we all know right away what to do."

Students suggested several feature changes to improve the PartoPen for student use. The suggestions include modifying the form itself to make the boxes larger and thus easier for entering data into (the PartoPen system used the standard WHO partograph form), and developing a flexible instruction-creation platform so that instructions can be easily modified to keep up with changes to WHO and Kenya Ministry of Health protocols. Several students also voiced concern that one unintended consequence of the

PartoPen might be a decrease in situational awareness, creating too great a reliance on the pen for instructions and decision support in an actual labor and delivery scenario. This concern was explored during the maternity ward study, and was not observed by the authors or reported by nurse midwife participants.

The results of the PartoPen study at the UoN suggests that using the PartoPen system in classrooms can improve students' ability to correctly complete a partograph form. These results also suggest that training on the PartoPen device does not significantly affect student performance on partograph completion tasks. The results of the teaching and training study support the hypothesis that a significant increase in partograph completion and accuracy can be achieved with little or no training on the device itself. We conclude that this benefit is due to the intuitive design, push-based functionality, and enhancement – rather than replacement – of the current paper-based system.

5 Maternity Ward Study

The second PartoPen study took place at Kenyatta National Hospital (KNH) and Pumwani Maternity Hospital (PMH). The goals of this study were to evaluate the PartoPen for usability in labor wards, determine if PartoPen use impacts partograph completion, and to investigate the broader impacts of the PartoPen on patient care and maternal health outcomes.

5.1 Methodology

The maternity ward study evaluated partograph completion rates for the month immediately prior to the introduction of the PartoPen, and for the following month, during which time the PartoPen was in use. "Completion" was measured using a partograph completion rubric previously developed by KNH staff for hospital administrative purposes. According to this rubric, a complete partograph has measurements for all of the partograph form sections, and a complete labor summary. A research assistant scanned the 369 partograph forms completed in the month prior to PartoPen introduction. During the month of PartoPen use, 457 partograph forms were initiated.

There were three phases in the introduction of the PartoPen system at KNH and PMH: (1) training nurses how to use the PartoPen system, (2) introducing the PartoPen system for use during 2–3 shifts per day, and (3) establishing sustainable infrastructure and gradually reducing researcher supervision of PartoPen use.

During the first phase, small groups of nurses received a 10–20 min introduction to the project and were trained on how to effectively use the system during their shift. Nurses were given a demonstration of the PartoPen functionality to introduce them to features of the system (reminders, audio decision-support, and additional instruction access), as well as a brief tutorial on exchanging pens during shift changes.

In phase two we introduced the PartoPen system in both KNH and PMH labor wards during the day shifts – approximately 7:30AM until 6:00PM. During the introduction week, PartoPen functionality, as well as the study design, were adjusted to account for observed environmental factors. Thee adjustments included modifying reminder sounds and text wording to account for noisy and busy environments, and simplifying the patient reminder ID system to allow nurses to create short, personalized identifiers for patients, rather than relying on the handwriting recognition in the pen to capture the patient's full name.

During the third phase, no changes were made to the code or the study design in order to keep study conditions consistent during data collection. Quantitative data was collected using a back-end logging system on the digital pens, which was downloaded every day at the beginning of the morning shift. Data logged by the pens included the following time-stamped variables: when audio prompts were played, which audio prompts were played when measurements were made, how many times instruction buttons were tapped, when the partograph form was started and completed, and which pen completed the form. Qualitative observations were also recorded during the three weeks of PartoPen use.

At the end of the three-week use period, nurses were asked to complete a survey on their experience before and during the PartoPen project (see Table 3).

Table 3. Survey questions that nurses were asked to answer after three weeks of using the PartoPen system.

Nurse survey questions
(1) **Before** the PartoPen project, how would you rate your level of expertise using the partograph form (on a scale of 1–10)?
(2) **After** the PartoPen project, how would you rate your level of expertise using the partograph form (on a scale of 1–10)?
(3) On average, how many patients in active labor do you care for during one **night** shift? (Circle one of the ranges below)
(4) On average, how many patients in active labor do you care for during one **day** shift? (Circle one of the ranges below)
(5) **Before** the PartoPen project, for what percentage of your patients did you complete a partograph? (Circle one of the ranges below)
(6) **After** the PartoPen project, for what percentage of your patients did you complete a partograph? (Circle one of the ranges below)
(7) On a scale of 1–10, rate your satisfaction with the PartoPen project in terms of usability (i.e., ease of use, functionality, instruction clarity, etc.)
(8) On a scale of 1–10, rate your satisfaction with the PartoPen project in terms of usefulness (i. e., level of patient care, level of job satisfaction, amount of time spent on tasks, etc.)

5.2   Partopen Software Implementation

In the PartoPen implementations at KNH and PMH, half-hour and four-hour reminders were enabled and activated by plotting a fetal heart rate measurement or a cervical dilation measurement, respectively. When a reminder would play, the patient's name and the type of measurement needed would scroll five times across the OLED display.

During the labor monitoring process, if a nurse plots a measurement indicating potential labor abnormality, the PartoPen decision support functionality is activated, a sound is played, and text scrolls across the OLED display indicating the available options for patient care.

The number of audio instructions for the maternity ward implementation was reduced during the first week of the study, because the nurses rarely used the audio instruction functionality, and the audio was unnecessarily using limited PartoPen memory. However, nurse midwives later began using the instruction buttons to teach students doing their clinical rotations how to use the partograph. Thus, in order to better facilitate nurses using the PartoPen to teach students, the full set of audio prompts were added back onto the pens. In the Appendix, the partograph that was used for the KNH maternity ward study is shown. The boxes (buttons) around the text on the left side of the form can be tapped repeatedly to access the use instructions mentioned above.

5.3   Results

Quantitative Results. During the maternity ward study three types of quantitative data were collected: the 369 scored patient partographs collected prior to PartoPen introduction, the 457 scored patient partographs collected during the period of PartoPen use, and survey responses from nurses who had completed the three-week usage period. This paper focuses on the survey results.

After three weeks of using the PartoPen system consistently on every shift, nurses were asked to fill out a short survey that captured demographic information about the participant, and gathered before-and-after information about PartoPen use. The survey consisted of eight Likert scale questions, and six free-form response questions.

On average, nurses self-reported an improvement of +2, on a scale of 1 to 10, in partograph expertise during the PartoPen project, a 9 out of 10 for usability of the PartoPen, and a 9.2 out of 10 for usefulness. Nurses also reported that the number of partographs they completed during the PartoPen study was, on average, 25 % more than they completed before the study. This increase in partograph completion rates is supported by initial data analysis of the completed partograph forms, and by anecdotal evidence by KNH staff and administration. In addition to the functionality provided by the PartoPen, which encouraged higher rates of partograph completion, the general increase in conversation and interest in the partograph due to the PartoPen study was also a likely contributing factor to the improved partograph completion rates.

Overall, the quantitative data gathered from the surveys suggest an increase in partograph knowledge among nurses, an increase in the number of partographs completed, and strongly positive perceptions of the PartoPen's usability and usefulness.

Qualitative Observations. During the first week of the PartoPen implementation, the first author was present in the labor wards from 7:30AM until 6:00PM to answer questions, facilitate PartoPen handoffs during shift changes, and to observe usage of the PartoPens. The most significant observations fall into two categories: digital pen design and PartoPen functionality.

Digital Pen Design. The nurses emphasized the necessity of a functional cap for the pens to keep ink from getting on their uniforms. One nurse, after getting pen ink on her uniform, remarked "Here, take it back, I won't use it unless there is a cap - or I'll bring you my laundry!" Caps for the pens were the distributed to the nurses, although the currently available cap for the Livescribe Echo pen was considered difficult to use.

A makeshift lanyard system was created to allow nurses to wear the pens around their necks, but a shirt clip or similar way to attach the pen to a pocket would be preferred. Other pen design improvement suggestions included having different colors of ink available, and making the pen thinner and lighter.

PartoPen Functionality Observations and Changes. During the first week of the study (the implementation and training phase), we observed nurses getting reminders from the pen, shaking their heads, and dismissing the reminder. Upon further investigation, we realized that the reminders nurses were receiving were for patients who had already delivered or had received a cesarean section. New functionality was added to the PartoPen that enabled a reminder ID system and a reminder cancelation system. The reminder ID system (pictured in the Appendix under the "Summary of Labor" section) was implemented to give nurses a way to create custom identifiers for patients that would scroll across the display when a reminder for that patient was triggered. Nurses write the identifier in one of the reminder ID boxes at the bottom of the form when a patient is admitted. The handwriting recognition engine in the pen interprets and stores this identifier and displays it for all future reminders for this patient. The reminder cancelation system addresses the issue of outstanding reminders for a patient that has already delivered or has been prescribed a cesarean section. A blue box at the top of the form (pictured in the Appendix in the top right-hand corner of the form) was created for nurses to sign their initials in once a patient has delivered or has been transferred for a cesarean section. The act of signing in the blue box cancels any existing reminders for that patient, and thus nurses will not receive unnecessary reminders.

The reminder and decision support functionality used in the maternity ward study relied on distinct pen tones and scrolling text on the pen. Nurses informed us that while this implementation did reduce the distractions associated with long audio prompts, they were unable to look at the OLED display to see which patient needed an exam if they were in the middle of another delivery. The text displayed for both reminders and decision support prompts is only scrolled five times before the display returns to showing the current clock time. Several modifications could be made to address this problem, including implementing a repeat button that would re-scroll the most recent text, continuing to scroll the text until the nurse uses the pen again, or implementing an audio based reminder system that uses an audio recording (made by the nurses themselves) of the patient's name, which is played back for that patient's reminders. The last solution is currently being developed, and will be tested in future PartoPen studies.

Displaying the time on the OLED display on the pens proved to be an important feature of the PartoPen system. Because measurements and exams are time-based, and each observation is associated with the time it is taken, nurses often ask each other for the clock time. Nurses had been using their mobile phones to get the time, but hospitals are increasingly restricting the use of personal phones during nurses' shifts to help reduce distractions and increase nurses' involvement with the patients. Nurses therefore began using the PartoPen to determine the exact time measurements were taken, increasing the accuracy of recorded data.

6 Conclusions and Future Work

The preliminary results of the nursing student study indicate that student performance on a partograph worksheet completion task improves when using the fully functional PartoPen system. A key finding of this study was that the PartoPen significantly improved student scores on the more complex patient case study, suggesting that reinforcement of existing knowledge, and real-time decision-making may be amplified and improved by using the PartoPen system. Based on the positive results from the student study at UoN, we are currently working with other Kenyan nursing schools to integrate the PartoPen into their existing nursing curricula. Additionally, the authors intend to examine how the PartoPen can be used to facilitate initial training on the partograph, as well as the transition from in-class partograph instruction to clinical use of the partograph form.

While the maternity ward study is on going, preliminary results suggest that using the PartoPen system increases partograph completion rates and increases nurses' accuracy when completing partographs. In addition, nurses were satisfied with both the usability and the usefulness of the PartoPen, suggesting that continued and sustainable use is possible in this environment.

After the three-week period of PartoPen use, twenty pens were left at KNH to continue being used by nurses in the labor wards. At the time of writing, these pens have been in use at KNH for a total of ten months.

Future work on the PartoPen project will focus on expanding the number of study sites to include clinics at various levels of healthcare, including rural health clinics, dispensaries, and district level facilities. Future research will also expand on the nursing student study data described in this paper, and evaluating the impact of long-term PartoPen use in the classroom, and how this affects performance among students during clinical rotations and evaluations.

The next step in determining the impact of the PartoPen system is to expand the goals of the study from looking solely at completion rates to include how partograph completion (or incompletion) affects patient outcomes. We are collaborating with a larger maternal health project based at KNH to study the effects of PartoPen use on maternal and fetal outcomes.

Acknowledgements. This research is funded by a Gates Grand Challenge in Global Health grant, a National Science Foundation Graduate Research Fellowship, and by the ATLAS Institute at the University of Colorado Boulder. We would like to thank the leadership of KNH, PMH, and the UoN, particularly Dr. John Ong'ech and Dr. Grace Omoni, for their support and cooperation during the PartoPen studies, and all of the students, nurses, and staff who participated in the studies. Research assistants Maya Appley, Addie Crawley, Sara Rosenblum, and Vincent Ochieng contributed significantly to this study.

Appendix: KNH Partograph Form

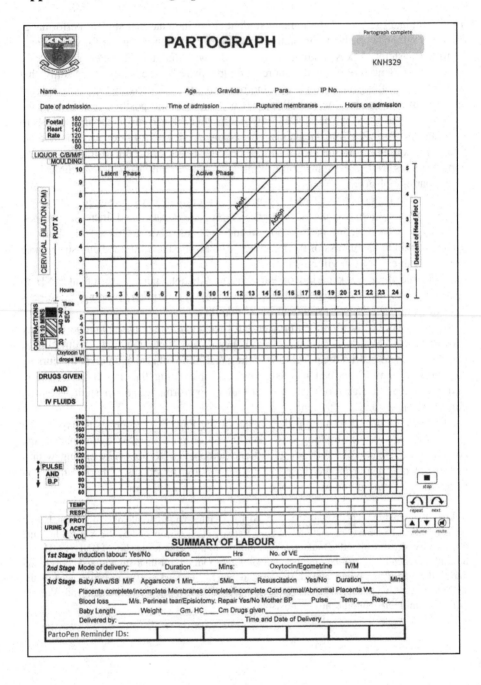

References

1. World Health Organization. Fact Sheet, World Health Organization Media Center for Maternal mortality. http://www.who.int/mediacentre/factsheets/fs348/en/index.html (2010). Retrieved 20 June 2011
2. Stanton, M.E.: A case for investment in maternal survival and health. Presentation at the Woodrow Wilson International Centre for Scholars, Washington, DC. http://www.wilsoncenter.org/events/docs/Mary%20Ellen%20Stanton%20Presentation.pdf (2010). Accessed 30 April 2012
3. Friedman, E.: The graphic analysis of labor. Am. J. Obstet. Gynecol. **68**(6), 1568–1575 (1954)
4. Mugerwa, K.Y., Namagembe, I., Ononge, S., Omoni, G., Mwuiva, M., Wasiche, J.: The use of Partographs in Public Health Facilities in Kenya. http://www.rcqhc.org/download/FP_DOCS/Final_paper_Kenya.pdf. Accessed 30 April 2012
5. Lawn, J., Kerber, K.: Opportunities for Africa's Newborns: Practical Data, Policy and Programmatic Support for Newborn Care in Africa. World Health Organization, Geneva (2006)
6. Levin, L.: Use of the partograph: effectiveness, training, modifications, and barriers: a literature review. Washington, DC, United States Agency for International Development, Fistula Care, EngenderHealth: 28 (2011)
7. Lavender, T., Omoni, G., Lee, K., Wakasiaka, S., Waitit, J., Mathai, M.: Students' experiences of using the partograph in Kenyan labour wards. Afr. J. Midwifery Women's Health **5**(3), 117–122 (2011)
8. Underwood, H.: Using a digital pen to improve labor monitoring and reinforce birth attendant training. University of Colorado at Boulder, ATLAS Institute. http://www.colorado.edu/atlas/technicalreports (2011) Retrieved 1 Aug. 2011
9. Underwood, H., Sterling, S.R., Bennett, J.: Improving maternal labor monitoring in Kenya using digital pen technology: a user evaluation. In: Proceedings of Global Humanitarian Technology Conference, 2012 (2012)
10. Hartung, C., Anowka, Y., Brunette, W., Lerer, A., Tseng, C., Borriello, G.: Open data kit: tools to build information services for developing regions. In: Proceedings of the ACM/IEEE Conference on Information and Communication Technology for Development, London, United Kingdom, 13–16 December 2010 (2010)
11. Grameen Foundation. Mobile Technology for Community Health. http://www.grameenfoundation.org/what-we-do/technology/mobile-health (2010). Retrieved 2 July 2011
12. Parikh, T.: CAM: a mobile interaction framework for digitizing paper processes in the developing world. In: Proceedings of ACM Symposium on User Interface Software and Technology (UIST), Seattle, Washington, 23–26 October 2005 (2005)
13. Sherwani, J., et al.: HealthLine: speech-based access to health information by low-literate users (2007)
14. Svoronos, T., et al.: CommCare: automated quality improvement to strengthen community-based health the need for quality improvement for CHWs. Health (San Francisco) (2010)
15. Derenzi, B., Mitchell, M., Schellenberg, D., Lesh, N., Sims, C., Maokola, W.: e-IMCI: improving pediatric health care in low-income countries (2008)
16. Dell, N., Breit, N., Crawford, J.: Digitizing paper forms with mobile imaging technologies. In: Second Annual Symposium on Computing for Development (2012)
17. Ratan, A.L., Chakraborty, S., Chitnis, P.V., Toyama, K., Ooi, K.S., Phiong, M., Koenig, M.: Managing microfinance with paper, pen and digital slate. Scientia **196**(36) (2010)

18. Yeh, R., Liao, C., et al.: ButterflyNet: a mobile capture and access system for field biology research. In: Proceedings of the SIGCHI Conference on Human Factors in Computing Systems, Montreal, Quebec, Canada, 22–27 April 2006 (2006)
19. Cowan, L.P., Griswold, W., Weibel, N., Hollan, J.: UbiSketch: bringing sketching out of the closet. La Jolla, University of California, San Diego: 10 (2011)
20. Song, H., Benko, H., et al.: Grips and gestures on a multi-touch pen. In: Proceedings of the 2011 Annual Conference on Human Factors in Computing Systems, Vancouver, BC, Canada, 7–12 May 2011 (2011)
21. Landau, S., Bourquin, G., van Schaack, A., Miele, J.: Demonstration of a universally accessible audio-haptic transit map built on a digital pen-based platform. In: 3rd International Haptic and Auditory interaction Design Workshop, Jyvaskyla, Finland, 15–16 September 2008 (2008)
22. Sarcevic, A.: TraumaPen: supporting documentation and situational awareness through real-time data capture and presentation in safety-critical work. Technical report, Computer Science Department, University of Colorado at Boulder (2010)
23. Jhpiego Corporation. E-Partogram. Saving Lives at Birth: A Grand Challenge for Development. http://www.savinglivesatbirth.net/summaries/35 (2011)
24. Mathai, M.: WHO Partograph E-learning Course. World Health Organization. http://streaming.jointokyo.org/viewerportal/vmc/player.do?eventContentId=995 (2010)

Author Index